Logical Dynamics of Information and Interaction

This book develops a new view of logic as a theory of information-driven agency and intelligent interaction between many agents – with conversation, argumentation, and games as guiding examples. It provides one uniform account of dynamic logics for acts of inference, observation, questions, and communication, that can handle both update of knowledge and revision of beliefs. It then extends the dynamic style of analysis to include changing preferences and goals, temporal processes, group action, and strategic interaction in games. Throughout, the book develops a mathematical theory unifying all these systems, and positioning them at the interface of logic, philosophy, computer science, and game theory. A series of further chapters explores repercussions of the 'dynamic stance' for these areas, as well as cognitive science.

JOHAN VAN BENTHEM is University Professor of Logic at the University of Amsterdam, Henry Waldgrave Stuart Professor of Philosophy at Stanford University, and Weilun Visiting Professor at Tsinghua University, Beijing. He is the author of *Language in Action* (1991) and *Exploring Logical Dynamics* (1996) and an editor of the handbooks of *Logic and Language* (1997), *Modal Logic* (2006), *Spatial Logic* (2007), and *Philosophy of Information* (2008).

Logical Dynamics of Information and Interaction

JOHAN VAN BENTHEM
The University of Amsterdam and Stanford University

CAMBRIDGE
UNIVERSITY PRESS

CAMBRIDGE
UNIVERSITY PRESS

University Printing House, Cambridge CB2 8BS, United Kingdom

Published in the United States of America by Cambridge University Press, New York

Cambridge University Press is part of the University of Cambridge.

It furthers the University's mission by disseminating knowledge in the pursuit of education, learning and research at the highest international levels of excellence.

www.cambridge.org
Information on this title: www.cambridge.org/9781107417175

First published 2011
First paperback edition 2014

A catalogue record for this publication is available from the British Library

Library of Congress Cataloguing in Publication data

Benthem, J. F. A. K. van, 1949–
Logical dynamics of information and interaction / Johan van Benthem.
p. cm.
Includes bibliographical references and index.
ISBN 978-0-521-76579-4 (Hardback)
1. Logic, Symbolic and mathematical. I. Title.
QA9.B3988 2011
511.3–dc23
2011030309

ISBN 978-0-521-76579-4 Hardback
ISBN 978-1-107-41717-5 Paperback

To Arthur and Lucas

Contents

Preface

This book is about Logical Dynamics, a theme that first gripped me in the late 1980s. The idea had many sources, but what it amounted to was this: make actions of language use and inference first-class citizens of logical theory, instead of studying just their products or data, such as sentences or proofs. My programme then became to explore the systematic repercussions of this 'dynamic turn'. It makes its first appearance in my book *Language in Action* (1991), where categorial grammars are linked to procedures of linguistic analysis using relational algebra – viewing natural language as a sort of cognitive programming language for transforming information. My next book *Exploring Logical Dynamics* (1996) continued with this perspective, linking it to modal logic and process theories in computer science: in particular, dynamic logic of programs. This added new themes like process invariances and definability, dynamic inference, and computational complexity of logics. In the meantime, my view of logical dynamics has evolved again. I now see it as a general theory of agents that produce, transform, and convey information – and in all this, their social interaction should be understood just as much as their individual powers. Just think of this: asking a question and giving an answer is just as logical as drawing a conclusion on your own. And likewise, I would see argumentation with different players as a key notion of logic, with proof just a single-agent projection. This stance is a radical break with current habits, and I hope it will gradually grow on the reader, the way it did on me.

The book presents a unified account of the resulting agenda, in terms of *dynamic-epistemic logic*, a framework developed around 2000 by several authors. Many of its originators are found in my references and acknowledgments, as are others who helped shape this book. In this setting, I develop a systematic way of describing actions and events that are crucial to agency, and show how it works uniformly for observation-based knowledge update, inference, questions, belief revision, and preference change, all the way up to complex social scenarios over time, such as games. In doing so, I am not claiming that this approach solves all problems of agency, or that logic is the sole guardian of intelligent interaction. Philosophy, computer science, probability theory, or game theory have important things to say as well. But I do claim that logic has a long-standing art of choosing abstraction levels that are sparse and yet revealing. The perspective offered here is simple, illuminating, and a useful tool to have in your arsenal when studying foundations of cognitive behaviour. Moreover, the logical view that we develop has a certain mathematical elegance that can be appreciated even when the grand perspective leaves you cold. And if that technical appeal does not work either, I would already be happy if I could convey that the dynamic stance throws fresh light on many old things, helps us see new ones – and that it is fun!

This book is based on lectures and papers since 1999, many co-authored. Chapter 1 explains the program, Chapter 2 gives background in epistemic logic, and Chapters 3–12 develop the logical theory of agency, with a baseline for readers who just wish to see the general picture, and extra topics for those who want more. Chapters 13–16, which can be read separately, explore repercussions of logical dynamics in other disciplines. Chapter 17 summarizes where we stand, and points at roads leading from here. In composing this story, I had to be selective, and the book does not cover every alley I have walked. Also, throughout, there are links to other areas of research, but I could not chart them all. Still, I would be happy if the viewpoints and techniques offered here would change received ideas about the scope of logic, and in particular, revitalize its interface with philosophy.

Acknowledgments

First of all, I want to thank my co-authors on papers that helped shape this book: Cédric Dégrémont, Jan van Eijck, Jelle Gerbrandy, Patrick Girard, Tomohiro Hoshi, Daisuke Ikegami, Barteld Kooi, Fenrong Liu, Maricarmen Martinez, Ştefan Minică, Siewert van Otterloo, Eric Pacuit, Olivier Roy, Darko Sarenac, and Fernando Velázquez Quesada. I also thank the students with whom I have interacted on topics close to this book: Marco Aiello, Guillaume Aucher, Harald Bastiaanse, Boudewijn de Bruin, Nina Gierasimczuk, Wes Holliday, Thomas Icard, Lena Kurzen, Minghui Ma, Marc Pauly, Ben Rodenhäuser, Floris Roelofsen, Ji Ruan, Joshua Sack, Tomasz Sadzik, Merlijn Sevenster, Josh Snyder, Yanjing Wang, Audrey Yap, Junhua Yu, and Jonathan Zvesper. Also, many colleagues gave comments, from occasional to extensive, that improved the manuscript: Krzysztof Apt, Giacomo Bonanno, Davide Grossi, Andreas Herzig, Wiebe van der Hoek, Hans Kamp, Larry Moss, Bryan Renne, Gabriel Sandu, Sebastian Sequoiah-Grayson, Yoav Shoham, Sonja Smets, Rineke Verbrugge, and Tomoyuki Yamada. I also profited from the readers' reports solicited by Cambridge University Press, though my gratitude must necessarily remain *de dicto*. Finally, I thank Hans van Ditmarsch and especially Alexandru Baltag for years of contacts on dynamic-epistemic logic and its many twists and turns.

1 Logical dynamics, agency, and intelligent interaction

1.1 Logical dynamics of information-driven agency

Human life is a history of millions of actions flowing along with a stream of information. We plan our trip to the hardware store, decide on marriage, rationalize our foolish behaviour last night, or prove an occasional theorem, all on the basis of what we know or believe. Moreover, this activity takes place in constant interaction with others, and it has been claimed that what makes humans so unique in the animal kingdom is not our physical strength, nor our powers of deduction, but rather our planning skills in social interaction – with the mammoth hunt as an early example, and legal and political debate as a late manifestation. It is this intricate cognitive world that I take to be the domain of logic, as the study of the invariants underlying these informational processes. In particular, my programme of *Logical Dynamics* (van Benthem 1991, 1996, 2001) calls for identification of a wide array of informational processes, and their explicit incorporation into logical theory, not as didactic background stories for the usual concepts and results, but as first-class citizens. One of the starting points in that programme was a pervasive ambiguity in our language between *products* and *activities* or processes. 'Dance' is an activity verb, but it also stands for the product of the activity: a waltz or a mambo. 'Argument' is a piece of a proof, but also an activity one can engage in, and so on. Logical systems as they stand are product-oriented, but Logical Dynamics says that both sides of the duality should be studied to get the complete picture. And this paradigm shift will send ripples all through our standard notions. For instance, natural language will now be, not a static description language for reality, but a dynamic programming language for changing cognitive states.

Recent trends have enriched the thrust of this action-oriented programme. 'Rational agency' stresses the transition from the paradigm of proof and computation performed by a single agent (or none at all) to agents with abilities, goals, and preferences plotting a meaningful course through life. This turn is also clear in computer science, which is no longer about lonely Turing Machines scribbling on tapes, but about complex intelligent communicating systems with goals and purposes. Another recent term, 'intelligent interaction', emphasizes what is perhaps the most striking feature here, the role of *others*. Cognitive powers show at their best in many-mind, rather than single-mind settings – just as physics only gets interesting, not with single bodies searching for their Aristotelian natural place, but on the Newtonian view of many bodies influencing each other, from nearby and far.

1.2 The research programme in a nutshell

What phenomena should logic study in order to carry out this ambitious programme? I will first describe these tasks in general terms, and then go over them more leisurely with a sequence of examples. A useful point of entry here is the notion of *rationality*. Indeed, the classical view of humans as 'rational animals' seems to refer to our reasoning powers:

To be rational is to reason *intelligently.*

These powers are often construed narrowly as deductive skills, making mathematical proof the paradigm of rationality. This book has no such bias. Our daily skills in the common sense world are just as admirable, and much richer than proof, including further varieties of reasoning such as justification, explanation, or planning. But even this variety is not yet what I am after. As our later examples will show, the essence of a rational agent is the ability to use information from many sources, of which reasoning is only one. Equally crucial information for our daily tasks comes from, in particular, observation and communication. I will elaborate this theme later, but right now, I cannot improve on the admirable brevity of the Mohist logicians in China around 500 BC (Zhang & Liu 2007):[1]

Zhi: Wen, Shuo, Qin 知 问 说 亲

knowledge arises through questions, inference, and observation.

[1] Somewhat anachronistically, I use modern simplified Chinese characters.

Thus, while I would subscribe to the above feature of rationality, its logic should be based on a study of all basic informational processes as well as their interplay.

But there is more to the notion of rationality as I understand it:

To be rational is to act *intelligently.*

We process information for a purpose, and that purpose is usually not contemplation, but action. And once we think of action for a purpose, another broad feature of rationality comes to light. We do not live in a bleak universe of *information*. Everything we do, say, or perceive is coloured by a second broad system of what may be called *evaluation*, determining our preferences, goals, decisions, and actions. While this is often considered alien to logic, and closer to emotion and fashion, I would rather embrace it. Rational agents deal intelligently with both information and evaluation, and logic should get this straight.

Finally, there is one more crucial aspect to rational agency, informational and evaluational, that goes back to the roots of logic in Antiquity:

To be rational is to interact *intelligently.*

Our powers unfold in communication, argumentation, or games: multi-agent activities over time. Thus, the rational quality of what we do resides also in how we interact with *others*: as rational as us, less, or more so. This, too, sets a broader task for logic, and we find links with new fields such as interactive epistemology, or agent studies in computer science.

I have now given rationality a very broad sense. If you object, I am happy to say instead that we are studying 'reasonable' agents, a term that includes all of the above. Still, there remains a sense in which mathematical deduction is crucial to the new research programme. We want to describe our broader agenda of phenomena with *logical systems*, following the methods that have proven so successful in the classical foundational phase of the discipline. Thus, at a meta-level, in terms of modelling methodology, throughout this book, the reader will encounter systems obeying the same technical standards as before. And meta-mathematical results are as relevant here as they have always been. That, to me, is in fact where the unity of the field of logic lies: not in a restricted agenda of 'consequence', or some particular minimal laws to hang on to, but in its methodology and modus operandi.

So much for grand aims. The following examples will illustrate what we are after, and each adds a detailed strand to our view of rational agency. We will then summarize the resulting research programme, followed by a brief description of the actual contents of this book.

1.3 Entanglement of logical tasks: inference, update, and information flow

The Amsterdam Science Museum *NEMO* (www.nemo-amsterdam.nl/) organizes regular 'Kids' Lectures on Science', for some sixty children aged around eight in a small amphitheatre. In February 2006, it was my turn to speak – and my first question was this:

> *The Restaurant* 'In a restaurant, your Father has ordered Fish, your Mother ordered Vegetarian, and you have Meat. Out of the kitchen comes some new person carrying the three plates. What will happen?'

The children got excited, many little hands were raised, and one said: 'He asks who has the Meat.' 'Sure enough', I said: 'He asks, hears the answer, and puts the plate on the table. What happens next?' The children said: 'He asks who has the Fish!' Then I asked once more what happens next? And now one could see the Light of Reason suddenly start shining in those little eyes. One girl shouted: 'He does not ask!' Now, *that* is logic ... After that, we played a long string of scenarios, including card games, Master Mind, Sudoku, and even card magic, and we discussed what best questions to ask and conclusions to draw.

Two logical tasks The Restaurant is about the simplest scenario of real information flow. And when the waiter places that third plate without asking, you see a logical inference in action. The information in the two answers allows the waiter to infer (implicitly, in a flash of the 'mind's eye') where the third plate must go. This can be expressed as a logical form

$$A \text{ or } B \text{ or } C, \text{ not-}A, \text{ not-}B \Rightarrow C$$

One can then tell the usual story about the power of valid inference in other settings. With this moral, the example goes back to Greek Antiquity. But the scenario is much richer. Let us look more closely: perhaps, appropriately, with the eyes of a child.

To me, the Restaurant cries out for a new look. There is a natural unity to the scenario. The waiter first obtains the right information by asking questions and understanding answers, acts of *communication* and perhaps *observation*, and once enough data have accumulated, he *infers* an explicit solution. Now on the traditional line, only the latter step of deductive elucidation is logic proper, while the former are at best pragmatics. But in my view, both informational processes are on a par, and both should be within the compass of logic. Asking a question and grasping an answer is just as logical as drawing an inference. And accordingly, logical systems should account for both of these, and perhaps others, as observation, communication, and inference occur entangled in most meaningful activities.

Information and computation And logic is up to this job, if we model the relevant actions appropriately. Here is how. To record the information changes in the Restaurant, a helpful metaphor is *computation*. During a conversation, information states of people – alone, and in groups – change over time, in a systematic way triggered by information-producing events. So we need a set of information states and transitions between them. And as soon as we do this, we will find some fundamental issues, even in the simplest scenarios.

Update of semantic information Consider the information flow in the Restaurant. The intuitive information states are sets of 'live options' at any stage, starting from the initial six ways of giving three plates to three people. There were two successive *update actions* on these states, triggered by the answers to the waiter's two questions. The first reduced the uncertainty from six to two options, and the second reduced it to *1*, i.e., just the actual situation. Here is a 'video sequence' of how the answers for Meat and Fish would work in case the original order was ***FMV*** (fish for the first person, meat for the second, vegetarian for the third):

This is the common sense process of semantic update for the current *information range*, where new information is produced by events that rule out possibilities. In Chapters 2 and later, we will call this elimination scenario a case of 'hard information', and typical events producing it are public announcements in communication, or public observations.

Inference and syntactic information The first two updates have zoomed in on the actual situation. This explains why no third question is needed. But then we have a problem. What is the *point* of drawing a logical conclusion if it adds no further information? Here, the common explanation is that inferences 'unpack' information that we may have only implicitly. We have reached the true world, and now we want to spell it out in a useful sort of code. This is where inference kicks in, elucidating by means of linguistic description what the world looks like:

(6) ***update 1 >*** (2) ***update 2 >*** (1) *inference to 'full arrangement'*

This sounds fine, but it makes sense only when we distinguish two different notions of information: one 'explicit', the other 'implicit' (van Benthem & Martinez 2008). Now, while there are elegant logics for semantic update of the latter, there is no consensus on how to model the explicit information produced by inference. Formats include syntactic accumulation of formulas, but more graphical ones also make sense. For instance, here is how propositional inferences drive stages in the solution of puzzles:

Example Take a simple 3×3 Sudoku diagram, produced by applying the two rules that 'each of the nine positions must have a digit', but 'no digit occurs twice on a row or column':

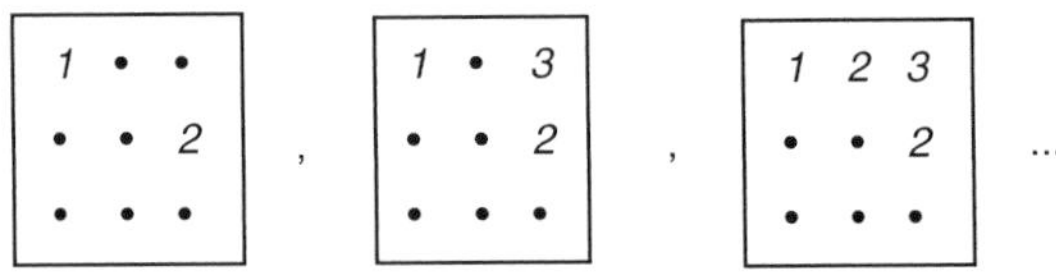

Each successive diagram displays a bit more about the unique solution (one world) determined by the initial placement of the digits *1*, *2*. Thus, explicit information is brought to light in logical inference in a process of what may be called deductive *elucidation*. Chapter 5 of this book will make a more systematic syntactic proposal for representing the dynamics of inference, that works in tandem with semantic update. For now, we just note that what happened at the Restaurant already involves a basic issue in the philosophy of logic (cf. Chapter 13): capturing and integrating different notions of information.

Putting things together, the *dynamics* of various kinds of informational actions becomes a target for logical theory. But to make this work, we must, and will, also give an account of the underlying *statics*: the information states that the actions work over. As a first step toward this programme, we have identified the first level of skills that rational agents have:

> *their powers of* inference, *and their powers* of observation, *resulting in information updates that change what they currently know.*

1.4 Information about others and public social dynamics

Another striking feature to the information flow in the Restaurant are the questions. Questions and answers typically involve more than one agent, and their dynamics is *social*, having to do also with what people come to know about each other. This higher-order knowledge about others is crucial to human communication and interaction in general.

Questions and answers Take just one simple 'Yes/No' question followed by a correct answer, a ubiquitous building block of interaction. Consider the following dialogue:

> *Me*: 'Is this Beihai Park?'
> *You*: 'Yes.'

This conveys facts about the current location. But much more is going on. By asking the question in a normal scenario (not, say, a competitive game), I indicate that I do not know the answer. And by asking you, I also indicate that I think you may know the answer, again under normal circumstances.[2] Moreover, your answer does not just transfer bare facts to me. It also achieves that you know that I know, I know that you know that I know, and in the limit of such iterations, it achieves *common knowledge* of the relevant facts in the group consisting of you and me. This common knowledge is not a by-product of the fact transfer. It rather forms the basis of our mutual expectations about future behaviour.[3] Keeping track of higher-order

[2] All such presuppositions are off in a classroom with a teacher questioning students. The logics that we will develop in this book can deal with a wide variety of such scenarios.

[3] If I find your pin code and bank account number, I may empty your account – if I know that you do not know that I know all this. But if I know that you know that I know,

information about others is crucial in many disciplines, from philosophy (interactive epistemology) and linguistics (communicative paradigms of meaning) to computer science (multi-agent systems) and cognitive psychology ('theory of mind').[4] Indeed, the ability to move through an informational space keeping track of what other participants know and do not know, including the crucial ability to switch and view things from other people's perspectives, seems characteristic of human intelligence.

So, logical activity is interactive, and its theory should reflect this. Some colleagues find this alarming, as social aspects are reminiscent of gossip, status, and Sartre's 'Hell is the Others'. The best way of dispelling such fears may be a concrete example. Here is one, using a card game, a useful normal form for studying information flow in logical terms. It is like the Restaurant in some ways, but with a further layer of higher-order knowledge.

The Cards (van Ditmarsch 2000) Three cards 'red', 'white', 'blue' are distributed over three players: *1, 2, 3*, who get one each. Each player sees her own card, but not the others. The real distribution over *1, 2, 3* is *red, white, blue.* Now a conversation takes place (this actually happened during the *NEMO* children session, on stage with three volunteers):

2 asks 1	*'Do you have the blue card?'*
1 answers truthfully	*'No.'*

Who knows what then, assuming the question is sincere? Here is the effect stated in words:

> Assuming the question is sincere, *2* indicates that she does not know the answer, and so she cannot have the blue card. This tells *1* at once what the deal was. But *3* did not learn, since he already knew that *2* does not have blue. When *1* says she does not have blue, this now tells *2* the deal. *3* still does not know the deal; but since he can perform the reasoning just given, he does know that the others know it.

We humans go through this sort of reasoning in many settings, with different knowledge for different agents. In Chapters 2, 3, we will analyse this information flow in detail.

I will not touch your account. Crime is triggered by fine iterated epistemic distinctions: that is why it is usually better left to experts.

[4] Cf. Hendricks (2005), Verbrugge (2009), van Rooij (2005), and many other sources.

These scenarios can be much more complex. Real games of 'who is the first to know' arise by restricting possible questions and answers, and we will consider game logics later on. Also, announcements raise the issue of the reliability of the speaker, as in logic puzzles with meetings of Liars and Truth-Tellers. Our systems will also be able to deal with these in a systematic way, though separating one agent type from another is often a subtle manner of design. Logic of communication is not easy, but it is about well-defined issues.

Thus, we have a second major aspect of rational agents in place as a challenge to logic:

their social powers *of mutual knowledge and communication.*

Actually, these powers involve more than pure information flow. Questions clearly have other uses than just conveying information: they define *issues* that give a purpose to a conversation or scientific investigation. This dynamics, too, can be studied per se, and Chapter 6 will show how to deal with 'issue management' within our general framework.

1.5 Partial observation and differential information

The social setting suggests a much broader agenda for logical analysis. Clearly, public announcement as we saw in the Restaurant or with the Cards is just one way of creating new information. The reality in many games, and most social situations, is that information flows differentially, with partial observation by agents. When I draw a card from the stack, I see which card I am getting. You do not, though you may know it is one of a certain set: getting *some* information. When you take a peek at my card, you learn something by cheating, degrading my knowledge of the current state of the game into mere belief. When you whisper in your neighbour's ear during my talk, this is a public announcement in a subgroup – where I and others need not catch what you are saying, and I may not even notice that any information is being passed at all.

Modelling such information flow is much more complicated than public announcement, and goes beyond existing logical systems. The first satisfactory proposals were made only in the late 1990s, as we shall see in Chapter 4. By now, we can model information flow in parlour games like

Clue, that have an intricate system of public and private moves. All this occurs in natural realities all around us, such as *electronic communication*:

> I send you an email, with the message '*P*': a public announcement in the group {*you, me*}. You reply with a message '*Q*' with a cc to others: a public announcement to a larger group. I respond with '*R*' using a 'reply-to-all' plus a *bcc* to some further agents.

In the third round, we have a partly hidden act again: my *bcc* made an announcement to some agents, while others do not know that these were included. The information flow in this quite common episode is not simple. After a few rounds of *bcc* messages to different groups, it becomes very hard to keep track of who is supposed to know what. And that makes sense: differential information flow is complex, and so is understanding social life.

There are intriguing thresholds here. Using *bcc* is not misleading to agents who know that it is a possible event in the system. A further step is *cheating*. But even judicious lies seem a crucial skill in civilized life. Our angelic children are not yet capable of that, but rational agents at full capacity can handle mixtures of lies and truths with elegance and ease.

Thus, we have a further twist to our account of the powers of rational agents:

> *different* observational access *and processing differential information flow.*

This may seem mere engineering. Who cares about the sordid realities of cheating, lying, and social manoeuvring? Well, differential information is a great good: we do not tell everyone everything, and this keeps things civilized and efficient. Indeed, most successful human activity is social, from hunting cave bears to mathematics. And a crucial feature of social life is organization, including new procedures for information flow. Even some philosophy departments now do exams on Skype, calling for new secret voting procedures on a public channel. What is truly amazing is how this fascinating informational reality has been such a low priority of mainstream logicians and epistemologists for so long.

1.6 Epistemic shocks: self-correction and belief revision

So far, we considered information flow and knowledge. It is time for a next step. Agents who correctly record information from their observations, and industriously draw correct conclusions from their evidence, may be rational in some Olympian sense. But they are still cold-blooded recording devices. But knowledge is scarce, and rationality does not reside in always being cautious, and continual correctness. Its peak moments occur with warm-blooded agents, who are opinionated, make mistakes, but who subsequently *correct* themselves.[5] Thus, rationality is about the dynamics of being wrong just as much as about that of being right: through belief revision, i.e., *learning* by giving up old beliefs.[6] Or maybe better, rationality is about a balance between two abilities: jumping to conclusions, and subsequent correction if the jump was over-ambitious.

Here, events become more delicate than with information flow through observation. Our knowledge can never be falsified by true new information, but beliefs can, when we learn new facts contradicting what we thought most plausible so far. Feeling that an earthquake is hitting the Stanford campus, I no longer believe that a short nocturnal bike ride will get me home in ten minutes. There is much for logic to keep straight in this area. For instance, the following nasty scenario has been discussed by computer scientists, philosophers, and economists in the 1990s. Even true beliefs can be sabotaged through true information:

> *Misleading with the Truth*
> You know that you have finished *3rd*, *2nd*, or *1st* in the election, and you find lower outcomes more plausible. You also know that being *2nd* makes your bargaining chances for getting some high office small ('dangerous heavy-weight'). In fact you were *1st*. I know this, but only say (truly) that you are not *3rd*: and you become unhappy. Why?

Initially, you find being *3rd* the most plausible outcome, and may believe you will get high office by way of compensation. So, this is a true belief of yours, but for the wrong reason. Now you learn the true fact that you are not *3rd*, and being *2nd* becomes the most plausible world. But then, you now believe, falsely, that you will not have any high office – something you would not believe if you knew that you won the election.

[5] Compare a lecture with a mathematician writing a proof on a blackboard to a research colloquium with people guessing, spotting problems, and then making brilliant recoveries ...

[6] In a concrete setting, revision comes to the fore in *conversation*. People contradict each other, and then something more spectacular is needed than update. Maybe one of them was wrong, maybe they all were, and they have to adjust. Modelling this involves making a further distinction between information coming from some source, and agents' various attitudes and responses to it.

Our logics of belief revision in Chapter 7 of this book can deal with such scenarios. They even include others with a softer touch, where incoming *soft information* merely makes certain worlds less or more plausible, without ever removing any world entirely from consideration. In this same line, Chapter 8 will show how dynamic logics can also incorporate *probability*, another major approach to beliefs of various strengths.

Thus, in addition to the earlier update that accumulates knowledge, we have identified another, more complex, but equally important feature of truly rational agents:

> *their capacity for hypothesizing, being wrong, and then* correcting themselves.

In many settings, these capacities seem the more crucial and admirable human ability. A perfectly healthy body is great, but passive, and the key to our biological performance is our immune system responding to cuts, bruises, and diseases. Likewise, I would say that flexibility in beliefs is essential: and logic is all about the immune system of the mind.

1.7 Planning for the longer term

So far, we have mostly discussed single moves that rational agents make in response to incoming information, whether knowledge update or belief revision. But in reality, these single steps make sense only as part of longer processes through time. Even the Restaurant involved a conversation, that is, a sequence of steps, each responding to earlier ones, and directed towards some goal, and the same is true for games and social activities in general. There is relevant structure at this level, too, and as usual, it is high-lighted by well-known puzzles. Here is an evergreen:

> *The Muddy Children* (Fagin *et al.* 1995):
> After having played outside, two of three children have got mud on their foreheads. They can only see the others, so they do not know their own status.[7] Now their Father comes along and says: 'At least one of you is dirty.' He then asks: "Does anyone know if he is dirty?" Children answer truthfully, and this is repeated round by round. As questions and answers repeat, what will happen?

[7] This observational access is the inverse of our earlier card games, but very similar.

One might think that nothing happens, since the father just tells the children something they already know – the way parents tend to do – viz. that there is at least one dirty child. But in reality, he does achieve something significant, making this fact into *common knowledge*. Compare the difference between every colleague knowing that your partner is unfaithful: no doubt unpleasant, but maybe still manageable, with this fact being common knowledge, including everyone knowing that the others know, etcetera: the shame at department meetings becomes unbearable. Keeping this in mind, here is what happens:

> Nobody knows in the first round. But in the next round, each muddy child reasons like this: 'If I were clean, the one dirty child I see would see only clean children, and so she would know that she was dirty. But she did not. So I must be dirty, too!'

Note that this scenario is about what happens in the long run: with more children, common knowledge of the muddy children arises after more rounds of ignorance announcement, after which, in the next step, the clean children will know that they are clean. There is a formal structure to this. The instruction to the children looks like a computer program:

> REPEAT (IF you don't know your status THEN say you don't know
> ELSE say you do).

This is no coincidence. Conversation involves *plans*, and plans have a control structure for actions also found in computer programs: choice, sequential composition, and iteration of actions. The muddy children even have *parallel composition* of actions, since they answer simultaneously. Thus, actions may be composed and structured to achieve long-term effects – and this, too, will be an aspect of our logics. But for the moment, we note this:

> *Rational agency involves planning in* longer-term scenarios, *and its quality also lies in the ways that agents compose their individual actions into larger wholes.*

We will study long-term perspectives on agent interaction in Chapter 11, with connections to temporal logics of branching time as the Grand Stage where human activity takes place: as in Jorge Luis Borges' famous story *The Garden of Forking Paths.*

1.8 Preferences, evaluation, and goals

Now we move to another phenomenon, that is crucial to understanding the driving force of much informational behaviour as discussed so far. Just answering 'a simple question' is rare. Behind every question, there lies a *why*-question: why does this person say this, what does she want, and what sort of scenario am I entering? Pure informational activities are rare, and they tend to live in an ether of preferences, and more generally, *evaluation* of situations and actions. This is not just greed or emotion. 'Making sense' of an interaction involves meaning and information, but also getting clear on the goals of everyone involved. This brings in another fundamental level of agent structure:

> *logic of rational agency involves* preferences *between situations and actions, and agents'* goals, *usually aligned with these preferences.*

Preferences determine actions, and knowing your preferences helps me make predictions about what will happen.[8] It is hard to separate information from evaluation, and this may reflect some deep evolutionary entanglement of our cognitive and emotional brain systems.

While preference has been studied extensively in decision theory and game theory, it has been more marginal in logic. In Chapter 9 of this book, we will incorporate preference logic, and show how it fits well with logical analysis of knowledge update and belief revision. Indeed, it might be said that this move provides the explanatory dynamics in the physicists' sense behind the 'kinematics' of knowledge and belief that we have emphasized so far.[9]

1.9 Games, strategies, and intelligent interaction

Temporal perspective and preference combine in the next crucial feature of rational agents that we noted earlier, their responding to others and mutually influencing them. Even a simple conversation involves choosing assertions depending on what others say. This interactive aspect means that dynamic logics must eventually come to turn with *games*:

[8] If you think this is just daily life, not science, think of how a referee will judge your paper on its interest rather than its mere truth, where 'interesting' depends on the preferences of a scientific community.

[9] I take this analogy seriously. *Rationality* (cf. Chapters 10, 15) links preference, belief, and action in ways reminiscent of Newton's Laws for the dynamics of moving bodies using force, mass and acceleration. We mix theoretical and observational terms in suitable laws, and then base explanations on them.

True interaction and games To sample the spirit of interaction, consider the following game played between a Student and a Teacher. The Student is located at position *S* in the following diagram, but wants to reach the position of escape *E* below, whereas the Teacher wants to prevent him from getting there. Each line segment is a path that can be travelled. At each round of the game, the Teacher cuts one connection, anywhere in the diagram, while the Student can, and must, travel one link still open to him at his current position:

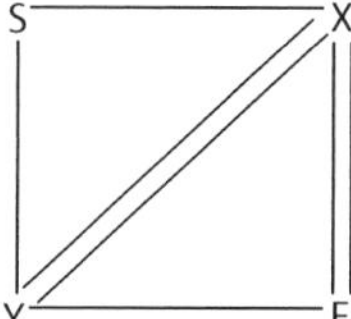

If Teacher is greedy, and starts by cutting a link *S*–*X* or *S*–*Y* right in front of Student, then Student can reach the escape *E*. However, Teacher does have a *winning strategy* for preventing Student from ever reaching *E*, by doing something else:

> first cut one line between *X* and *E*, and then let further cutting be guided in a straightforward manner by where Student goes next.

Here *strategies* for players are rules telling them what to do in every eventuality. Solving games like this can be complex, emphasizing the non-trivial nature of interaction.[10]

Digression: learning Formal Learning Theory (Kelly 1996) concentrates on single-agent settings where a student forms hypotheses on the basis of some input stream of evidence: there is a Student, but no Teacher, unless we think of Nature as a disillusioned teacher doing a minimum of presentation without adjustment. But the realities of teaching and learning

[10] Rohde (2005) shows that solving 'sabotage graph games' like this is *Pspace*-complete, a high degree of complexity. The reader will get an even better feel for the complexity of interaction by considering a variant. This time, the Teacher wants to force the Student to *end up in E* without any possibility of escape. Who of the two has the winning strategy this time, in the same graph?

are social, with Students and Teachers responding to each other – and learning is a social process. We even learn at two levels: 'knowledge that', and know-how or skills.[11]

Logic and game theory These multi-agent scenarios are close to game theory (Osborne & Rubinstein 1994 is an excellent introduction whose style also speaks to logicians) where information, evaluation, and strategic interaction are entangled. For a start, *Zermelo's Theorem* says that extensive two-player games of finite depth with perfect information and zero-sum outcomes are *determined*: that is, one of the players has a winning strategy. In our teaching game, this explains why Student or Teacher has a winning strategy.[12] Real game theory arises when players have preferences and evaluate outcomes. The reasoning extending Zermelo's is *Backward Induction*. Starting from values on leaves, nodes get evaluated up the tree, representing players' intermediate beliefs as to expected outcomes, given that both are acting 'rationally'. Here is an example, with nodes indicating the turns of two players *A*, *E*, while branching indicates different available moves. End nodes mark utility values for players in the order ('value for *A*', 'value for *E*'):

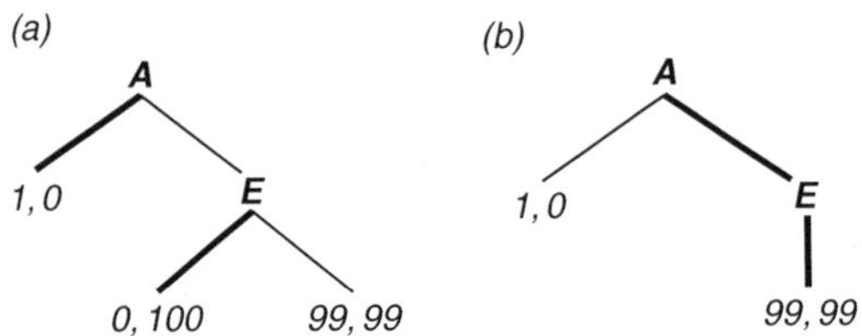

The thick black lines in Tree (a) indicate the backward induction moves of 'rational' players who choose those actions whose outcomes they believe to be best for themselves. Interestingly, this is an equilibrium with a socially undesirable outcome, as *(1, 0)* makes both players worse off than *(99, 99)*. Thus, we need to reassess the assumptions behind the usual solution procedures for games. Dynamic logics of communication help here, with new takes. Think of *promises* that change a game by announcement of intentions.

[11] For the general importance of know-how in epistemology, cf. Gochet (2006). In a related vein, Chapter 14 discusses epistemic plans merging knowledge and action, a tradition that goes back to the pioneering study Moore (1985).

[12] For details of this and the next examples, cf. Chapter 10.

E might promise that she will not go left, changing game (a) to game (b) – and the new equilibrium *(99, 99)* results, making both players better off by restricting the freedom of one.[13]

Finally, games are not just an analytical tool. They are also a ubiquitous human activity across cultures, serving needs from gentle elegant wastes of time to training crucial skills:

> *a full logical understanding of rational agency and intelligent interaction requires a logical study of* games, *as a crucial model for human behaviour.*

This theme is mostly the subject of van Benthem (to appearA) but Chapters 10, 11, 15 take it up in some detail, including games with partial observation and imperfect information.

1.10 Groups, social structure, and collective agency

Single agents need not just interact on their own: typically, they also form *groups* and other collective agents, whose behaviour does not reduce to that of individual members. For instance, groups of players in games can form coalitions, and social choice theory is about groups creating group preferences on the basis of individual preferences of their members. We saw some of this in the notion of common knowledge, which is about the degree of being informed inside a group. But there are more themes that concern group agency, and in Chapter 12 we will show how our dynamic logics interface with group behaviour, and even may help provide a 'micro-theory' of information-based rational deliberation.

1.11 The programme of Logical Dynamics in a nutshell

Rational agency involves information flow with many entangled activities: inference, observation, communication, and evaluation, all over time. Logical Dynamics makes all of these first-class citizens, and says that logical theory should treat them on a par. The resulting dynamic logics add subtlety and scope to classical systems, going beyond agent-free proof and

[13] Van Benthem (2007f) proposes alternatives to Backward Induction in more history-oriented games, where players constantly remind themselves of the *legitimate rights of others*, or of past favours received.

computation.[14] Thus, we get new interdisciplinary links beyond the old friends of mathematics, philosophy,[15] and linguistics, including computer science and economics. While reclaiming a broader agenda, with formal systems a means but not the end, logic also becomes a central part of academic life, overflowing the usual disciplinary boundaries.

A historical pedigree Is this new-fangled tinkering with the core values of logic? I do not think so. The ideas put forward here are ancient. We already mentioned broader Chinese views from Mohist logic. Likewise, traditional Indian logic stressed three ways of getting information. The easiest route is to observe, when that is possible. The next method is inference, in case observation is impossible or dangerous – the example being a coiled object in a room where we cannot see if it is a piece of rope, or a cobra. And if these two methods fail, we can resort to communication, and ask an expert. And also in the Western tradition, the social interactive aspect of information flow was there from the start. While many see Euclid's *Elements* as the paradigm for logic, with its crystalline mathematical proofs and eternal truths, the true origin may be closer to Plato's *Dialogues*, an argumentative practice. It has been claimed that logic arose out of legal, philosophical, and political debate in all its three main traditions.[16] And this multi-agent interactive view has emerged anew in modern times. A beautiful case is the *dialogue games* of Lorenzen (1955), that explained logical validity pragmatically in terms of a winning strategy for a proponent arguing a conclusion against an opponent granting the premises. Similar views occur in Hintikka (1973), another pioneering work on games and informational activities inside logic.

But in the end, I see no opposition between Platonic and Euclidean images. This book is about a broad range of logical activities, but still pursued by mathematical means. Logical dynamics has no quarrel with classical standards of explicitness and precision.

[14] Moreover, we find a side benefit to the Dynamic Turn. Chapter 13 shows how to replace 'non-standard logics', that have sprung up in droves in recent years, by perfectly classical systems, once we identify the right information-changing events, and make them an explicit part of the logic.

[15] In particular, I see strong connections with (social) *epistemology*, cf. Goldman (1999).

[16] For instance, the Mohists in early China discuss the Law of Non-Contradiction as a principle of conversation: 'resolve contradictions with others', 'avoid contradicting yourself'. Cf. Zhang & Liu (2007).

The promised land? The area staked out here may be the logician's Promised Land, but it is hardly virgin territory. Like Canaan in the Old Testament, it is densely settled by other nations, such as philosophers, computer scientists, or economists, worshipping other gods such as probability or game theory. Can we just dispossess them? Indeed we must not. Rational agency is a deep subject, calling for all the help we can. I think logic has fresh insights to offer, beyond what is already there. But it has no favoured position: polytheism is a civilized idea. I do predict new fruitful liaisons between logic and its neighbours.[17]

A bridge too far? On top of the logical micro-structure and discrete temporal processes that we will study, there is emergent statistical behaviour of large groups over a long time. That is the realm of probability, evolutionary games, and *dynamical systems*. The intriguing interface of dynamic logics and dynamical systems is beyond the scope of this book.

1.12 The chapters explained

Some students find it helpful to think of the whole programme presented here pictorially, in stages:

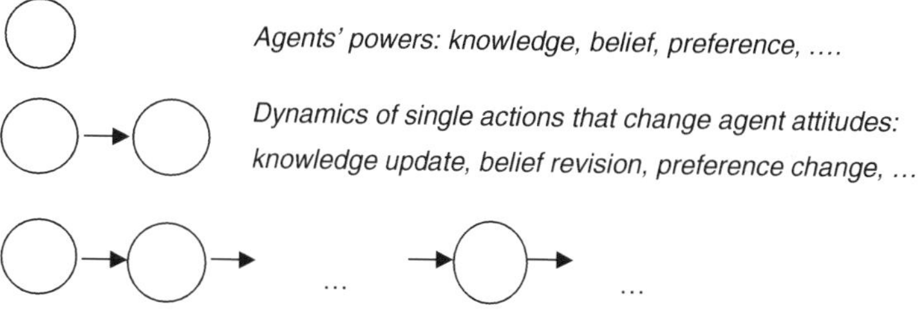
Agents' powers: knowledge, belief, preference, ...
Dynamics of single actions that change agent attitudes:
knowledge update, belief revision, preference change, ...
...
...
Long-term phenomena and interactive processes[18]

Our chapters follow these lines, highlighting activities and powers of agents. We start with semantic information and update, taking knowledge in the relaxed sense of what is true according to the agent's hard information. Our tools are epistemic logic over possible worlds models in a concrete sense (Chapter 2), its dynamified version of public announcement logic (*PAL*; Chapter 3) for communication and observation, and the more

[17] For a wonderful sample of combining disciplines, cf. Shoham & Leyton-Brown (2008) on multi-agent systems in terms of computer science, game theory, and logic.

[18] One might add a second dimension of *group size*, but it would overload the picture.

sophisticated dynamic-epistemic logic (*DEL*) of Chapter 4 with private scenarios and many agents. These systems are our paradigm for the analysis of definable changes in current models of the agents' information. This methodology is applied in Chapter 5 to deal with inference, and other actions turning implicit into explicit knowledge. Chapter 6 shows how this also works for questions and issue management. Next, we turn to belief revision in Chapter 7, using changes in plausibility orders to model learning systematically. Chapter 8 is a digression, showing how similar ideas work for probabilistic update, with a richer quantitative view of learning mechanisms. Then we move to agents' evaluation of worlds and actions, and show how our techniques for plausibility change also apply to preferences and ways of changing them, providing a unified account of information, belief, and goals (Chapter 9). Next, moving beyond one-step dynamics, Chapter 10 is about longer-term multi-agent interaction, with a special emphasis on games. A still wider perspective is that of Chapter 11, with embeddings of our dynamic logics in epistemic temporal logics of branching time, and logics of protocols that add 'procedural information' to our study of agency. Completing the development of our theory, Chapter 12 considers groups as new logical agents, showing how the earlier systems lift to this setting, including new phenomena such as belief merge, as well as links with social choice theory.

Advice to the reader These chapters are all arranged more or less as follows. First comes the motivation, then the basic system, then its core theory, followed by a conclusion explaining how one more building block of our logic of agency has been put in place. Readers could opt out at this stage, moving on to the next topic in the chapter sequence. What follows in a chapter is usually a logician's pleasure garden with further technical themes, open problems, and a brief view of key literature. Our open problems are both technical and conceptual – and non-logicians, too, may find some of them worthwhile.

The remaining Chapters 13–16 show how the logical dynamics developed here applies to a range of disciplines. Chapter 13 is concerned with philosophy, putting many old issues in a fresh light, and in that same light, adding new themes. Chapters 14–16 extend the interface to computer science, game theory, and cognitive science. These chapters can be read independently: there is no sequence, and they are different in style and level of technicality. Finally, Chapter 17 states our main conclusions and recommendations.

2 Epistemic logic and semantic information

Our first topic in the study of agency is the intuitive notion of information as a *semantic range of possibilities*, widespread in science, but also a common sense view. For this purpose, we use epistemic logic, proposed originally for analysing the philosophical notion of knowledge (Hintikka 1962). While the latter use remains controversial, we will take the system in a neutral manner as a logic of semantic information. More precisely, the *hard information* that an agent currently has is a set of possible worlds, and what it 'knows' is that which is true in all worlds of that range (van Benthem 2005c). This picture makes knowledge a standard universal modality. We present some basics in this chapter, stressing points of method that will recur. We refer to Blackburn, de Rijke & Venema (2000), Blackburn, van Benthem & Wolter (2006) for all details in what follows.[1] We add a few special themes and open problems, setting a pattern that will return in later chapters.

The topics to come reflect different aspects of a logical system. Its language and semantics provide a way of describing situations, evaluating formulas for truth or falsity, engaging in communication, and the like. This leads to issues of definition and *expressive power*. Next there is the *calculus of valid reasoning*, often with completeness theorems tying this to the semantics. In this book, this theme will be less dominant. To me, modal logics are not primarily about inferential life-styles like *K*, *KD45*, or *S5*, but about describing agency. Finally, there is *computational complexity*. We want to strike a balance between expressive power and complexity of logical tasks (cf. van Benthem & Blackburn 2006). This comes out well in a procedural perspective on meaning and proof in the form of 'logic games' – an interesting case of interactive

[1] Epistemic logic is flourishing today in computer science, game theory, and other areas beyond its original habitat: cf. Fagin *et al.* (1995), van der Hoek & Meijer (1995), and van der Hoek & Pauly (2007).

methods entering logic. Our presentation is not a textbook for epistemic logic,[2] but a showcase of a standard logical system that we will 'dynamify' in this book – not to replace it with something else, but to make it do even further things than before.

At this point, readers who know their epistemic logic might want to skip to the next chapters.

2.1 The basic language

We start with the simplest epistemic language, describing knowledge of individual agents, taken from some set I of agents that are relevant to the application at hand. But let us first illustrate what sort of situation this language is typically supposed to describe.

Example Questions and Answers.
A stranger approaches you in Beijing, and asks

Q 'Is this Beihai Park?'

As a well-informed and helpful Chinese citizen, you answer truly

A 'Yes.'

As we noted in Chapter 1, this involves a mix of factual information and higher-order information about information of others. In particular, the answer produces *common knowledge* between the agents *Q* and *A* of the relevant topographical fact. ■

It is worth having a logic that gets clear on these matters, and epistemic logic, first created to describe cogitation by lonesome thinkers, is just right for this interactive purpose. Thus, we can deal in a natural manner with the social aspects of agency that we noted before.

Definition Basic epistemic language.
The *basic epistemic language EL* has a standard propositional base with proposition letters and Boolean operators ('not', 'and', 'or', 'if then'), plus modal operators $K_i\varphi$ ('*i knows that* φ'), for each agent i in some set I that we fix in

[2] A new textbook in the current spirit is van Benthem (to appearB).

specific applications, and also $C_G\varphi$: 'φ is common knowledge in the group G' (with $G\subseteq I$). The inductive syntax rule is as follows, where the 'p' stands for any choice of atoms from some set of proposition letters:

$$p \mid \neg\varphi \mid \varphi \wedge \psi \mid K_i\varphi \mid C_G\varphi$$

We write $<i>\varphi$ for the existential modality $\neg K_i\neg\varphi$ saying intuitively that 'agent i considers φ possible'. Likewise, the existential dual modality of $C_G\varphi$ is written as $< C_G>\varphi$.[3,4] ■

Through its agent-indexed modalities, this basic language describes the preceding scenario.

Example Questions and Answers, continued.
Let Q ask a factual question 'φ?', to which A answers truly: 'Yes.' The presupposition for a normal truthful answer is that A knows that φ: written as $K_A\varphi$. The question itself, if it is a normal cooperative one, conveys at least the presuppositions (i) $\neg K_Q\varphi \wedge \neg K_Q\neg\varphi$ (Q does not know if φ) and (ii) $<Q>(K_A\varphi \vee K_A\neg\varphi)$ (Q thinks that A may know the answer). After the whole two-step communication episode, φ is known to both agents: $K_A\varphi \wedge K_Q\varphi$, while they also know this about each other: $K_QK_A\varphi \wedge K_AK_Q\varphi$, up to any depth of iteration. Indeed, they achieve the limit notion of common knowledge, written as $C_{\{Q,\ A\}}\varphi$. ■

Common knowledge is a group notion whose importance has long been recognized in many areas: from philosophy to game theory, computer science, linguistics, and psychology.

2.2 Models and semantics

The preceding assertions only make precise sense when backed up by a formal semantics. Here is the formal version of the earlier intuitive idea of information as range.

DEFINITION Epistemic models.
Models **M** for the epistemic language are triples $(W, \{\to_i \mid i\in I\}, V)$, where W is a set of *worlds*, the $\to_i$ are binary *accessibility relations* between worlds, and V is

[3] Individual modalities are essentially first-order universal quantifiers, but common knowledge is a higher-order notion. In this initial phase of our presentation, this technical difference will not matter.

[4] We will also use other connectives $\vee$, $\to$, $\leftrightarrow$ as convenient, defined in the usual way.

a *valuation* assigning truth values to proposition letters at worlds. In what follows, our primary semantic objects are *pointed models* (**M**, *s*) where *s* is the actual world representing the true state of affairs. ■

These models stand for collective information states of a group of agents. We impose no general conditions on the accessibility relations, such as reflexivity or transitivity – leaving a 'degree of freedom' for a modeller using the system. Many of our examples work well with equivalence relations (reflexive, symmetric, and transitive) – and such special settings may help the reader. But in some chapters, we will need arbitrary accessibility relations allowing end-points without successors, to leave room for false or misleading information.

Important notes on notation In some contexts, the difference between the semantic accessibility symbol $\rightarrow_i$ and a syntactic implication $\rightarrow$ is hard to see. To avoid confusion, we will often also use the notation $\sim_i$ for epistemic accessibility. The latter usually stands for an equivalence relation, while the arrow has complete generality. Also, we mostly write **M**, *s* for pointed models, using the brackets (**M**, *s*) only to remove ambiguity.

Example Setting up realistic epistemic models.
We will mainly use simple models to make our points. Even so, a real feel for the sweep of the approach only comes from the 'art of modelling' for real scenarios. Doing so also dispels delusions of grandeur about possible worlds. Consider this game from Chapter 1. Three cards 'red', 'white', 'blue' were given to three players: *1, 2, 3*, one each. Each player can see her own card, but not that of the others. The real distribution over the players *1, 2, 3* is *red, white, blue* (written as **rwb**). Here is the resulting information state:

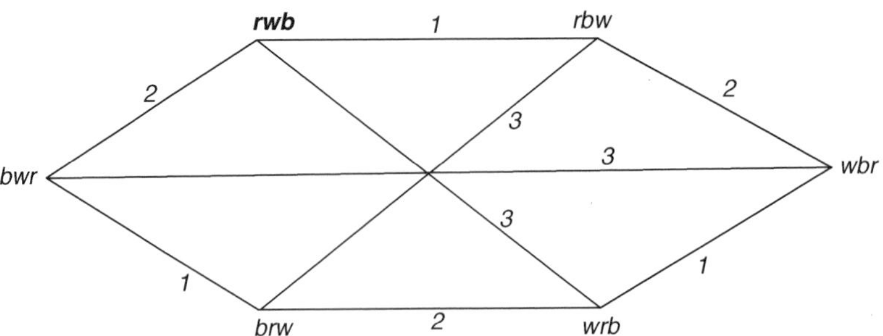

This pictures the six relevant states of the world (the hands, or distributions of the cards), with the appropriate accessibilities (equivalence relations in this case) pictured by the uncertainty lines between hands. E.g., the single *1*-line between *rwb* and *rbw* indicates that player *1* cannot distinguish these situations as candidates for the real world, while *2* and *3* can (they have different cards in them). Thus, the diagram says the following. Though they are in ***rwb*** (as an outside observer can see), no player knows this. Of course, the game itself is a dynamic process yielding still further information, as we will see in Chapter 3. ■

Over epistemic models, that may often be pictured concretely as information diagrams of the preceding sort, we now interpret the epistemic language:

DEFINITION Truth conditions.

$\mathbf{M}, s \models p$	iff	*V makes p true at s*
$\mathbf{M}, s \models \neg\varphi$	iff	*not* $\mathbf{M}, s \models \varphi$
$\mathbf{M}, s \models \varphi \wedge \psi$	iff	$\mathbf{M}, s \models \varphi$ *and* $\mathbf{M}, s \models \psi$
$\mathbf{M}, s \models K_i\varphi$	iff	*for all t with* $s \rightarrow_i t$: $\mathbf{M}, t \models \varphi$
$\mathbf{M}, s \models C_G\varphi$	iff	*for all t that are reachable from s by some finite sequence of* $\rightarrow_i$ *steps* ($i \in G$): $\mathbf{M}, t \models \varphi$[5] ■

Example A model for a question/answer scenario.
Here is how a question/answer episode might start (this is just one of many possible initial situations!). In the following diagram, reflexive arrows are presupposed, but not drawn (reflexivity represents the usual assumption that knowledge is truthful). Intuitively, agent *Q* does not know whether *p*, but *A* is fully informed about it:

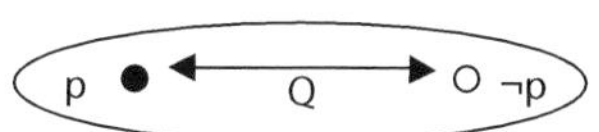

In the black world to the left, the following formulas are true:

$$p,\ K_A p,\ \neg K_Q p \wedge \neg K_Q \neg p,\ K_Q(K_A p \vee K_A \neg p),$$

$$C_{\{Q,A\}}(\neg K_Q p \wedge \neg K_Q \neg p),\ C_{\{Q,A\}}(K_A p \vee K_A \neg p)$$

This is an excellent situation for *Q* to ask *A* whether *p* is the case: he even knows that she knows the answer. Once the answer 'Yes' has been given,

[5] Thus, we quantify over all sequences like $\rightarrow_1 \rightarrow_2 \rightarrow_3 \rightarrow_1$, etc. For the cognoscenti, in this way, common knowledge acts as a *dynamic logic* modality $[(\cup_{i \in G} \rightarrow_i)^*]\varphi$.

intuitively, this model changes to the following one-point model where maximal information has been achieved:

Now, of course $C_{\{Q,\ A\}}p$ holds at the black world. ■

Over our models, an epistemic language sharpens distinctions. For instance, that everyone in a group knows φ is not yet common knowledge, but so-called *universal knowledge* $E_G\varphi$, being the conjunction of all formulas $K_i\varphi$ for all $i \in G$. Now let us broaden things still further.

The above scenarios can also be seen as getting information from any source: e.g., Nature in the case of *observation*. Observation or general learning are congenial to epistemic logic. We just emphasized the conversational setting because it is lively and intuitive.

From semantics to language design There is something conservative to 'giving a semantics': a language is given, and we fit it to some model class. But epistemic models are appealing in their own right as geometrical information structures. And then, there is an issue of what language best describes these structures, perhaps changing the given one. To illustrate this design issue, we return to the *social character* of group information.

Example From implicit to explicit group knowledge.
Consider a setting where both agents have information that the other lacks, say as follows:

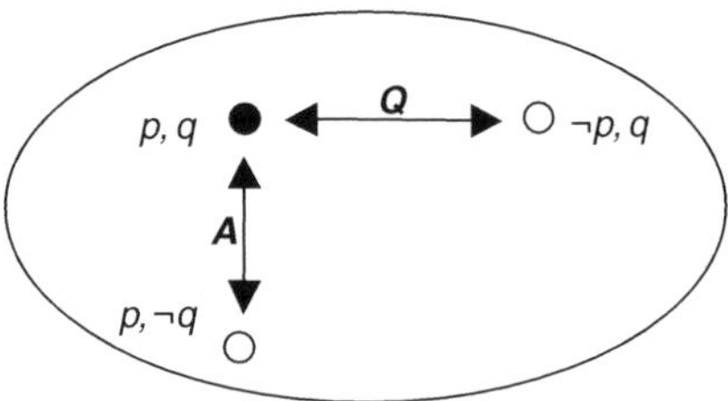

Here, the black dot on the upper left is the actual world. The most cooperative scenario here is for Q to tell A that q is the case (after all, this is something he knows), while A can tell him that p is the case. Intuitively, this reduces the initial three-point model to the one-point model where $p \wedge q$ is common knowledge. Other three-world examples model other interesting, sometimes surprising conversational settings (cf. van Benthem 2006b). ■

Another way of saying what happens here is that, when Q, *A* maximally inform each other, they will cut things down to the *intersection* of their individual accessibility relations. This suggests a new natural notion for groups, beyond the earlier common knowledge:

DEFINITION Distributed knowledge.
Intuitively, a formula φ is *implicit* or *distributed knowledge* in a group, written $D_G\varphi$, when agents could come to see it by pooling their information. More technically, extending our language and the above truth definition, this involves intersection of accessibility relations:

$$\mathbf{M}, s \models D_G\varphi \quad \text{iff} \quad \text{for all } t \text{ with } s\cap_{i\in G} \to_i t\text{: } \mathbf{M}, t \models \varphi$$ [6] ■

Intuitively, groups of agents can turn their implicit knowledge into common knowledge (modulo some technicalities) by communication. We will pursue this phenomenon in Chapters 3, 12.

As we proceed, we will use other extensions of the basic language as needed, including a *universal modality* over all worlds (accessible or not) and *nominals* defining single worlds.

Digression: epistemic models reformulated Though our logic works over arbitrary models, in practice, equivalence relations are common. Our card examples were naturally described as follows: each agent has local states it can be in (say, the card deals it may receive), and worlds are global states consisting of vectors *X*, *Y* with each agent in some local state. Then the natural accessibility relation is component-wise equality:

$$X \to_i Y \;\; \textit{iff} \;\; (X)_i = (Y)_i$$

Another case are the models for games in Chapters 10, 14, where worlds are vectors of strategies for players. This is a semantic 'normal form' (Fagin *et al.* 1995, van Benthem 1996):

FACT Each epistemic model with equivalence relations for its accessibilities is isomorphic to a submodel of a vector model.

6 Other definitions of implicit knowledge exist, preserving our later bisimulation invariance. Cf. Roelofsen (2006), van der Hoek, van Linder & Meijer (1999).

Proof Give each agent as its states the equivalence classes of its accessibility relation. This yields an isomorphism since $\rightarrow_i$-equivalence classes of worlds s, t are equal iff $s \rightarrow_i t$. ■

2.3 Validity and axiomatic systems

Validity of formulas φ in epistemic logic is defined as usual in semantic terms, as truth of φ in all models at all worlds. Consequence may then be defined through validity of conditionals.

Minimal logic The following completeness result says that the validities over arbitrary models may be described purely syntactically by the following calculus of deduction:[7]

THEOREM The valid formulas are precisely the theorems of the *minimal epistemic logic* axiomatized by (a) all valid principles of propositional logic, (b) the definition $\Diamond\varphi \leftrightarrow \neg K\neg\varphi$, (c) modal distribution $K(\varphi \rightarrow \psi) \rightarrow (K\varphi \rightarrow K\psi)$, together with the inference rules of (d) Modus Ponens ('from $\varphi \rightarrow \psi$ and φ, infer ψ') and (e) Necessitation ('if φ is a theorem, then so is $K\varphi$').

Proof The proof for this basic result can be found in any good textbook. Modern versions employ Henkin-style constructions with maximally consistent sets over some finite set of formulas only, producing a finite counter-model for a given non-derivable formula. ■

One axiom of this simple calculus has sparked continuing debate, viz. the epistemic distribution law

$$K(\varphi \rightarrow \psi) \rightarrow (K\varphi \rightarrow K\psi)$$

This seems to say that agents' knowledge is closed under logical inferences, and this 'omniscience' seems unrealistic. At stake here is our earlier distinction between semantic and inferential information. Semantically, with processes of observation, an agent who has the hard information that $\varphi \rightarrow \psi$ and that φ also has the hard information that ψ. But in a more finely grained perspective of syntactic inferential information, geared toward processes of elucidation, Distribution need not hold. We will not join the fray here,

[7] We will often drop agent subscripts for K-operators when they play no essential role.

but Chapter 5 shows one way of introducing inferential information into basic epistemic logic and Chapter 13 contains more philosophical discussion.

Internal versus external Another point to note is this. Deductive systems may be used in two modes. They can describe agents' own reasoning inside scenarios, or outside reasoning by theorists about them. In some settings, the difference will not matter – but sometimes, it may. In this book, we will not distinguish 'first person' and 'third person' perspectives on epistemic logic, as our systems accommodate both.[8]

Stronger epistemic logics and frame correspondence On top of this minimal deductive system, two further steps go hand in hand: helping ourselves to stronger axioms endowing agents with further features, and imposing further structural conditions on accessibility in our models. For instance, here are three more axioms with vivid epistemic interpretations:

$K\varphi \rightarrow \varphi$	*Veridicality*
$K\varphi \rightarrow KK\varphi$	*Positive Introspection*
$\neg K\varphi \rightarrow K\neg K\varphi$	*Negative Introspection*

The former seems uncontroversial (knowledge is in synch with reality), but the latter two have been much discussed, since they assume that, in addition to their logical omniscience, agents now also have capacities of unlimited introspection into their own epistemic states. Formally, these axioms correspond to the following structural conditions on accessibility:

$K\varphi \rightarrow \varphi$	*reflexivity*	$\forall x\colon x \rightarrow x$
$K\varphi \rightarrow KK\varphi$	*transitivity*	$\forall xyz\colon (x \rightarrow y \wedge y \rightarrow z) \Rightarrow x \rightarrow z$
$\neg K\varphi \rightarrow K\neg K\varphi$	*euclidity*	$\forall xyz\colon (x \rightarrow y \wedge x \rightarrow z) \Rightarrow y \rightarrow z$

The term correspondence can be made precise using *frame truth* of modal formulas under all valuations on a model (cf. van Benthem 1984). Powerful results exist matching up axioms with relational conditions: first-order like here, or in higher-order languages. We will occasionally refer to such techniques, but refer to the literature for details.

The complete deductive system with all the above axioms is called *S5*, or *multi-S5* when we have more than one agent. Taking the preceding conditions

[8] Aucher (2008) is a complete reworking of dynamic-epistemic logic in the internal mode.

together, it is the logic of equivalence relations over possible worlds. What may seem surprising is that this logic has no interaction axioms relating different modalities K_i, K_j. But none are plausible:

Example Your knowledge and mine do not commute.
The following model provides a counter-example to the putative implication $K_1\, K_2\, p \rightarrow K_2\, K_1\, p$. Its antecedent is true in the black world to the left, but its consequent is false:

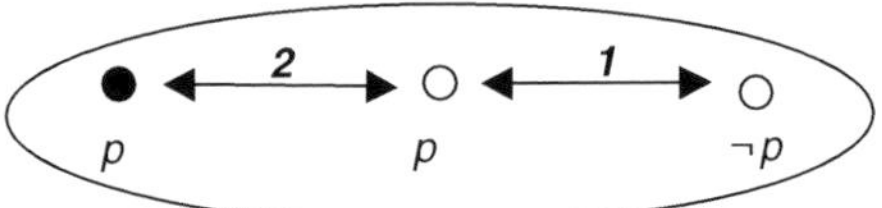

Such implications only hold in a scenario when agents have special informational relationships.[9] ■

In between the minimal logic and S5, many other logics live, such as *KD45* for belief. In this book, we use examples from the extremes, though our results apply more generally.

Group logic We conclude with a typical operator for groups of agents:

THEOREM The complete epistemic logic with common knowledge is axiomatized by adding the following two principles to the minimal epistemic logic, where E_G is the earlier modality for 'everybody in the group knows':

$C_G\varphi \leftrightarrow (\varphi \wedge E_G\, C_G\, \varphi)$	*Fixed-Point Axiom*
$(\varphi \wedge C_G\, (\varphi \rightarrow E_G\varphi)) \rightarrow C_G\varphi$	*Induction Axiom*

These laws are of wider interest. The Fixed-Point Axiom expresses reflexive equilibrium: common knowledge of φ is a proposition p implying φ while every group member knows that p is true. The Induction Axiom says that common knowledge is not just any such proposition, but the largest: technically, a 'greatest fixed-point' (cf. Chapters 3, 4, 10).

[9] In the combined dynamic-epistemic logics of Chapter 3 and subsequent ones, some operator commutation principles will hold, but then between epistemic modalities and action modalities.

This completes our tour of the basics of epistemic logic. Next, we survey some technical themes that will play a role later in this book. Our treatment will be light, and we refer to the literature for details (Blackburn, de Rijke & Venema 2000; van Benthem, to appearB).

2.4 Bisimulation invariance and expressive power

Given the importance of languages in this book, we first elaborate on expressive power and invariance. Expressive strength of a language is often measured by its power of telling models apart by definable properties, or by invariance relations measuring what it cannot distinguish. One basic invariance is *isomorphism*: structure-preserving bijection between models. First-order formulas $\varphi(\boldsymbol{a})$ cannot distinguish between a tuple of objects $\boldsymbol{a}$ in one model $\boldsymbol{M}$, and its image $f(\mathbf{a})$ in another model $\boldsymbol{N}$ linked to $\boldsymbol{M}$ by an isomorphism f.[10] This style of analysis also applies to epistemic logic. But first we make a technical connection:

First-order translation Viewed as a description of epistemic models, our language is weaker than a first-order logic whose variables range over worlds:

FACT There exists an effective translation from the basic epistemic language into first-order logic yielding equivalent formulas.

Proof This is the well-known Standard Translation. For instance, the epistemic formula $p \wedge K\Diamond q$ goes to an equivalent $Px \wedge \forall y(Rxy \rightarrow \exists z(Ryz \wedge Qz))$, with R a binary predicate symbol for the accessibility relation, and P, Q unary predicates for proposition letters. The first-order formula is a straightforward transcription of the truth conditions for the epistemic one. ■

These modal translations have only special bounded or *guarded* quantifiers over accessible worlds, making them a special subclass of first-order logic (van Benthem 2005B). This shows in powers of distinction. The full first-order language can distinguish a one-point reflexive cycle from an irreflexive

[10] Another candidate is *potential isomorphism*: van Benthem (2002b) has some extensive discussion. Again we refer to the cited literature for more details in what follows.

2-cycle (two non-isomorphic models) – but it should be intuitively clear that these models verify the same epistemic formulas everywhere.

Bisimulation and information equivalence Here is a notion of semantic invariance that fits the modal language like a Dior gown:

DEFINITION A *bisimulation* between two models **M**, **N** is a binary relation $\equiv$ between their states s, t such that, whenever $x \equiv y$, then (a) x, y satisfy the same proposition letters, (b1) if $x \, R \, z$, then there exists a world u with $y \, R \, u$ and $z \equiv u$, and (b2) the same 'zigzag' or 'back-and-forth clause' holds in the opposite direction. The following diagram shows this:

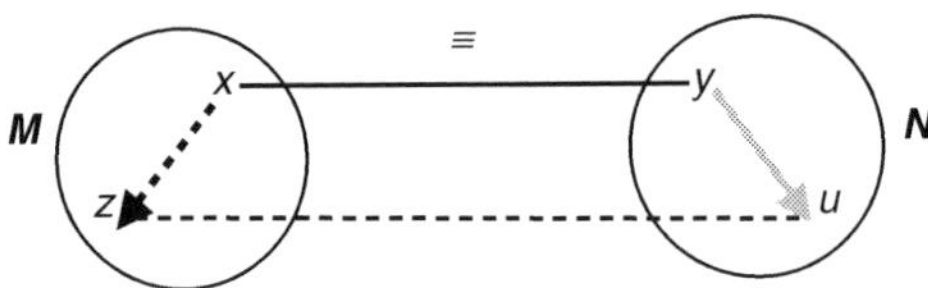

■

Clause (1) expresses 'local harmony', the zigzag clauses (2) the dynamics of simulation. This is often given a procedural spin: bisimulation identifies processes that run through similar states with similar local choices. Thus, it answers a fundamental question about computation: 'When are two processes the same?'[11] Likewise, we can ask:

When are two information models the same?

Example Bisimulation-invariant information models.
Bisimulation occurs naturally in epistemic updates changing a current model. Suppose that the initial model is like this, with the actual world indicated by the black dot:

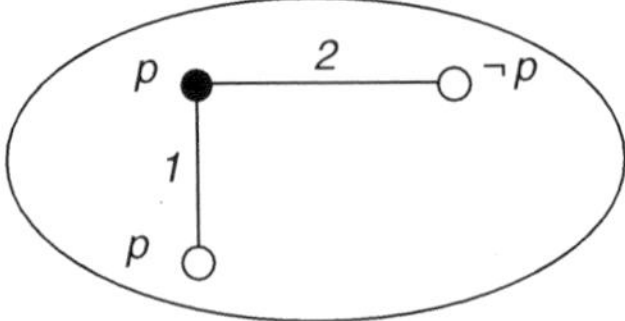

[11] There are different natural notions of identity between processes (cf. van Benthem 1996) or games (cf. van Benthem 2002a and Chapter 10). This diversity in answers is sometimes called Clinton's Principle: *It all depends on what you mean by 'is'*.

All three worlds satisfy different epistemic formulas. Now, despite her uncertainty, in the actual world, agent *1* does know that *p*, and can say this – updating to the model

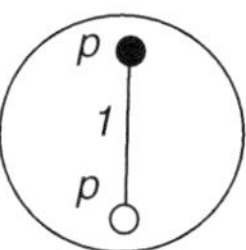

But here the two worlds are intuitively redundant, and indeed this information state for the two agents has an obvious bisimulation to just the one-point model

■

Each model has a smallest 'bisimulation contraction' satisfying the same modal formulas. But bisimulation can also make models larger, turning them into *trees* through the method of 'unravelling', providing a nice geometrical normal form (used in Chapter 11):

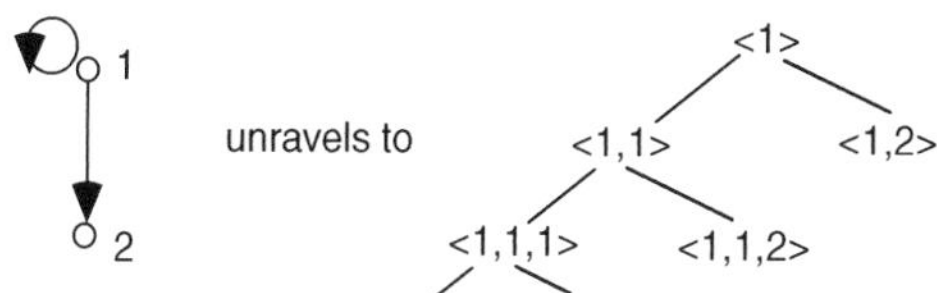

Thus, informational equivalence of models can work both ways, contracting or expanding, depending on what we find useful.

Invariance and definability Some model-theoretic results tie bisimulation closely to truth of modal formulas. Just for convenience, we restrict attention to *finite models* – but this is easily generalized in the theory of modal logic. We formulate results for general relations:

INVARIANCE LEMMA The following two assertions are equivalent:

(a) **M**, *s* and **N**, *t* are connected by a bisimulation,

(b) **M**, *s* and **N**, *t* satisfy the same epistemic formulas.

Proof The direction from (a) to (b) is by induction on epistemic formulas. In the opposite direction, for finite models, we argue as follows. Let *xZy* be the relation which holds if the worlds *x* in **M** and *y* in **N** satisfy the same epistemic formulas. For a start, then, $s\,Z\,t$. Clearly also, Z-related worlds satisfy the same proposition letters. Now, assume that $x\,Z\,y$, and let $x \rightarrow x'$. Suppose for contradiction that there is no world y' with $y \rightarrow y'$ satisfying the same modal formulas. Then there is a modal formula $\alpha_{y'}$ true in x' and false in y'. Taking the conjunction α of all these formulas $\alpha_{y'}$ for all successors y' of y, we see that α is true in x', and so $\Diamond\alpha$ is true in world x, whence $\Diamond\alpha$ is also true in y, as $x\,Z\,y$. But then there must be a successor y^* of y satisfying α: and this contradicts the construction of α.[12] ■

Semantic versus syntactic information states The epistemic language and bisimulation-invariant structure are essentially two sides of one coin. This is relevant to our view of information states. Looking explicitly, the *epistemic theory* of a world s in a model $\boldsymbol{M}$ is the set of all formulas that are true internally at s about the facts, agents' knowledge of these, and their knowledge of what others know. By contrast, the models $\boldsymbol{M}, s$ themselves locate the same information implicitly in the local valuation of a world plus its pattern of interaction with other worlds.[13] The next result says that these two views of information are equivalent:

State Definition Lemma For each model $\boldsymbol{M}, s$ there is an epistemic formula β (involving common knowledge) such that the following are equivalent:

(a) $\boldsymbol{N}, t \models \beta$,

(b) $\boldsymbol{N}, t$ has a bisimulation $\equiv$ with $\boldsymbol{M}, s$ such that $s \equiv t$.

Proof This result is from Barwise & Moss (1996) (our version follows van Benthem 1997). We sketch the proof for equivalence relations $\sim_a$ with existential modalities $<a>$, but it works for arbitrary relations. First, note that any finite multi-S5 model $\boldsymbol{M}, s$ falls into maximal zones of worlds that satisfy the same epistemic formulas in our language.

Claim 1 There is a finite set of formulas φ_i $(1 \leq i \leq k)$ defining a partition of the model such that any two worlds satisfying the same φ_i agree on all epistemic formulas.

[12] The lemma even holds for arbitrary epistemic models, provided we take epistemic formulas from a language with arbitrary *infinite* conjunctions and disjunctions.

[13] Compare the way in which category theorists describe a mathematical structure externally by its *connections to other objects* in a category through the available morphisms.

To see this, take any world s, and take difference formulas $\delta^{s,\,t}$ between it and any t not satisfying the same epistemic formulas: say, s satisfies $\delta^{s,\,t}$ and t does not. The conjunction of all $\delta^{s,\,t}$ is a formula φ_i true only in s and all worlds sharing its epistemic theory. We may assume φ_i also lists all information about proposition letters true and false throughout its zone. We also make a quick observation about uncertainty links between these zones:

If any world satisfying φ_i is $\sim_a$-linked to a world satisfying φ_j, then all worlds satisfying φ_i also satisfy $<a>\varphi_j$.

Next take the following description $\beta_{\mathbf{M},\,s}$ of **M**, s:

(a) all proposition letters and their negations true at s, plus the unique φ_i true at **M**, s,
(b) common knowledge of (b1) the disjunction of all the zone formulas φ_i, (b2) all negations of conjunctions $\varphi_i \wedge \varphi_j$ $(i \neq j)$, (b3) all true implications $\varphi_i \rightarrow <a>\varphi_j$ for which situation # occurs, and (b4) all true implications $\varphi_i \rightarrow [a]\vee\varphi_j$, with a disjunction $\vee$ over all cases enumerated in (b3).

Claim 2 **M**, $s \models \beta_{\mathbf{M},\,s}$

Claim 3 If **N**, $t \models \beta_{\mathbf{M},\,s}$, then there is a bisimulation between **N**, t and **M**, s.

To prove Claim 3, let **N**, t be any model for $\beta_{\mathbf{M},\,s}$. The φ_i partition **N** into disjoint zones Z_i of worlds satisfying these formulas. Now relate all worlds in such a zone to all worlds that satisfy φ_i in the model **M**. In particular, t gets connected to s. We check that this gives a bisimulation. The atomic clause is clear by construction. But also, the zigzag clauses follow from the given description: (a) Any $\sim_a$-successor step in **M** has been encoded in a formula $\varphi_i \rightarrow <a>\varphi_j$ that holds everywhere in **N**, producing the required successor there. (b) Conversely, if there is no $\sim_a$-successor in **M**, this shows up in the limitative formula $\varphi_i \rightarrow [a]\vee\varphi_j$, which also holds in the model **N**, so that there is no excess successor there either. ■

The Invariance Lemma says that bisimulation has a good fit with the modal language. The State Definition Lemma strengthens this to say that each semantic state is captured by one formula. Again this extends to arbitrary models with an infinitary epistemic language.

Example Defining a model up to bisimulation.
Consider the two-world model for our earlier basic question/answer episode.

Here is an epistemic formula that defines its φ-state up to bisimulation:

$$\varphi \wedge C_{\{Q,A\}}((K_A\varphi \vee K_A\neg\varphi) \wedge \neg K_Q\varphi \wedge \neg K_Q\neg\varphi)$$ ■

Thus we can always switch between syntactic and semantic information states. The latter view will dominate this book, but the reader may want to keep this duality in mind.

2.5 Computation and the complexity profile of a logic

While derivability and definability are the main pillars of logic, issues of *task complexity* form a natural complement. Given that information has to be recognized or extracted to be of use to us, it is natural to ask how complex such extraction processes really are.

Decidability In traditional modal logic, the interest in complexity has gone no further than just asking the following question. Validity in first-order logic is *undecidable*, validity in propositional logic is *decidable*: what about modal logic, which sits in between?

THEOREM Validity in the minimal modal logic is decidable. So is validity for multi-*S5*.

There are many proofs, exploiting special features of modal languages, especially their bounded local quantifiers. One method uses the *effective finite model property*: each satisfiable modal formula φ has a finite model whose size can be computed effectively from the length of φ.[14] Validity is decidable for many further modal and epistemic logics.

But things can change rapidly when we consider logics with combinations of modalities that satisfy what look like natural commutation properties:

THEOREM The minimal logic of two modalities *[1]*, *[2]* satisfying the axiom $[1][2]\varphi \rightarrow [2][1]\varphi$, plus a universal modality U over all worlds is undecidable.

The technical reason is that such logics encode undecidable *tiling problems* on the structure $IN \times IN$ (Harel 1985, Marx 2006). By frame correspondence, the commutation axiom defines a *grid structure* satisfying the following first-order convergence property:[15]

[14] This property typically fails for the full language of first-order logic.

[15] The universal modality is needed in the encoding: cf. van Benthem (to appearB).

$\forall xyz$: $(xR_1y \wedge yR_2z) \rightarrow \exists u$: $(xR_2u \wedge uR_1z)$

Here is a diagram picturing this, creating a cell of the grid:

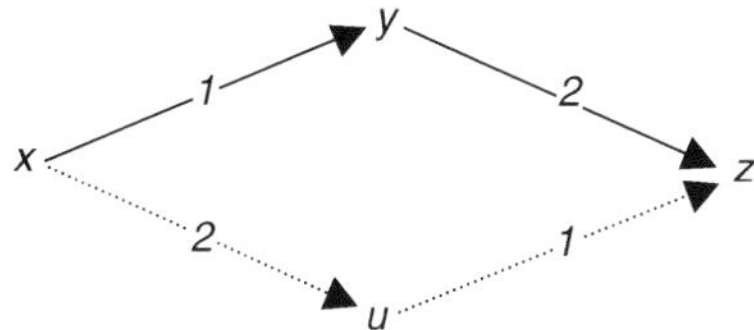

This complexity danger is general. Modal logics of trees are harmless, but modal logics of grids are dangerous!

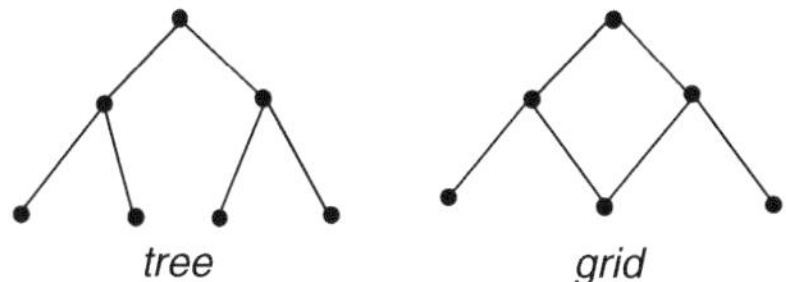

Modal logics with such commutation axioms occur with agency (Chapter 10, 11), and undecidability is around the corner. But in this chapter, we first look at the fine-structure of decidable tasks.

Computational complexity Complexity theory studies computation time or memory space needed for tasks as a function of input size. Inside the decidable problems, it distinguishes feasible polynomial rates of growth (linear, quadratic, ...): forming the time complexity class ***P***, from non-feasible non-deterministic polynomial time ***NP***, polynomial space (***Pspace***), exponential time (***Exptime***), and ever higher up. For details, see Papadimitriou (1994).

Complexity profile of a logic To really understand how a logical system works, it helps to check the complexity for some basic tasks that it is used for. Determining validity, or equivalently, *testing for satisfiability* of given formulas, is one of these:

Given a formula φ, determine whether φ has a model.

But there are other, equally important tasks, such as *model checking*:

Given a formula φ and a finite model *(M, s)*, check whether **M**, $s \models \varphi$.

Here is a third key task, that of *testing for model equivalence*:

Given two finite models *(M, s)*, *(N, t)*, check if they satisfy the same formulas.

Here is a table of the complexity profiles for two well-known logics. Entries mean that the problems are in the class indicated, and no lower or higher:

	Model Checking	Satisfiability	Model Comparison
Propositional logic	*linear time (P)*	*NP*	*linear time (P)*
First-order logic	***Pspace***	*undecidable*	*NP*

Where does the basic modal language fit? Model checking is efficient. While first-order evaluation has exponential growth via quantifier nesting, there are fast modal algorithms. Next, close-reading decidability arguments helps us locate satisfiability. Finally, testing for modal equivalence, or for the existence of a bisimulation, turns out efficient, too:

FACT The complexity profile for the minimal modal logic is as follows:

Model Checking	Satisfiability	Model Comparison
P	***Pspace***	*P*

For epistemic logic, results are similar. But *S5* has a difference. Single-agent satisfiability is in ***NP***, as *S5* allows for a normal form without iterated modalities. But with two agents, satisfiability jumps back to ***Pspace***: social life is more complicated than being alone.

Complexity results are affected by the expressive power of a language. When we add our common knowledge modality $C_G\varphi$ to epistemic logic, the above profile changes as follows:

Model Checking	Satisfiability	Model Comparison
P	***Exptime***	*P*

The Balance: expressive power versus computational effort Logic has a Golden Rule: what you gain in one desirable dimension, you lose in another. Expressive strength means high complexity. Thus, it is all about striking a balance. First-order logic is weaker than second-order logic in defining mathematical notions. But its poverty has a reward, viz. the axiomatizability of valid consequence, and the emergence of useful model-theoretic properties such as the Compactness Theorem. Many modal logics are good compromises lower down this road. They become decidable, or they have more perspicuous proof systems.

2.6 Games for logical tasks

Computation is not just a routine chore: procedures are a fundamental theme in their own right. This comes out well with *game versions* of logical tasks (van Benthem 2007d):

Evaluation games We cast the process of evaluating modal formula φ in model $(\mathbf{M}, s)$ as a *two-person game* between a Verifier, claiming that φ is true, and a Falsifier, claiming that φ is false. The game starts at some world s. Each move is dictated by the main operator of the formula at hand (the total length of the game is bounded by its modal depth):

DEFINITION The modal evaluation game *game* $(\mathbf{M}, s, \varphi)$.
The following rules of the games endow the logical operators with an interactive dynamic meaning:

atom p	test p at s in $\mathbf{M}$: if true, then V wins – otherwise, F wins
disjunction	V *chooses* a disjunct, and play continues with that
conjunction	F chooses a conjunct, and play continues with that
$\Diamond\varphi$	V picks an R-successor t of the current world; play continues with φ at t
$\Box\varphi$	F picks an R-successor t of the current world; play continues with φ at t.

A player also loses when (s)he must pick a successor, but cannot do so. ■

Example A complete game tree.
We give an illustration in general modal logic over relational models. Here is the complete game tree for the modal formula $\Box(\Diamond p \vee \Box\Diamond p)$ played starting at state *1* in the following model:

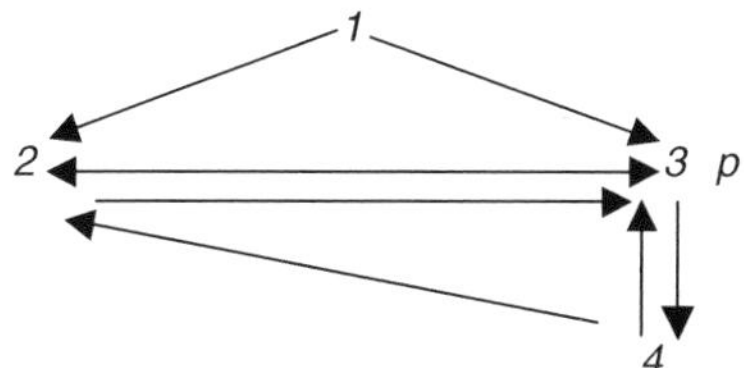

We draw game nodes, plus the player which is to move, plus the relevant formula. Bottom leaves have a marker 'win' if Verifier wins (the atom is true there), 'lose' otherwise:

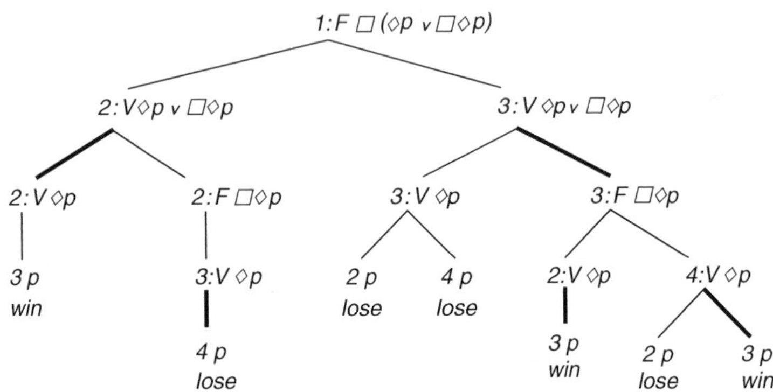

Each player has three winning runs, but the advantage is for Verifier, who can always play to win, whatever Falsifier does. Her *winning strategy* is marked by the bolder lines. ■

A strategy can encode subtle behaviour: Verifier must hand the initiative to Falsifier at state 3 on the right if she is to win. Some further background is found in Chapter 10. The reason why Verifier has a winning strategy is that she is defending a true statement:

FACT For any evaluation game, the following assertions are equivalent:
(a) modal formula φ is true in model **M** at world *s*,
(b) player *V* has a winning strategy in *game* (**M**, *s*, φ).

Proof The proof is a straightforward induction on the formula φ. This is a useful exercise to get a feel for the workings of strategies, and the game dynamics of the modalities. ■

Model comparison games The next game provides fine-structure for bisimulation between models (**M**, *s*), (**N**, *t*). It involves a Duplicator (claiming that there is an analogy) and a Spoiler (claiming a difference), playing over pairs of worlds (*x*, *y*), one from each model.

DEFINITION Modal comparison games.
In each round of the game, Spoiler chooses a model, and a world *u* that is a successor of *x* or *y*, and Duplicator responds with a corresponding successor *v*

in the other model. Spoiler wins if *u, v* differ in their atomic properties, or Duplicator cannot find a successor in 'her' model. The game continues over some finite number of rounds, or infinitely. ■

There is a tight connection between these games and modal properties of the two models:

FACT (a) Spoiler's winning strategies in a *k*-round game between **M**, *s* and **N**, *t* match exactly the modal formulas of *operator depth k* on which the worlds *s, t* disagree.

(b) Duplicator's winning strategies in the *infinite* round game between **M**, *s* and **N**, *t* match the bisimulations between **M**, **N** linking *s* to *t*.

It is not hard to prove this result, but we merely give some illustrations.

Example 'Choosing now or later'.
Consider the game between the following models starting at their roots:

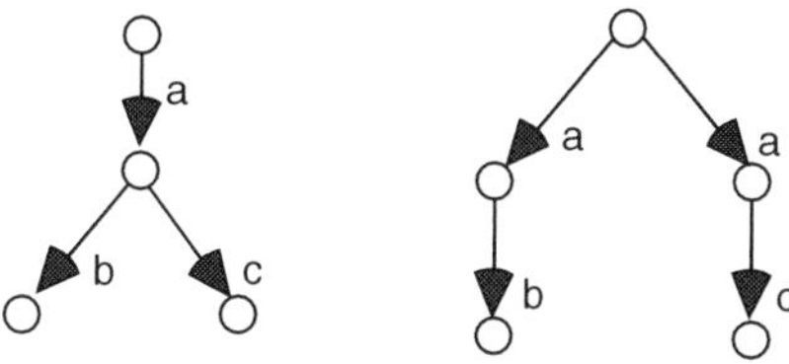

Spoiler can win the comparison game in two rounds, with several strategies. One stays inside the same model, exploiting the modal difference formula *<a>(<b>T ∧ <c>T)* of depth *2*.[16] Another winning strategy for Spoiler switches models, using the modal formula *[a]<b>T* that contains only two operators in all. The switch is signalled by the change in modalities. ■

This concludes our sketch of games that show logic *itself* in a dynamic interactive light.[17] We discuss this issue of logics as dynamic procedures at greater length in van Benthem (to appearA). For here, note also that the above game trees are themselves modal models. This suggests applications of modal logic to game theory, as we will see in Chapter 10.

[16] Here '*T*' stands for the *always true* formula.

[17] Other logic games do tasks like model construction or proof (van Benthem 2007d).

2.7 Conclusion

We have presented epistemic logic in a new manner. First, we shook off some dust, and emphasized the themes of hard semantic information and modelling multi-agent scenarios. In doing so, we dropped the claim that the *K*-operator stands for knowledge in any philosophical sense, making it a statement about current semantic information instead. This is a radical shift, and the issue of what real knowledge is now shifts to the analysis of further aspects of that notion such as explicit awareness (Chapter 5), belief (Chapter 7), and dynamic stability under new information (cf. the epistemological passages in Chapter 13).

Next we presented the usual basics of the formal system, plus some newer special topics. Together, these illustrated three key themes: (a) expressive power of definition, (b) inferential power of deduction, and (c) computational task performance, leading to a suggestive interactive embodiment of the logic itself in dynamic games.

2.8 Further directions and open problems

There are many further issues about knowledge, information, and epistemic logic. A few have been mentioned, such as connections with dynamic logics, or the contrast between first and third person perspectives. We mention a few more, with pointers to the literature.

Topological models Relational possible worlds models are a special case of a more general and older semantics for modal logic. Let **M** be a topological space $(X, \mathbb{O}, V)$ with points X, a topology of open sets $\mathbb{O}$, and a valuation V. This supports a modal logic:

Definition Topological models.
φ is true at point s in **M**, written **M**, $s \models \Box\varphi$, if s is in the *topological interior* of the set $[[\varphi]]^{\mathbf{M}}$: the points satisfying φ in **M**. Formally, this can be written as follows: $\exists O \in \mathbb{O}$: $s \in \mathbb{O}$ & $\forall t \in O$: $\mathbf{M}, t \models \varphi$. ■

Typical topologies are metric spaces like the real numbers, or trees with sets closed in the tree order as opens (essentially, our relational models). Modal axioms state properties of interior: $\Box\varphi \to \varphi$ is inclusion, $\Box\Box\varphi \leftrightarrow \Box\varphi$ idempotence, and $\Box(\varphi \wedge \psi) \leftrightarrow \Box\varphi \wedge \Box\psi$ closure of opens under intersections. But infinitary distribution, relationally valid, fails:

Example $\Box \wedge_{i \in I} p_i \leftrightarrow \wedge_{i \in I} \Box p_i$ fails on metric spaces.
Interpret p_i as the open interval $(-1/i, +1/i)$, for all $i \in N$. The $\Box p_i$ denote the same interval, and the intersection of all these intervals is $\{0\}$. But the expression $\Box \wedge_{i \in I} p_i$ denotes the topological interior of the singleton set $\{0\}$, which is the empty set $\emptyset$. ■

All earlier techniques generalize, including a topological bisimulation related to continuous maps, plus a vivid matching game. The handbook Aiello, Pratt & van Benthem (2007) has chapters on the resulting theory. Van Benthem & Sarenac (2005) use these models for epistemic logic, with multi-agent families of topologies closed under operations of group formation. They exploit the failure of infinitary distribution to separate iterative and fixed-point views of common knowledge (cf. Barwise 1985):

THEOREM On topological models, common knowledge defined through countable iteration is strictly weaker than common knowledge defined as a greatest fixed-point for the above equation $\mathbf{p} \leftrightarrow \varphi \wedge E_G\, \mathbf{p}$.

In fact, the standard product topology on topological spaces models a form of group knowledge stronger than both, as 'having a shared situation'. Given the epistemic interest in such richer models, there is an open problem of extending the logics of this book from relational to topological models.

Neighbourhood models A further generalization of dynamic epistemic logic extends the topological setting to abstract *neighbourhood models* where worlds have a family of subsets as their 'neighbourhoods'. Then a box modality $\Box\varphi$ is true at a world w if at least one of w's neighbourhoods is contained in $[[\varphi]]$ (cf. van Benthem, to appearB; Hansen, Kupke & Pacuit 2008). The resulting modal base logic has $\Box$ upward monotone, but not distributive over $\wedge$ or $\vee$. The epistemic update rules of Chapters 3, 4 extend to neighbourhood models, as long as the set of relevant epistemic events is finite (a generalization of the standard product topology works; cf. the seminar paper Leal 2006). Zvesper (2010) has further developments, motivated by an analysis of some key results in game theory (see Chapter 10). But an elegant general approach remains a challenge. Intuitively, neighbourhood models record the underlying information or 'evidence' that combines to create an over-all epistemic range.[18]

[18] Van Benthem & Pacuit (2011) explores this richer framework in detail.

Techniques for such more fine-grained epistemic representations will return at various places in this book.

Tandem View: translation and correspondence This book will mostly use modal logics. But there is the option of translating these to first-order of higher-order logics. Then, our whole theory becomes embedded in *fragments of classical logical systems*. Which ones?

Internal versus external logical views of knowledge Epistemic logic builds on classical propositional logic, adding explicit knowledge operators. By contrast, *intuitionistic logic* treats knowledge by 'epistemic loading' of the interpretation of standard logical constants like negation and implication. Van Benthem (1993, 2009e) discuss the contrast between the two systems, and draw comparisons. As will be shown in Chapter 13, the dynamic content of intuitionistic logic involves observational update (Chapter 3), awareness-raising actions (Chapter 5), as well as procedural information in the sense of Chapter 11. There is a general question of an intuitionistic version for the information dynamics explored in this book.

From propositional to predicate logic We have surveyed propositional epistemic logic. But many issues of knowledge have to do with information about objects, such as knowing the location of the treasure, or knowing a method for breaking into a house. This book will not develop *quantificational counterparts*, but it is an obvious next stage throughout.

3 Dynamic logic of public observation

Having laid the groundwork of semantic information and agents' knowledge in Chapter 2, we will now study dynamic scenarios where information flows and knowledge changes by acts of observation or communication. This chapter develops the simplest logic in this realm, that of public announcement or observation of hard information that we can trust absolutely. This simple pilot system raises a surprising number of issues, and its design will be a paradigm for all that follows in the book. We start with motivating examples, then define the basic logic, explore its basic properties, and state the general methodology coming out of this. We then sum up where we stand. At this stage, the reader could skip to Chapter 4 where richer systems start. What follows are further technical themes, open problems (mathematical, conceptual, and descriptive), and a brief view of key sources.

3.1 Intuitive scenarios and information videos

Recall the Restaurant scenario of Chapter 1. The waiter with the three plates had a range of six possibilities for a start, this got reduced to two by the answer to his first question, and then to one by the answer to the second. Information flow of this kind means stepwise range reduction. This ubiquitous view can even be seen in propositional logic:

Throwing a party The following device has long been used in my courses since 1970. You want to throw a party, respecting incompatibilities. You know that (a) John comes if Mary or Ann does, (b) Ann comes if Mary does not come, (c) If Ann comes, John does not. Can you invite people under these constraints? Logical inference might work as follows:

> By (c), if Ann comes, John does not. But by (a), if Ann comes, John does. This is a contradiction, so Ann does not come. But then, by (b), Mary comes. So, by (a) once more, John must come. Indeed a party {John, Mary} satisfies all three requirements.

This shows the usual propositional rules at work, and the power of inference to get to a goal. But here is a dynamic take on the informational role of the premises that takes things more slowly. At the start, no information was present, and all eight options remained:

(1) *{MAJ, MA-J, M-AJ, M-A-J, -MAJ, -MA-J, -M-AJ, -M-A-J}* *8 items*

Now the three given premises *update* this initial information state, by removing options incompatible with them. In successive steps, (a), (b), (c) give the following reductions:

(a) ***(M or A)*** $\rightarrow$ ***J***	new state	*{MAJ, M-AJ, -MAJ, -M-AJ, -M-A-J}*	*5 items*
(b) **not-M** $\rightarrow$ *A*	new state	*{MAJ, M-AJ, -MAJ}*	*3 items*
(c) *A* $\rightarrow$ **not-J**	new state	*{M-AJ}*	*1 item*

This resembles the information flow in games like Master Mind (van Benthem 1996), where information about some arrangement of coloured pegs comes in round by round.

Card games The same simple mechanism works in multi-agent settings like *card games*, where it performs sophisticated updates involving what agents know about each other. Recall the 'Three Cards' from Chapter 2, where the following scenario was played during the *NEMO* Science lecture. Cards *red, white, blue* are dealt to players: *1, 2, 3*, one for each. Each player sees his own card only. The real distribution over *1, 2, 3* is red, white, blue (***rwb***). This was the epistemic information model:

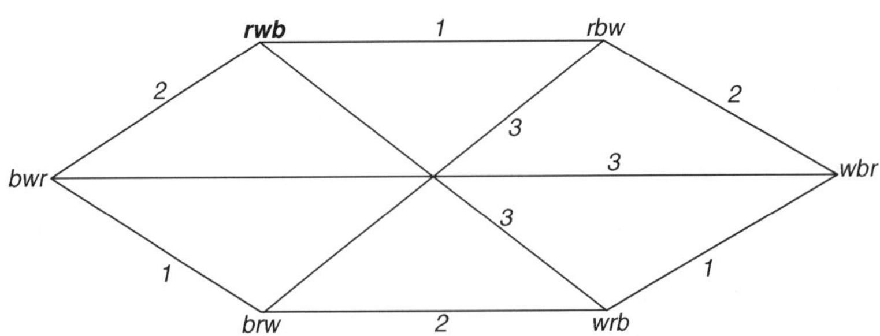

Now the following two conversational moves take place:

2 asks 1	'Do you have the blue card?'
1 answers truthfully	'No.'

Who knows what then? Here is the effect in words:

> Assuming the question is sincere, *2* indicates that she does not know the answer, and so she cannot have the blue card. This tells *1* at once what the deal was. But *3* does not learn, since he already knew that *2* does not have blue. When *1* says she does not have blue, this now tells *2* the deal, but *3* still does not know even then.

We now give the updates in the diagram, making all these considerations geometrically transparent. Here is a concrete 'update video' of the successive information states:

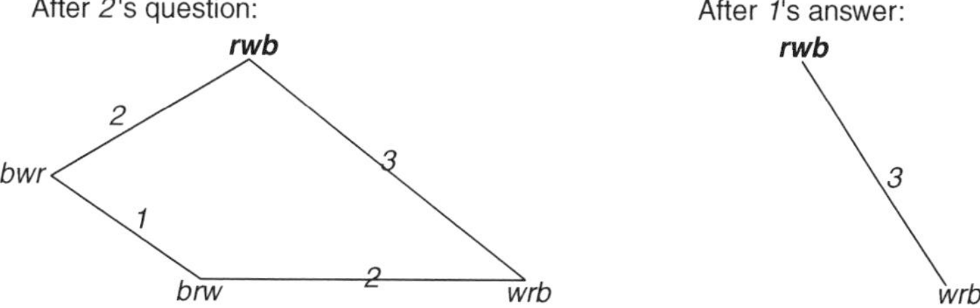

We see at once in the final diagram that players *1*, *2* know the initial deal now, as they have no uncertainty lines left. But *3* still does not know, given her remaining line, but she does know that *1*, *2* know – and in fact, the latter is common knowledge.

Similar analyses exist by now for other conversation scenarios, and for a wide variety of puzzles and games. In particular, one can also model the scenario where *2*'s question is not informative, with an alternative update video whose description we leave to the reader.

3.2 Modelling informative actions: update by hard information

Our task is now to 'dynamify' the notions of Chapter 2 to deal with these examples. We already have the statics in place, in the form of our earlier epistemic models, described by the epistemic language. Now we need an explicit account of the actions that transform these states, and an extension of the epistemic language to define these explicitly, and reason about them. But what are these basic actions of information flow? In what follows, we will mostly call them *public announcements*, as this term has become widespread. But this conversational phrasing reflects just one way of thinking about these basic acts, and equally good cases are *public observation* (the same for every agent), or learning in a more general sense. Indeed, even 'action' may be too specialized a term, and my preference is to think of informational *events*, whether or not with an actor involved. Now, such events come with different force, stronger or weaker. In this chapter, we study only the simplest phenomenon: events of *hard information* producing totally trustworthy facts.

Semantic update for hard information Here is what seems a folklore view of information flow: new information eliminates possibilities from a current range. More technically, public announcements or observations *!P* of true propositions *P* yield 'hard information' that changes the current model irrevocably, discarding worlds that fail to satisfy *P*:

DEFINITION Updating via definable submodels.
For any epistemic model **M**, world *s*, and formula *P* true at *s*, the model (**M**|*P*, *s*) (**M** *relativized to P at s*) is the submodel of **M** whose domain is the set $\{t \in \mathbf{M} \mid \mathbf{M}, t \models P\}$. ■

Drawn in a simple picture, such an update step *!P* goes

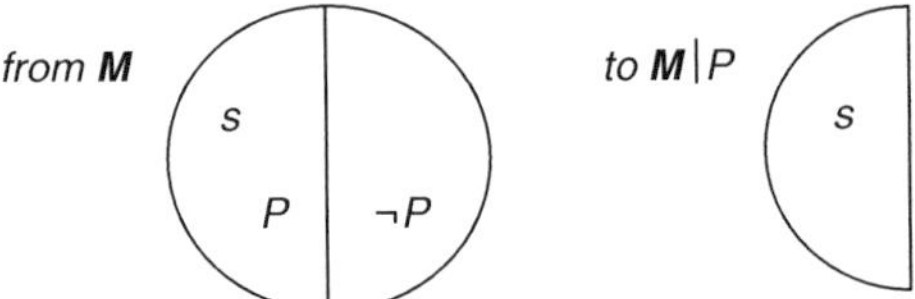

These diagrams of a jump from one model to another are useful in visualizing arguments about validity of logical principles in this setting.[1] These principles are not entirely obvious. Crucially, truth values of formulas may change in the update depicted here:[2] most notably, since agents who did not know that *P* now do after the announcement. This makes reasoning about information

[1] Note that an update action *!P* as described here may be seen as a *partial function* on epistemic models: it is only executable when **M**, $s \models P$, and in that case, it produces a unique new value.

[2] That is why we did not write '*P*' under the remaining zone in the model to the right.

flow more subtle than just a simple conditionalization. The best way of getting clear on such issues is, of course, the introduction of a logical system.

The Muddy Children This simple update mechanism explains the workings of many knowledge puzzles, one of which has become an evergreen in the area, as it packs many key topics into one simple story. It occurred in Chapter 1, but we repeat it here. A complete discussion of all its features must wait until our subsequent chapters.

Example Muddy Children (Fagin *et al.* 1995; Geanakoplos 1992).
After playing outside, two of three children have mud on their foreheads. They can only see the others, so they do not know their own status. (This is an inverse of our card games.) Now their Father says: 'At least one of you is dirty.' He then asks: 'Does anyone know if he is dirty?' Children answer truthfully. As questions and answers repeat, what happens?

> Nobody knows in the first round. But in the next round, each muddy child can reason like this: 'If I were clean, the one dirty child that I see would have seen only clean children, and so she would have known that she was dirty at once. But she did not. So I must be dirty, too!' ■

In the initial model, eight possible worlds assign *D* or *C* to each child. A child knows about the others' faces, not her own, as reflected in the accessibility lines in the diagrams below. Now, the successive assertions made in the scenario update this information:

Example, continued Updates for the muddy children.
Updates start with the Father's public announcement that at least one child is dirty. This simple communicative action merely eliminates those worlds from the initial model where the stated proposition is false. I.e., *CCC* disappears:

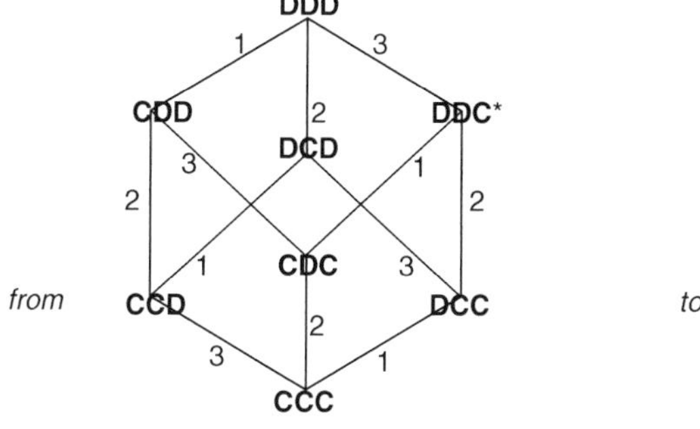

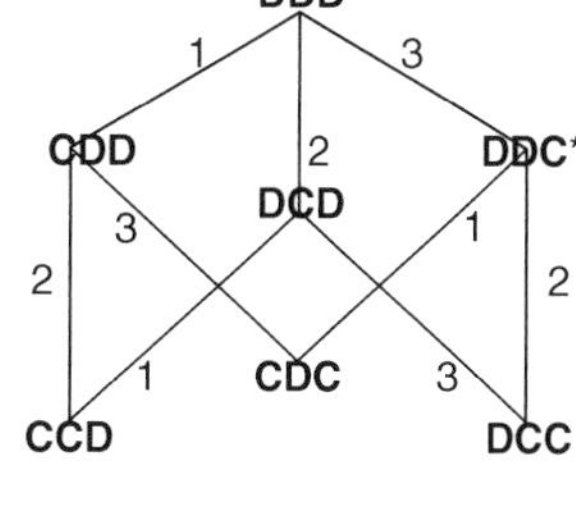

When no one knows his status, the bottom worlds disappear:

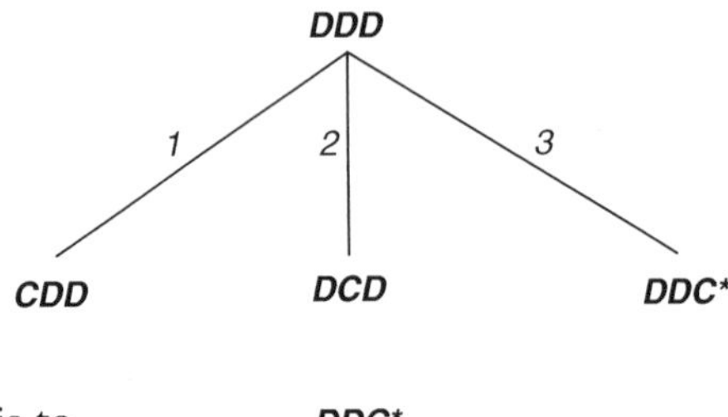

The final update is to *DDC** ■

In this model sequence, domain size decreases stepwise: *8, 7, 4, 1.* More generally, with *k* muddy children, *k* rounds of stating the same simultaneous ignorance assertion 'I do not know my status' by everyone yield common knowledge about which children are dirty. A few more such assertions by those who now know achieve common knowledge of the complete actual distribution of the **D** and **C** for the whole group.

The same setting also analyses the effects of changes in the procedure. For instance, a typical feature of dynamic actions is that their order of execution matters to the effects:

Example, still continued Dynamic effects of speaking order.
Moving from simultaneous to sequential announcement, the update sequence is quite different if the children speak in turn. The first update is as before, *CCC* disappears:

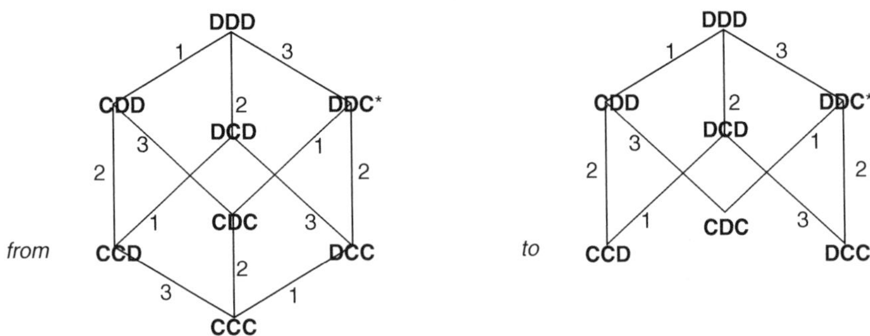

When the first child says it does not know, *DCC* is eliminated. Then the second child knows its status. Saying this takes out all worlds but *DDC, CDC*. In the final model, it is common knowledge that *2, 3* know, but *1* never finds out through epistemic assertions. ■

This is much more than an amusing puzzle. It raises deep issues of information flow that will return in this book. Just consider the following features just below the surface:

Enabling actions. The procedure is jump-started by the Father's initial announcement. No update happens if the further procedure is run on the initial cube-like 8-world model. Thus, internal communication only reaches the goal of common knowledge after some external information has *broken the symmetry* of the initial diagram. How general is this?

Self-refuting assertions. Children truly state their ignorance, but the last announcement of this fact leads to knowledge of their status, reversing its own truth. How can that be?

Iteration and update evolution. The Father gives an instruction that gets repeated to some limit. We will see that this repetition is a very common scenario, also in games (cf. Chapter 15), and thus, it is of interest to explore universes of evolution as updates get repeated.

Program structures. The scenario involves many program constructions: children follow a rule 'If you know your status, say so, else, say you do not' (*IF THEN ELSE*), there is sequential composition of rounds (;), we already noted iterations (*WHILE DO*), and the fact that the children speak simultaneously even adds a notion of *parallel composition*.

This combination of examples and deeper issues shows that public announcement of hard information is an update mechanism with hidden depths, and we now proceed to describe it more precisely in a logical system that can speak explicitly about the relevant events.

3.3 Dynamic logic of public announcement: language, semantics, axioms

First we must bring the dynamics of these successive update steps into a suitable extension of our static epistemic logic. Here is how:[3]

Definition Language and semantics of public announcement.
The language of *public announcement logic PAL* is the epistemic language with added action expressions, as well as dynamic modalities for these, defined by the following syntax rules:

[3] Van Benthem (2006b, c) are more extensive surveys of *PAL* and its technical properties.

Formulas	P: $p \mid \neg\varphi \mid \varphi\wedge\psi \mid K_i\varphi \mid C_G\varphi \mid [A]\varphi$[4]
Action expressions	A: $!P$

The language is interpreted as before in Chapter 2, while the semantic clause for the new dynamic action modality is forward-looking among models as follows:

$$\mathbf{M}, s \models [!P]\varphi \quad \text{iff} \quad \textit{if} \quad \mathbf{M}, s \models P, \textit{ then } \mathbf{M}|P, s \models \varphi$$ ■

This language allows us to make typical assertions about knowledge change such as

$$[!P]K_i\varphi$$

that states what an agent i will know after having received the hard information that P. This one formula of dynamified epistemic logic neatly highlights the combination of ideas from diverse fields that come together here. The study of speech acts $!P$ was initiated in linguistics and philosophy, that of knowledge assertions $K_i\varphi$ in philosophical logic and economics. And the dynamic effect modality $[\,]$ combining these actions and assertions into a new formal language comes from program logics in computer science.[5]

Axioms and completeness Reasoning about information flow in public update revolves around the formula $[!P]K_i\varphi$. In particular, we need to analyse the dynamic *recursion equation* driving the informational process, that relates the new knowledge to the old knowledge the agent had before the update took place. Here is the relevant principle:

FACT The following equivalence is valid for *PAL*: $[!P]K_i\varphi \leftrightarrow (P \rightarrow K_i(P \rightarrow [!P]\varphi))$.

Proof This can be verified using the above truth clauses, with the above diagrams for concreteness. Compare the models ($\mathbf{M}$, s) and ($\mathbf{M}|P$, s) before and after the update:[6]

[5] *A point of notation*. I write $[!P]\varphi$ to stress the different roles of the announced proposition P and the postcondition φ. Since P and φ are from the same language, the more popular notation $[!\psi]\varphi$ will be used as well. Some dynamic-epistemic literature even has a variant $[\psi]\varphi$ suppressing the action marker $!$.

[6] Note that an update action $!P$ as described here may be seen as a *partial function* on epistemic models: it is only executable when $\mathbf{M}$, $s \models P$, and in that case, it produces a unique new value.

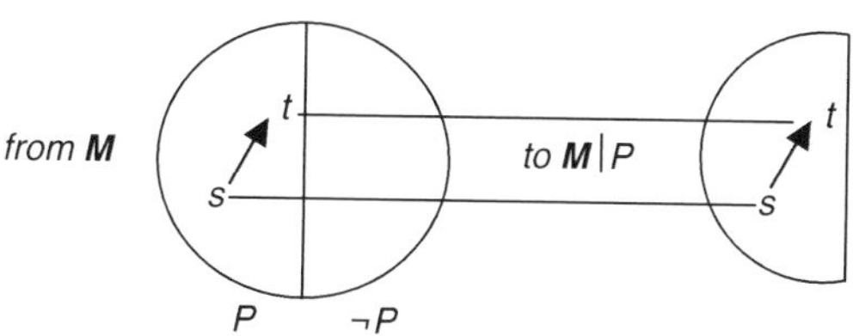

The formula $[!P]K_i\varphi$ says that, in $M|P$, all worlds $t \sim_i$-accessible from s satisfy φ. The corresponding worlds t in M are those $\sim_i$-accessible from s *that satisfy P*. As truth values of formulas may change in an update step, the right description of these worlds in M is not that they satisfy φ (which they do in $M|P$), but rather $[!P]\varphi$: they *become* φ after the update. Finally, $!P$ is a partial function: P must be true for its public announcement to be executable. Thus, we make our assertion on the right provided that $!P$ is executable, i.e., P is true. Putting this together, the formula $[!P]K_i\varphi$ says the same as $P \rightarrow K_i(P \rightarrow [!P]\varphi)$.[7] ■

We will discuss this key principle of information flow in more detail later. Here is how it functions in a complete calculus of public announcement. We state a result with a degree of freedom, since we do not care about the precise underlying epistemic base logic:

THEOREM *PAL* without common knowledge is axiomatized completely by the laws of epistemic logic over our static model class, now applied to the whole language, plus the following *recursion axioms*:[8]

$[!P]q$	$\leftrightarrow$	$P \rightarrow q$	*for atomic facts q*
$[!P]\neg\varphi$	$\leftrightarrow$	$P \rightarrow \neg[!P]\varphi$	
$[!P](\varphi \wedge \psi)$	$\leftrightarrow$	$[!P]\varphi \wedge [!P]\psi$	
$[!P]K_i\varphi$	$\leftrightarrow$	$P \rightarrow K_i(P \rightarrow [!P]\varphi)$[9]	

7 The consequent may be simplified to the equivalent formula $P \rightarrow K_i[!P]\varphi$.

8 In particular, for all dynamic logics in this book, we assume a *Necessitation Rule* for all modalities, as well as *Replacement of Equivalents*: if $\vdash \alpha \leftrightarrow \beta$, then $\vdash \varphi(\alpha) \leftrightarrow \varphi(\beta)$.

9 Some readers find equivalent formulations with existential dynamic modalities $<!P>$ easier to read. In this book, we switch occasionally to the latter for convenience.

Proof First, consider soundness. The first axiom says that update actions do not change the ground facts about worlds. The negation axiom interchanging $[\,]\neg$ and $\neg[\,]$ is a special law of modal logic expressing that update is a partial function. The conjunction axiom is always valid. And we have discussed the crucial knowledge axiom already. As for our inference rules, in particular, Replacement of Equivalents is sound.

Next, we turn to completeness. Suppose that some formula φ of *PAL* is valid. Start with some innermost occurrence of a dynamic modality in a subformula $[!P]\psi$ in φ. Now the axioms allow us to push this modality $[!P]$ through Boolean and epistemic operators in ψ until it attaches only to atoms, where it disappears completely thanks to the base axiom. Thus, we get a provably equivalent formula where $[!P]\psi$ has been replaced by a purely epistemic formula. Repeating this process until all dynamic modalities have disappeared, one obtains a formula φ' provably equivalent to φ. Since φ', too, is valid, it is provable in the base logic, which is complete by assumption, and hence, so is the formula φ itself. ■

Example Announcing an atomic fact makes it known.
Indeed, $[!q]Kq \leftrightarrow (q \rightarrow K(q \rightarrow [!q]q)) \leftrightarrow (q \rightarrow K(q \rightarrow (q \rightarrow q))) \leftrightarrow (q \rightarrow KT) \leftrightarrow T$.[10] ■

Example Diamonds and boxes are close.
For a partial function, a modal diamond and a modal box state almost the same. Here is how this shows in *PAL*: $\langle !P\rangle\varphi \leftrightarrow \neg[!P]\neg\varphi \leftrightarrow \neg(P \rightarrow \neg[!P]\varphi) \leftrightarrow P \wedge [!P]\varphi$. ■

This concludes the introduction of the first dynamic logic of this book. *PAL* is a natural extension of epistemic logic for semantic information that agents have, but it can also talk about events that change this information, and the resulting changes in knowledge. So we have made good on one promise in our programme. We now explore things a bit further.

3.4 A first exploration of planet *PAL*

The simple *PAL* system is remarkable in several ways, both technical and conceptual – and it is surprising how many issues of general interest attach to the above calculus.

[10] Like earlier, 'T' stands for the *always true* proposition.

Descriptive scope *PAL* describes what single agents learn in puzzles like Master Mind, but also multi-agent settings like the Three Cards. It also works in sophisticated scenarios like the Muddy Children that will return in Chapter 11, and the analysis of solution procedures in game theory (Chapter 15). Further examples are speech act theories in philosophy, linguistics, and agent theory in computer science.[11] Of course, *PAL* also has its limits. To mention just one, puzzles or conversations may refer explicitly to the *epistemic past*, as in saying 'What you said just now, I knew already.' This calls for a past-looking version that does not eliminate worlds for good, but keeps a record of the update history until now (see below, and Chapter 11). Moreover, many realistic scenarios involve partial observation and different information flow for different agents, which will be the topic of our next chapter. Our interest in *PAL* is because it is the first word, not the last.

A lens for new phenomena: Moore sentences, learning, and self-refutation *PAL* also helps us see new phenomena. Public announcement of atomic facts p makes them common knowledge. This is often thought to be just the point of speech acts of public assertion. But what guarantees that this is so? Indeed, one must be careful. It is easy to see that the following principle is valid for purely *factual formulas* φ without epistemic operators:

$$[!\varphi]C_G\varphi$$

But announcements $!\varphi$ of epistemic truths need not result in common knowledge of φ. A simple counter-example are *Moore-type sentences* that can be true, but never known. For instance, in one of our question–answer scenarios in Chapter 2, let the answerer A say truly

$p \wedge \neg K_Q p$ 'p, but you don't know it'

This very utterance removes the questioner Q's lack of knowledge about the fact p, and thus makes its own content false. Hence, announcing Moore sentences leads to knowledge of their negation. This switch may seem outlandish, but with the Muddy Children, repeated assertions of ignorance eventually led to knowledge in a last round. Similar beneficial reversals are known from game theory (Dégrémont & Roy 2009).

[11] Speech act theory (Searle & Vanderveken 1985) has specifications for successful assertions, questions, or commands. These can help apply *PAL* in real settings.

These switching cases have philosophical relevance. In Chapter 13, we will discuss the Fitch Paradox in Verificationism, the view that every true assertion can become known to us. This thesis must be qualified, as Moore examples show, and following van Benthem (2004b), we will show how dynamic epistemic logic meets epistemology here.[12,13]

Making time explicit These examples also highlight a temporal peculiarity of our system. Saying that *PAL* is about learning that φ is ambiguous between: (a) φ *was* the case, before the announcement, and (b) φ *is* the case after the announcement. For worlds surviving in the smaller updated model, factual properties do not change, but epistemic properties may. Making this temporal aspect implicit, and helping ourselves to an ad-hoc *past operator* $Y\varphi$ for 'φ was true at the preceding stage' (Yap 2006; Sack 2008), here is a principle about knowledge, and indeed common knowledge, that does hold in general for all assertions:

$$[!\varphi]C_G Y\varphi$$

It always becomes common knowledge that φ *was* true at the time of the announcement.[14]

Iterated assertions or observations In what sense is *PAL* also a logic of conversation or experimental procedure? We will discuss this issue in Chapters 10 and 11, but here is a first pointer. It may have seemed to the reader that the *PAL* axioms omitted one recursive case. Why is there no axiom for a combination of assertions

$$[!P][!Q]\varphi?$$

The reason is that our reduction algorithm started from innermost occurrences of dynamic modalities, always avoiding this case. Still, we have the

12 For a taxonomy of 'self-fulfilling' and 'self-refuting' statements, see Chapter 15.

13 A broader issue here is *learnability*: how can we come to know assertions by any means? Define a new modality $M, s \models \langle learn \rangle\varphi$ iff there is an formula P with $M, s \models \langle !P \rangle\varphi$. Prefixing a negation, this can also express that we *cannot* come to know a proposition. Van Benthem (2004b) asked if *PAL* plus the modal operator $\langle learn \rangle$ is decidable. Balbiani *et al.* (2007) proved that it is axiomatizable, but French & van Ditmarsch (2008) then showed that it is undecidable.

14 Factual formulas φ satisfy an implication $Y\varphi \rightarrow \varphi$ from 'then' to 'now'. More complex epistemic ones need not.

following observation that tells us how to achieve the effect of saying two consecutive things by saying just one:

FACT The formula $[!P][!Q]\varphi \leftrightarrow [!(P \wedge [!P]Q)]\varphi$ is valid.[15]

The proof is immediate if you think about what this says. The embedding inside the box uses the mutual recursion between actions and formulas in the inductive *PAL* syntax.

The agents behind PAL: observation and memory Static epistemic logic highlights assumptions that have sparked debate. As we saw in Chapter 2, distribution $K(\varphi \rightarrow \psi) \rightarrow (K\varphi \rightarrow K\psi)$ has been read in terms of agents' powers of *inference*, and the *.4* axiom $K\varphi \rightarrow KK\varphi$ as positive *introspection*. In Chapters 5, 13 we will question both for explicit knowledge, but this chapter sticks with semantic information. Even so, *PAL* does put up a new principle for scrutiny that has been much less discussed, viz. the recursion axiom

$[!P]K\varphi \leftrightarrow (P \rightarrow K[!P]\varphi)$ *Knowledge Gain*

What is remarkable about this principle is how it interchanges knowledge after an event with knowledge before that event. What does this mean in terms of agent powers? This is slightly easier to see when we adopt a more abstract stance. Consider the formula

$K[a]\varphi \leftrightarrow [a]K\varphi$

This says that I know now that action a will produce effect φ if and only if, after action a has occurred, I know that φ. For many actions a and assertions φ, this equivalence is fine. But there are counter-examples. For instance, I know now already that, after drinking, I get boring. But alas, after drinking, I do not know that I am boring.[16] At stake here are two new features of agents: their powers of *memory*, and of *observation* of relevant events. Drinking impairs these,[17] while public announcements respect both. We will discuss issues of Perfect Recall more fully in Chapter 11, but my point is that powers of observation and memory are just as crucial to epistemology as powers of deduction or introspection.

[15] Cf. van Benthem (1999b). This is like a colleague who used to monopolize faculty meetings by saying 'I say *P*, and then you will say *Q* to that, and then I will say *R* to *Q*', etcetera.

[16] In the converse direction, it may be true that after the exam I know that I have failed, but it still need not be the case that I know right now that after the exam I have failed.

[17] It does of course have intriguing epistemic features of its own, like forgetting or avoiding information.

'Tell All': optimal communication While much of epistemology is about single agents, *PAL* offers a much richer world of social scenarios where agents communicate, like games or dialogues. These usually constrain what can be said. Life is civilized just because we do not 'tell it like it is'. Even so, what can a group achieve by optimal communication? Consider two agents in a model ***M***, *s*. They can tell each other what they know, cutting ***M*** down to smaller parts. If they are cooperative, what is the best information they can give via successive updates – and what does the final collective information state look like?

Example The best agents can do by internal communication.
What is the best that can be achieved in this model (van Benthem 2002b)?

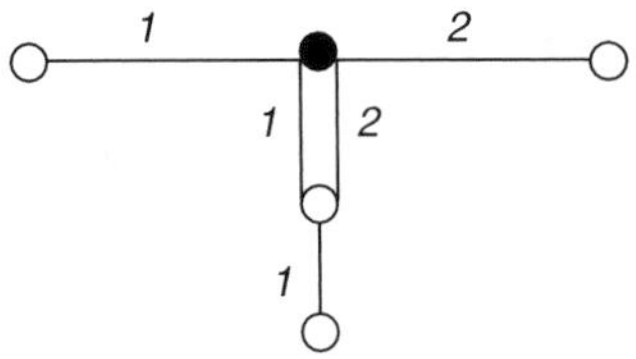

Geometrical intuition suggests that this must be:

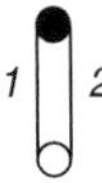

This is correct, and we leave a description of the moves to the reader. ■

In Chapters 12, 15, we will look at such scenarios in more detail, as they model game solution or social deliberation. In particular, in the finite case, agents will normally reach a minimal submodel where further true announcements have no effect. In a sense that can be made precise, they have then converted their *factual distributed knowledge* into *common knowledge*. With two agents, it can be shown that just two announcements suffice.

Example Optimal communication in two existentialist steps.
A two-step solution to the preceding example is the following conversation:

1 sighs: 'I don't know what the real world is'
2 sighs: 'I don't know either'

It does not matter if you forget details, because the scenario also works in the opposite order. ■

3.5 The dynamic methodology

Having seen some of the interesting issues that *PAL* brings within the compass of logic, let us now try to tease out the general methodology behind public announcement logic.

Compositional analysis Together, the *PAL* axioms analyse the effects of new information compositionally, breaking down the 'post-conditions' behind the dynamic modalities. The key principle here is the recursion axiom for knowledge gain after informational events, and it provides a model of analysis for all further dynamic phenomena that we will study in this book: from belief revision to preference change or group formation.

Static reduction and complexity The recursive character of the *PAL* axioms drives an effective reduction of all assertions with dynamic action modalities to static epistemic statements. This allowed us to establish completeness via a completeness theorem for the static language, while also, a decidable base logic gets a *decidable* dynamic extension. In particular, public announcement logic is decidable because basic epistemic logic is. Thus, in a sense, the dynamics comes for free: we just see more by making it explicit.[18]

Fitting a dynamic superstructure to a static base The general picture here is as follows, and it will recur in this book. First, one chooses a static language and matching models that represent information states for groups of agents, the snapshots or stills for the dynamic process we are after. Then we analyse the relevant informative events as updates changing these models. Next, these events are defined explicitly in a dynamic extension of the base language, that can also state effects of events by propositions true after their occurrence. This adds a dynamic superstructure over an existing logical system with a two-tier set-up:

At the static level, one gets a complete axiom system for whatever models one has chosen. On top of that, one seeks dynamic *recursion axioms* that describe static effects of events. In cases where this works, every formula is equivalent to a static one – and the above benefits follow. This design of dynamic-epistemic logics is modular, and independent from specific

[18] In a little while, we will also see reasons for going beyond this reductionist approach.

properties of static models. In particular, the *PAL* axioms make no assumptions about accessibility relations. Hence our completeness theorem holds just as well on arbitrary models for the minimal logic *K*, serving as some minimal logic of *belief*.[19,20]

Pre-encoding and language design Our recursive mechanism needs a static base language with operators that can do conditionalization, as in the clause '$K(P \rightarrow$' of the Knowledge Gain Axiom. We call this *pre-encoding*: a static model already has all information about what happens after informative events take place. Pre-encoding is ubiquitous, witness the key role of conditionals in belief revision (Chapter 7). If the base language lacks this power, we must *redesign* it to carry the weight of its natural dynamics:

Extending PAL with common knowledge Public announcement logic was designed to reason about what people tell each other, and it is quite successful in that. Nevertheless, its basic axioms have no special interaction principles relating different agents. The social character of *PAL* only shows in formulas with iterated epistemic modalities. But what about the acme of that: common knowledge? So far, we have not analysed this at all. As it turns out, there just is no *PAL* reduction axiom for $[!P]C_G\varphi$. This is strange, as one main purpose of a public announcement was producing the latter group knowledge.[21] The solution turns out to be the following redesign (van Benthem 1999b). We need to enrich the standard language of epistemic logic in Chapter 2 with a new operator:

[19] Indeed, this is how some core texts on *PAL* and general *DEL* set up things from the start.

[20] *Preserving frame conditions.* Some interplay between statics and dynamics can occur after all. Suppose we have a static special condition on models, say, accessibility is an equivalence relation. Then a constraint arises on the update mechanism: it should *preserve these conditions*. There is no guarantee here, and one must look case by case. Here is a general fact about *PAL*, given how its update rule works: *PAL* respects any condition on relations that is preserved under submodels. This includes all frame conditions definable by *universal first-order sentences*. But *PAL* update need not preserve existential properties, like every world having a successor. Things are more fragile with the *DEL* product update of Chapter 4, that outputs submodels of direct products of epistemic models. In that case, the only first-order frame conditions automatically preserved are *universal Horn sentences*. Reflexivity, symmetry, and transitivity are of that form, but a non-Horn universal frame condition like linearity of accessibility is not, and hence it may be lost in update.

[21] Baltag, Moss & Solecki (1998) do axiomatize *PAL* with C_G, using special inference rules.

DEFINITION Conditional common knowledge.
Conditional common knowledge $C_G^P\varphi$ says that φ is true in all worlds reachable from the current one by a finite path of accessibilities running only through worlds satisfying P. ■

Plain $C_G\varphi$ is just $C_G^T\varphi$. Van Benthem, van Eijck & Kooi (2006) show that $C_G^P\varphi$ cannot be defined in epistemic logic with just common knowledge. The new operator is natural. It is bisimulation-invariant, and also, standard completeness proofs (cf. Fagin *et al.* 1995) carry over to the natural axioms for $C_G^P\varphi$. On top of these static principles, here is a valid recursion axiom for common knowledge in *PAL*:

FACT The following equivalence is valid: $[!P]C_G\varphi \leftrightarrow (P \leftrightarrow C_G^P[!P]\varphi)$.

Proof Just check with a picture like with the Knowledge Gain Axiom. ■

But we have a richer base language now, so we need recursion axioms that work for $C_G^P\varphi$ itself, not just $C_G\varphi$. The next and final axiom shows that the hierarchy stops here:

THEOREM *PAL* with conditional common knowledge is axiomatized completely by adding the valid reduction law $[!P]C_G^\varphi\psi \leftrightarrow (P \rightarrow C_G^{P\wedge[!P]\varphi}[!P])\psi$.

Example Atomic announcements produce common knowledge.
Indeed, $[!q]C_Gq \leftrightarrow (q \rightarrow C_G^q[!q]q) \leftrightarrow (q \rightarrow C_G^qT) \leftrightarrow (q \rightarrow T) \leftrightarrow T$. ■

By now, we have teased out most general features of our pilot system.

We conclude with a possible objection, and a further twist to our methodology.

Reduction means redundancy? If *PAL* reduces every dynamic formula to an equivalent static one, does it not shoot itself in the foot, since the dynamics is essentially redundant? Here are a few thoughts. First, the *PAL* axioms are of intrinsic interest, raising many issues that no one had thought of before. The fact that they drive a reduction is secondary – and reductions in science often say much less anyway than it seems at first sight. The heart of the information dynamics is the recursive reduction process *itself*.[22] And its

[22] The arithmetic of + is not 'just that of the successor function': the recursion doing the work is crucial. Extensions of theories by definitions are often highly informative, not in a semantic sense, but in a fine-grained intensional sense of information (cf. the syntactic approach of Chapter 5).

dynamic language brings out phenomena that lay hidden in the epistemic base. But I even like the brute reduction, since it gives substance to my earlier slogan that 'logic can be more than it is'. Existing systems of logic have a much broader sweep than one might think: and this is seen by adding a dynamic superstructure, and then eliciting insights about actions.

Protocols and relaxing the reduction programme Even so, there are better reasons for relaxing the orthodox version of the dynamics programme, and we shall do so in Chapter 11. Here is the consideration (cf. van Benthem *et al.* 2007). Single steps of communication or observation, as described by *PAL*, make best sense inside longer-term scenarios of conversation or learning. And such scenarios may *constrain* the available sequences of assertions or observations. Not everything that is true can be said or observed.

Now *PAL* itself has no such constraints. It lives in a 'Supermodel' of all epistemic models related by all possible updates, where all histories of true assertions are possible:[23]

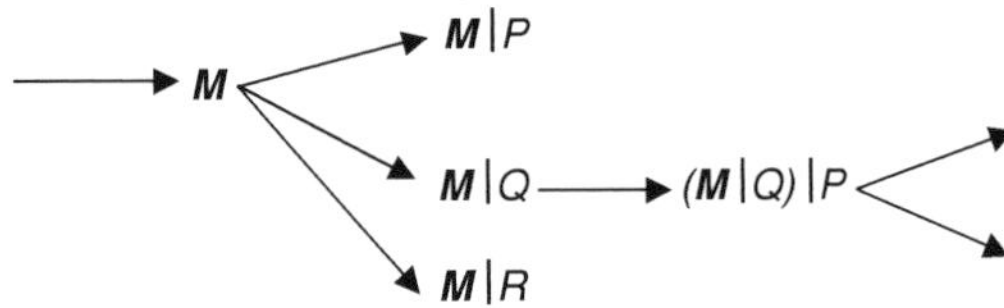

Once we accept constraints on histories in this full information process (cf. the *protocols* of Chapter 11), a simple but characteristic axiom of *PAL* so far must be abandoned, viz. the equivalence between the truth of a proposition and its 'executability' (we use the dual existential dynamic modality now, since it fits the current issue better):

$$\varphi \leftrightarrow \langle !\varphi \rangle T$$

Only the implication $\langle !\varphi \rangle T \rightarrow \varphi$ remains valid: announcability implies truth. This change suggests a fundamental conceptual distinction (van Benthem 2009f; Hoshi 2009; cf. Chapter 13) between *epistemic information* about facts in the world and what others know, and *procedural information* about the total process creating the epistemic information. The system *PAL* only describes the former kind of information, not the latter. While *PAL* with protocols still has

[23] Van Benthem (2003d) has representation results for the *PAL* language that use only small 'corners' of this huge class, consisting of an initial model **M** plus further submodels.

recursion axioms for knowledge, they no longer reduce formulas to exclusively epistemic ones. The procedural information that they contain may be essential.

3.6 Conclusion

The logic *PAL* of public announcement, or public observation, is a proof-of-concept: dynamic logics dealing with information flow look just like systems that we know, and can be developed maintaining the same technical standards. But on top of that, these systems raise a host of new issues, while suggesting a larger logical program for describing the dynamics of rational agency with much more sophisticated informational events than public announcements. We will turn to these in the following chapters.

At first sight so demure and almost trivial, *PAL* generates many interesting questions and open problems. We will now discuss a number of these, though what follows can be skipped without loss of continuity. This is the only chapter in the book where we discuss so many technical issues, and we apologize for its inordinate length. The reason is that *PAL* is a nice pilot system where things are simple to state. But in principle, every point raised here also makes sense for our later logics of belief, preference, or games.

3.7 Model theory of learning

First, we continue with the earlier semantic analysis for events of public announcement.

Bisimulation invariance Recall the notion of *bisimulation* from Chapter 2, the basic semantic invariance notion for the epistemic language. *PAL* still fits in this framework:

FACT All formulas of *PAL* are invariant for bisimulation.

Proof This may be shown via the earlier reduction to purely epistemic formulas, that were invariant for bisimulations. But a more informative proof suggesting generalizations goes inductively via an interesting property of update, viewed as an operation *O* on models. We say that such an operation O *respects bisimulation* if, whenever two models **M**, s and **N**, t are bisimilar, then so are their values $O(\mathbf{M}, s)$ and $O(\mathbf{N}, t)$:

FACT Public announcement update respects bisimulation.

Proof Let $\equiv$ be a bisimulation between **M**, s and **N**, t. Consider the submodels $\mathbf{M}|\varphi$, s and $\mathbf{N}|\varphi$, t after public update with φ. The point is that the *restriction* of $\equiv$ to these is still a bisimulation. Here is a proof of the zigzag clause. Suppose that some world w in $\mathbf{M}|\varphi$ with a $\equiv$-matching world w' in $\mathbf{N}|\varphi$ has an $\sim_i$-successor v in $\mathbf{M}|\varphi$. The original bisimulation then gives an $\sim_i$-successor v' for w' *in* **N** for which $v \equiv v'$. Now the world v satisfies φ in **M**, and hence, by the Invariance Lemma for the bisimulation $\equiv$, the matching v' satisfies φ as well. But that means that v' is inside the updated model $\mathbf{N}|\varphi$, as required. ■

Many other update mechanisms respect bisimulation, witness Chapter 4.[24]

Persistence under update We have seen that not all public events $!\varphi$ result in (common) knowledge that φ. Purely factual formulas φ have this property, and so do some epistemic ones, like Kp. Even $\neg Kp$ is preserved (by a simple argument involving reflexivity), but the Moore formula $p \wedge \neg Kp$ was a counter-example. This raises a general, and as yet unsolved, issue of persistence under update, sometimes called the Learning Problem:

Open Problem Characterize the syntax of the formulas φ that remain true when the $\neg\varphi$-worlds are eliminated from a model **M**, **s** where φ holds.

Here is a relevant result from modal logic:

THEOREM The epistemic formulas in the language without common knowledge that are preserved under submodels are precisely those definable using literals p, $\neg p$, conjunction, disjunction, and K-operators.

Proof Compare the universal formulas in first-order logic, that are just those preserved under submodels. Andréka, van Benthem & Németi (1998) have the technical proof for modal languages. ■

24 Cf. van Benthem (1996) and Hollenberg (1998) on analogous issues in process algebra.

A conjecture for the full language might add arbitrary formulas $C_G\varphi$ as persistent forms. Van Ditmarsch and Kooi (2006) discuss the case with dynamic modalities. The problem may be hard since lifting first-order model theory to non-first-order modal fixed-point logics (for the modality C_G) seems non-trivial, even on a universe of finite models.[25]

In any case, what we need in the *PAL* setting is not full submodel preservation, but rather preservation under 'self-defined submodels'. When we *restrict* a model to those of its worlds that satisfy φ, then φ should hold in that model, or in terms of an elegant validity:

$$\varphi \to (\varphi)^{\varphi}$$

Open Problem Which epistemic formulas imply their self-relativization?

Digression: enforcing persistence? Some people find non-persistence a mere side effect of infelicitous wording. E.g., when *A* said '*p*, but you don't know it', she should just have said '*p*', keeping her mouth shut about my mental state. Can we rephrase any message to make the non-persistence go away? An epistemic assertion φ defines a set of worlds in the current model **M**. Can we always find an equivalent persistent definition? This would be easy if each world has a simple unique factual description, like hands in card games. But there is a more general method that works in more epistemic settings, at least locally:

FACT In each finite epistemic *S5*-model, every public announcement has a persistent equivalent.[26]

[25] Recent development (spring 2010): Holliday and Icard (2010) have just solved the single-agent case, while classifying learning phenomena in a systematic manner.

[26] *Proof* Assume that models **M** are bisimulation-contracted (cf. Chapter 2) and connected, with no isolated accessibility zones. Now *j* publicly announces φ, going to the submodel $\mathbf{M}|\varphi$ with domain $\varphi^* = \{x \in \mathbf{M} \mid \mathbf{M}, x \models \varphi\}$. If this is still **M** itself, the persistent announcement "True" works. Now suppose φ^* is not all of **M**. Our equivalent persistent assertion has two disjuncts: $\Delta \vee \Sigma$. Using the proof of the State Definition Lemma in Chapter 2, Δ is an epistemic definition for φ^* in **M** describing each world in it up to bisimulation, and taking the disjunction. Now for Σ. Using the same proof, take a formula describing the model $\mathbf{M}|\varphi$ up to bisimulation, with common knowledge over an epistemic formula describing a pattern of zones and links in $\mathbf{M}|\varphi$ (but no specific world description added). Here is how the new statement works. First, $\Delta \vee \Sigma$ is common knowledge in $\mathbf{M}|\varphi$, because Σ is. But it also picks out the right worlds in **M**. Clearly, any world in φ^* satisfies its own disjunct of Δ. Conversely, let world *t* in **M** satisfy $\Delta \vee \Sigma$. If it satisfies some disjunct of Δ, then *t* is in φ^* by bisimulation-minimality of **M**. Otherwise, **M**, *t* satisfies Σ. But then by

Of course, the recipe for rephrasing your assertions in a persistent manner is ugly, and not recommended! Moreover, it is local to one model, and does not work uniformly.[27]

The test of language extensions Finally, how sensitive is this theory to generalization? Sometimes we want to express more than what is available in the basic epistemic language – and this poses an interesting challenge to the scope of *PAL*. An example in Chapter 2 was the notion D_G of distributed group knowledge, that crosses a clear semantic threshold:

FACT The modality $D_G\varphi$ for *intersection* of epistemic accessibility relations[28] is not invariant for bisimulation.

Proof The following models have $D_{\{a,\ b\}}\neg p$ true in world *1* on the right, but not on the left:

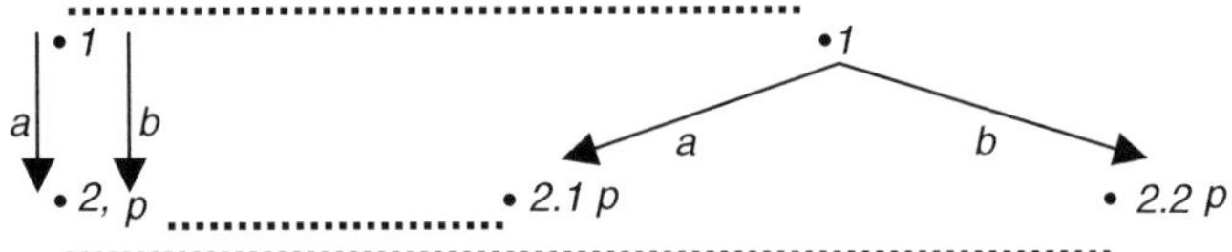

There is an obvious bisimulation with respect to the separate relations R_a, R_b – but zigzag fails for $R_a \cap R_b$. ■

Does *PAL* extend to this non-bisimulation invariant generalization? The answer is positive:

THEOREM *PAL* with distributed knowledge is axiomatized by the earlier principles plus the recursion axiom $[!P]D_G\varphi \leftrightarrow (P \rightarrow D_G[!P]\varphi)$.

Similar observations hold for other extensions of the epistemic base language. For instance, later on, we will have occasion to also use a global *universal modality* $U\varphi$ saying that φ is true in all worlds of the current model, epistemically accessible or not:

connectedness, every world in **M** satisfies Σ, and by the construction of Σ, there is a bisimulation between **M** and **M**$|\varphi$. But this contradicts the assumption that the latter update was proper.

27 Van Benthem (2006b) shows that no uniform solution is possible over all models.

28 On the other hand, adding the modality D_G does keep the epistemic logic decidable.

THEOREM *PAL* with the universal modality is axiomatized by the earlier principles plus the recursion axiom $[!P]U\varphi \leftrightarrow (P \rightarrow U(P \rightarrow [!P]\varphi))$.

So the dynamic extension methodology proposed here extends well beyond the basic epistemic language.

3.8 Abstract postulates on update and frame correspondence

Let us now step back from our update mechanism. *PAL* revolves around one concrete way of taking incoming hard information. But why should we accept this? What if we reversed the perspective, and asked ourselves which general postulates look plausible a priori for hard information update, and then see *which operations on models* would validate it?

One answer is via a modal frame correspondence argument (van Benthem 2007b), switching perspectives by taking the *PAL* axioms as abstract postulates. We consider only a simple version. Let abstract model-changing operations ♥p take pointed epistemic models $\boldsymbol{M}$, s with a set of worlds p in $\boldsymbol{M}$ to new epistemic models $\boldsymbol{M}$♥p, s – where the domain of worlds remains the same. For still more simplicity in what follows, we also assume that $s \in p$. Now, by *link elimination*, we mean the operation on epistemic models that changes accessibility by cutting all epistemic links between p-worlds and $\neg p$-ones.[29]

THEOREM Link elimination is the only model-changing operation that satisfies the equivalence [♥p]$Kq \leftrightarrow (p \rightarrow K(p \rightarrow$ [♥p]$q))$ for all propositions q.

Proof We identify propositions with sets of worlds. Start from any pointed model $\boldsymbol{M}$, s. From left to right, take q to be the set of worlds $\sim$-accessible from s in the model $\boldsymbol{M}$♥p. This makes [♥p]Kq true at s. Then the right-hand side says

29 What follows is a bit informal. More precise formulations would use the earlier Supermodel perspective, now with smaller families of epistemic models related by various transformations. We hope that the reader can supply formal details after a first pass through the correspondence proof. The general theme of logics for model transformations gets more interesting with the system *DEL* in Chapter 4, since we then also have constructions that can increase model size.

that $K(p \to [\heartsuit p]q)$ holds, i.e., all p-worlds that are $\sim$-accessible from s in $\mathbf{M}$ are in q. Thus, the operation $\heartsuit p$ preserves all existing $\sim$-arrows from s in $\mathbf{M}$ that went into the set p. In the converse direction, let q be the set of all p-worlds that are $\sim$-accessible from s in $\mathbf{M}$. This makes $K(p \to [\heartsuit p]q)$ true at s, and hence $[\heartsuit p]Kq$ must be true. But the latter says that all worlds that are $\sim$-accessible from s in $\mathbf{M}\heartsuit p$ are in q: that is, they can only have been earlier $\sim$-accessible worlds in $\mathbf{M}$. The two inclusions describe precisely the link-cutting version of epistemic update. ■

This is just one version in a sequence. Here is a next step. Assume now that also the domain of models may change from $\mathbf{M}$ to $\mathbf{M}\heartsuit p$. To zoom in on *PAL*-style eliminative updates, we now also need an *existential modality* $E\varphi$ (dual to the universal modality $U\varphi$):

THEOREM Eliminative update is the only model-changing operation that satisfies the following three principles: (a) $<!p>T \leftrightarrow p$, (b) $<\heartsuit p>Eq \leftrightarrow p \wedge E<\heartsuit p>q$, and (c) $[\heartsuit p]Kq \leftrightarrow (p \to K(p \to [\heartsuit p]q))$.

Proof sketch (a) $<!p>T \leftrightarrow p$ makes sure that inside a given model $\mathbf{M}$, the only worlds surviving into $\mathbf{M}\heartsuit p$ are those in the set denoted by p. Next, axiom (b) $<\heartsuit p>Eq \leftrightarrow p \wedge E<\heartsuit p>q$ holds for eliminative update, being dual to the earlier one for U. What it enforces in our abstract setting is that the domain of $\mathbf{M}\heartsuit p$ contain no objects beyond the set p in $\mathbf{M}$. Finally, the above axiom (c) for knowledge ensures once more that the epistemic relations are the same in $\mathbf{M}$ and $\mathbf{M}\heartsuit p$, so that our update operation really takes a *submodel*. ■

These results may suffice to show how an old modal technique acquires a new meaning with epistemic update, reversing the earlier direction of conceptual analysis.

3.9 Update versus dynamic inference

This chapter is about the information dynamics of observation, ignoring the dynamics of inference over richer syntactic information states, that will be touched upon in Chapter 5. Even so, *PAL* also suggests a local dynamic notion of inference summarizing the effects of successive updates (Veltman 1996; van Benthem 1996) that is worth a closer look:

DEFINITION Dynamic inference.
Premises $P_1, \ldots, P_k$ *dynamically imply* conclusion φ if after updating any information state with public announcements of the successive premises, all worlds in the end state satisfy φ. Stated in *PAL* terms, the following implication must be valid: $[!P_1] \ldots [!P_k]C\varphi$. ■

This notion behaves quite differently from standard logic in its style of premise management:

FACT All structural rules for classical consequence fail.

Proof (a) Order of announcement matters. Conclusions from *A, B* need not be the same as from *B, A*: witness the Moore-style sequences $\neg Kp\ ;\ p$ (consistent) versus $p\ ;\ \neg Kp$ (inconsistent). (b) Multiplicity of occurrence matters, and Contraction fails. $\neg Kp \wedge p$ has different update effects from the inconsistent $(\neg Kp \wedge p); (\neg Kp \wedge p)$. (c) Non-monotonicity. Adding premises can disturb conclusions: $\neg Kp$ implies $\neg Kp$, but $\neg Kp, p$ does not imply $\neg Kp$. (d) Similar dynamic counter-examples work for the Cut Rule. ■

These failures all reflect the essential dynamic character of getting epistemic information. Still, modified structural rules remain (van Benthem 1996):

FACT The following three structural rules are valid for dynamic inference:

Left Monotonicity	$X \Rightarrow A$ implies $B, X \Rightarrow A$
Cautious Monotonicity	$X \Rightarrow A$ and $X, Y \Rightarrow B$ imply $X, A, Y \Rightarrow B$
Left Cut	$X \Rightarrow A$ and $X, A, Y \Rightarrow B$ imply $X, Y \Rightarrow B$

THEOREM The structural properties of dynamic inference are axiomatized completely by Left Monotonicity, Cautious Monotonicity, and Left Cut.[30]

Thus, *PAL* suggests new notions of consequence in addition to classical logic, and as such, it may be profitably compared to existing *substructural logics*. Van Benthem (2008d) has a comparison with the programme of Logical Pluralism, a topic taken up in Chapter 13.

[30] A proof is in van Benthem (2003d). One uses an abstract representation of dynamic inference from van Benthem (1996), plus a bisimulation taking any finite tree for modal logic to a model where (a) worlds w go to a family of epistemic models $\mathbf{M}_w$, (b) basic actions a go to epistemic announcements $!\varphi_a$.

3.10 The complexity balance

The third main theme in Chapter 2 was computational complexity of logics.

THEOREM Validity in *PAL* is decidable.

Proof This may be shown by the same effective reduction that gave us the completeness, since the underlying epistemic base logic is decidable. ■

This does not settle the computational *complexity* – as translation via the axioms may increase the length of formulas exponentially.[31] Lutz (2006) uses a non-meaning-preserving but polynomial-time *SAT*-reduction to show that *PAL* sides with epistemic logic:

THEOREM The complexity of satisfiability in *PAL* is *Pspace-complete*.

As for the other two important items in the complexity profile of our system, we have the following result (van Benthem, van Eijck & Kooi 2006):

THEOREM In *PAL*, both model checking and model comparison take *Ptime*.

Both these tasks have interesting two-player game versions, extending those for epistemic logic in Chapter 2. In all then, from a technical point of view, *PAL* presents about the same good balance between expressive power and complexity as its static base logic *EL*.[32]

3.11 The long run: programs and repeated announcements

As we saw with the Muddy Children, to make *PAL* into a logic of conversation or longer-term informational processes, we need program structure for complex assertions and plans. An obvious extension uses *propositional dynamic logic* (*PDL*), an important system for describing sequential actions, which we take for granted in this book (cf. also Chapters 4, 10). The reader may consult Blackburn, de Rijke & Venema (2000), Harel, Kozen & Tiuryn (2000), Bradfield & Stirling (2006), or van Benthem (to appearB).

[31] Putting this more positively, dynamic modalities provide very *succinct notation*.

[32] Still, to really understand the complexity profile of a dynamic logic like this, we may have to add *new basic tasks* concerning the complexity of communication.

Program structure in conversation: PAL plus PDL We already noted that repeated assertions can be contracted to one, given the validity of $[!P][!Q]\varphi \leftrightarrow [!(P\wedge[!P]Q)]\varphi$. But despite this reduction, real conversation is driven by complex instructions. If you want to persuade your dean to give money to your ever-hungry institute, you praise him for making things go well if he looks happy, and criticize his opponents if he looks unhappy. And you apply this treatment until he is in a good mood, and then bring up the money issue. This recipe involves all major constructions of imperative programming. Sequential order of assertions is crucial (asking money first does not work), what you say depends on conditions, and flattery involves iteration: it is applied as long as needed to achieve the desired effect. Likewise, the Muddy Children puzzle involved program constructions of

(a) *sequential composition* ;
(b) *guarded choice* *IF ... THEN... ELSE...*
(c) *guarded iteration* *WHILE... DO...*

PAL plus *PDL* program operations on its actions is like *PAL* in that formulas remain invariant for epistemic bisimulation. Still, adding this long-term perspective crosses a dangerous threshold in terms of the complexity of validity:

THEOREM *PAL* with all *PDL* program operations added to the action part of the language is undecidable, and even non-axiomatizable.

Miller & Moss (2005) prove this using tiling methods (cf. Chapter 2). We will explain what is going on from a general point of view in the epistemic-temporal setting of Chapter 11.

Iterated announcements in the limit Muddy Children was a limit scenario where we keep making the same assertion as long as it is true. Such iterations create intriguing new phenomena. The muddy children repeat an assertion φ of ignorance until it can no longer be made truly. Their statement is *self-defeating*: when repeated iteratively, it reaches a sub-model of the initial model where it is false everywhere (ignorance turned into knowledge). The other extreme is *self-fulfilling* formulas φ whose iterated announcement eventually makes them common knowledge. The latter occur in solution procedures for games, announcing 'rationality' for all players (cf. Chapters 10, 15 for such *PAL* views of games).

Technically, iterated announcement of φ starting in an initial model $\boldsymbol{M}$ always reaches a *fixed point*.[33] This generalizes our earlier discussion of epistemic formulas φ whose one-step announcement led to truth of $C_G\varphi$ (say, factual assertions), or $C_G\neg\varphi$ (Moore-type formulas). The logical system behind these phenomena is an *epistemic* μ*-calculus* with operators for smallest and greatest fixed-points (Bradfield & Stirling 2006). Chapter 15 has details, and states mathematical properties of *PAL* with iteration in this setting.

3.12 Further directions and open problems

We end with some topics in the aftermath of this chapter that invite further investigation – ranging from practical modelling to mathematical theory.

Art of modelling Fitting epistemic logic and *PAL* to specific applications involves a choice of models to work with. In this book, this art of modelling will not be studied in detail, even though it is crucial to the reach of a framework. It raises general issues that have not yet been studied systematically in our logics. For instance, with public announcements in conversation, there is the issue of *assertoric force* of saying that φ. Normal cooperative speakers only utter statements that they know to be true, generating an event $!K\varphi$. They may also convey that they think the addressee does not know that φ. Such preconditions can be dealt with in the dynamic-epistemic logic of Chapter 4 and the protocols of Chapter 11. Weaker preconditions involving beliefs occur in Chapter 7, and others are discussed in our analysis of questions (Chapter 6). But there are many further features. For instance, there is an issue of language choice and *language change* in setting up the range of epistemic possibilities. Speech acts may make new proposition letters relevant, changing the representation space. When these phenomena are systematic enough, they may themselves enter the dynamic logic: the product update mechanism of Chapter 4 is one step towards a systematic view of model construction.

From analysing to planning *PAL* has been used so far to analyse given assertions, but it can also help plan assertions meeting specifications. Here is an example from a Russian mathematical Olympiad (cf. van Ditmarsch 2003): '7 cards are given to persons: *A* gets 3, *B* gets 3, C gets 1. How should *A*, *B*

[33] Repeated announcement in infinite models takes *intersections* at limit ordinals. In each model, each epistemic formula is then either self-defeating or self-fulfilling.

communicate publicly, in hearing of C, to find out the distribution of the cards while C does not?' Solutions depend on the number of cards. More generally, how can a subgroup communicate, keeping the rest in the dark? Normally, this needs 'hiding' (cf. Chapter 4), but some tasks can be done in public (cf. van Ditmarsch & Kooi 2008 on reachability from one model to another via announcements). A relevant analogy here comes from program logics with *correctness assertions*

$$\varphi \rightarrow [!P]\psi$$

saying that, if precondition φ holds, then public observation of P always leads to a model where postcondition ψ holds (cf. van Benthem 1996). Given *!P*, one can analyse its pre- and postconditions. In fact, $[!P]\psi$ defines the 'weakest precondition' for *!P* to produce effect ψ. Or given *!P* plus precondition φ, we can look for the 'strongest postcondition' ψ. E.g., with $\varphi = True$, and p atomic, it is easy to see that the strongest postcondition is common knowledge of p.[34] This is program analysis. But there is also *program synthesis*. Given precondition φ and postcondition ψ (say, specifying which agents are to learn what) we can look for an assertion *!P* guaranteeing that transition.

Agents and diversity We have seen how *PAL* makes idealizing assumptions about agents' powers of memory and observation. Liu (2005, 2008) discuss *diversity* in powers, and the need for *PAL* and other dynamic-epistemic logics to model agents with different powers interacting successfully. For this, we need to parametrize update rules (Chapter 4 gives an option). Indeed, *PAL* does not have an explicit notion of *agent* – and maybe it should.[35]

The secret of PAL: relativization Here is the ultimate technical explanation for the *PAL* axioms (van Benthem 1999b). Updating with P is *semantic relativization* of a model **M**, s to a definable submodel **M**|P, s. Now standard logic has a well-known duality here. One can either evaluate epistemic formulas in the new model **M**|P, or translate them into formulas in the old model **M**, s by means of *syntactic relativization*:

Relativization Lemma $\quad M|P, s \models \varphi \quad$ iff $\quad M, s \models (\varphi)^P$

[34] In general, epistemic post-conditions are not definable in *PAL*: cf. Chapter 11.

[35] Agents *themselves* might change by learning. But then our recursion axiom $[!P]K_i\varphi \leftrightarrow (P \rightarrow K_i(P \rightarrow [!P]\varphi))$ is too simplistic. Agent *i* before update changes into agent $i+!P$, and we must modify the axiom – perhaps to an equivalence like $[!P]K_{i+!P}\varphi \leftrightarrow (P \rightarrow K_i(P \rightarrow [!P]\varphi))$.

The recursion axioms of *PAL* are just the inductive clauses in syntactic relativization. And our 'pre-encoding' becomes the issue if the base language is closed under relativization. *EL* with common knowledge was not,[36] and conditional common knowledge solved this. In this light, *PAL* is a logic of relativization, related to logics of other basic operations such as *substitution*, that can be performed both semantically and syntactically.[37]

Schematic validity Next comes an open problem that I find slightly annoying. Unlike with most logical systems, the axioms of *PAL* are not closed under substitution of arbitrary formulas for atoms. For instance, the base axiom fails in the form $[!P]Kq \leftrightarrow (P \rightarrow Kq)$. With $P = q$, $[!q]Kq$ is always true, while the right-hand side $q \rightarrow Kq$ is not valid. Still, except for the atomic case, all *PAL* axioms are *schematically valid*: each substitution of formulas for Greek letters preserves validity. Even in the base, we have schematic validity of $<!\varphi>T \leftrightarrow \varphi$. Another nice example was our earlier law for repeated announcements, showing that there are interesting schematic validities that are not immediately apparent from the *PAL* axioms. The schematically valid principles are the truly general algebraic laws of public announcement. But it is not clear how to describe them well. For instance, our completeness proof does not show that schematically valid laws are derivable *as such*: we have only shown that each of their concrete instances is derivable.

Open Problem Are the schematic validities of *PAL* axiomatizable?[38]

Model theory of iteration, learning, and fixed-points There are many open problems in dynamic logics and μ-calculi extending *PAL* with program structure. Just ask yourself this: how do updates for related formulas compare? Suppose that φ implies ψ. Will the iterated $!\varphi$-sequence always go faster than that for $!\psi$? Are their limits included? Chapter 15 shows that neither needs to be the case, and asks further questions (cf. also van Benthem 2007f). But

[36] A related non-relativizing system is propositional dynamic logic *without tests*. In contrast, propositional dynamic logic with tests is closed under relativization.

[37] What is the complete logic of relativization $(\varphi)^A$ in first-order logic? Van Benthem (1999b) notes that the schematically valid iteration law of *PAL* here becomes a principle of *Associativity* $((A)^B)^C \leftrightarrow {}_A((B)^C)$.

[38] An answer is not obvious since schematic validity quantifies over all formulas of the language. The 'Stanford Logical Dynamics Lab' has recently announced decidability for the single-agent case: see W. Holliday, T. Hoshi & Th. Icard, 'Decidability of the PAL Substitution Core', CSLI, Stanford University.

mysteries abound. For instance, the *PDL* extension of *PAL* looks simple, but Baltag & Venema (p.c.) have shown that the formula $<(!P)^*>C_Gq$ is not even definable in the modal μ-calculus, as it lacks the finite model property. Intuitively, formulas $[(!P)^*]\varphi$ are still definable with greatest fixed-point operators of some sort, but we lack a good view of the best formalisms. Finally, to return to a concrete issue, recall the Learning Problem of determining just which *PAL*-formulas φ become common knowledge upon announcement: i.e., $[!\varphi]C_G\varphi$ is valid. Generalized issues of this sort arise with iteration. Say, when is a formula φ self-confirming or self-refuting in the announcement limit?

3.13 Literature

In this final section, I give a few key references as markers to the field, and the same sort of bibliography will conclude later chapters. My aim is not a survey of contributions, nor an official record of credits, though I try to be fair. Update of semantic information as reduction in a range of possible situations occurs all across science (think of probabilistic conditioning), and also in the world of common sense. There have been attempts to patent this idea to particular authors, but these are misguided. Systematic logics of update have been proposed in the semantics and pragmatics of natural language (Stalnaker 1978; Veltman 1996). The *PAL* style logic of model-changing announcements is due to Plaza (1989), though this paper was unknown and had no direct influence. It was rediscovered independently by students at ILLC Amsterdam. Gerbrandy & Groeneveld (1997) documents a first stage; Gerbrandy (1999a) goes further. Baltag, Moss & Solecki (1998) axiomatize *PAL* with common knowledge. Van Benthem (2006b) adds new themes and results that occur in this chapter, obtained since the late 1990s. Van Ditmarsch, van der Hoek & Kooi (2007) is a textbook on general dynamic-epistemic logic, but including much material on *PAL*. Van Benthem, van Eijck & Kooi (2006) has a streamlined version with recursion axioms for common knowledge, and definability results for *PAL* are proved uniformly with model-theoretic techniques like Ehrenfeucht games. Miller & Moss (2005) is a sophisticated mathematical study of the complexity of *PAL* with program constructions. The first significant philosophical uses of *PAL* are in Gerbrandy (1999a, b) and van Benthem (2004b).

4 Multi-agent dynamic-epistemic logic

The public announcements or observations *!P* studied in Chapter 3 are an easily described way of conveying information. The only difficulty may be finding good static epistemic models for the initial situation where updates start. But information flow gets much more challenging once we give up the uniformity in this scenario. In conversation, in games, or at work, agents need not have the same access to the events currently taking place. When I take a new card from the stack in our current game, I see which one I get, but you do not. When I overhear what you are telling your friend, you may not know that I am getting that information. At the website of your bank, you have an encrypted private conversation which as few people as possible should learn about. The most enjoyable games are all about different information for different players, and so on. In all these cases, the dynamics itself poses a challenge, and elimination of worlds is definitely not the right mechanism. This chapter is about a significant extension of *PAL* that can deal with partially private information, reflecting different *observational access* of agents to the event taking place. We will present motivating examples, develop the system *DEL*, and explore some of its technical properties. As usual, the chapter ends with a number of further directions and open problems, from practical and philosophical to more mathematical. *DEL* is the true paradigm of dynamic-epistemic logic, and its ideas will return over and over again in this book.

4.1 Multi-agent information flow

Here is what may be the simplest scenario that goes beyond Chapter 3, from a brochure for my Spinoza Award project in 1998, written to explain dynamic logic to university officials:

Example Two envelopes.

> We have both drawn a closed envelope. It is common knowledge between us that one holds an invitation to a lecture on logic, the other to a wild night out in Amsterdam. Clearly, we are both ignorant of which is what. Now I open my envelope, and read the contents, without showing them to you. Yours remains closed. Which information has passed exactly through my action? I know now which fate is in store for me. But you have also learnt something, viz. that I know – though not what I know. Likewise, I did not just learn what is in my envelope. I also learnt something about you, viz. that you know that I know. The latter fact has even become common knowledge between us.

What is a general principle behind all this information flow? Intuitively, we need to *eliminate links* here instead of worlds. The initial information state is one of collective ignorance, and the update removes my uncertainty link – with both worlds still needed to model your ignorance:

In the resulting model, all observations from our story come out right. ■

This kind of update occurs in many games (cf. van Ditmarsch 2000).

Example Partial showing in the Three-Cards game.
The three players have drawn their single card each, as in the earlier diagram:

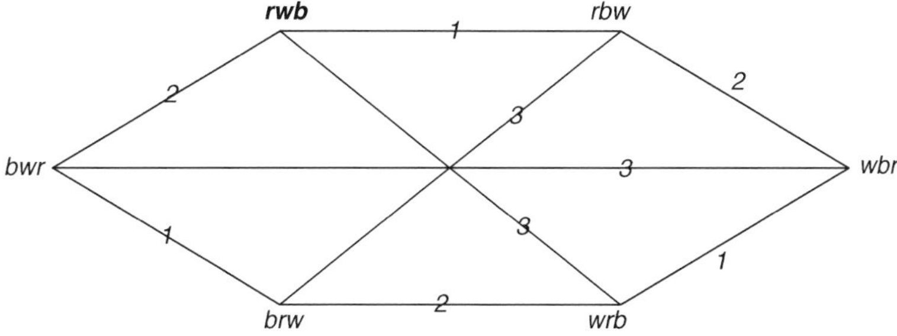

Now player *2* shows his card in public, but with its face *only to player 1*. The resulting update publicly removes the uncertainty lines for player *1*, but not for players *2* or *3*:

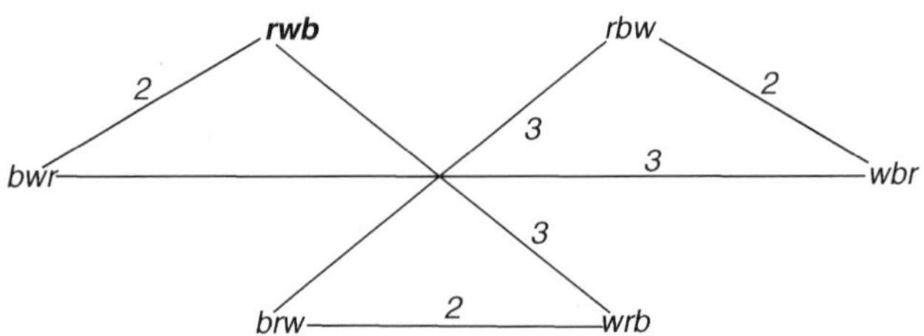

■

Games may not look serious, but they provide a *normal form* for many informational scenarios. Here is another phenomenon all around us:

Example *cc* and *bcc* in email.
Sending an email message to someone with a list of people under the *cc* button is like making a public announcement in the total group, assuming the computers work perfectly. But what does it mean to send a message with a *bcc* list? Now these people receive the message, but the others do not realize that, or, if they understand the system, they may think that there might be undisclosed recipients, but they cannot be sure. Scenarios like this do not work with link cutting, as it may be necessary to *add new worlds*. ■

Here is a simplest situation where a similar phenomenon happens. Initially, we are both ignorant whether some proposition p is true. A simplest model for this will again look as follows:

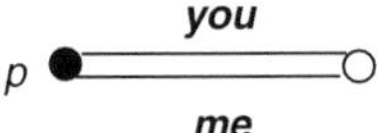

Now you hear a public announcement that p, but you are not sure whether I heard it, or I just thought it was a meaningless noise. In this case, intuitively, we need to keep two things around: one copy of the model where you think that nothing has happened, and one updated copy for the information that I received. This requires at least three worlds, and hence we need a new update mechanism that can even increase the number of worlds.

Most communication has private aspects, inadvertently or with deliberate hiding. Social life is full of procedures where information flows in restricted ways. Such diversity is highlighted in games that manipulate information flow and are an enjoyable training ground for social skills.[1] These phenomena

[1] The system of this chapter has been applied to fully describe parlour games like *Clue*.

pose real modelling challenges. It is not easy to write epistemic models with the right features by hand, and we need a systematic style of thinking. The usual practical art of modelling has to become a bit of a science.

It is serious! Before proceeding, I want to address a concern. Some colleagues see games and communication on a par with small-talk and gossip,[2] forced upon us by living in a crowd on this planet, but devoid of the lonely depth and beauty of truth and proof. In contrast, I think that multi-agent differences in information and observation are so crucial to human life, and the emerging social structures so essential, that it is *this* we want to understand in greater depth. It is easy to be rational when alone, but the greatest feats of the mind unfold in interaction. Social structure is not a curse, but constitutive of who we are as humans.

We will now analyse differential observational access, making epistemic logic much livelier. Other key logical features of multi-agency will unfold in subsequent chapters.

4.2 Event models and product update

The general idea in what follows is that agents learn from observing events to which they may have different access. An important step toward such a calculus was taken in Gerbrandy (1999a), using non-well-founded sets to model private announcement in *subgroups*: a subgroup gets informed while the rest learns nothing. Baltag, Moss & Solecki (1998) added the crucial idea of observational access via 'event models' to create the much more general system in use today.[3] A textbook is van Ditmarsch, van der Hoek & Kooi (2007), but our presentation will be self-contained, with some deviant aspects. The resulting system of *dynamic-epistemic logic (DEL)* provides a general account of multi-agent update of epistemic models, with some new features that are of interest in their own right.

For a start, the information models provided by epistemic logic have a natural companion, when we look at the events involved in scenarios of communication or interaction. Epistemic events can be treated much like epistemic states:

[2] But see Dunbar (1998) on the beauties and positive power of gossip.

[3] For an up-to-date survey of themes, cf. Baltag, van Ditmarsch & Moss (2008).

DEFINITION Event models.
An *epistemic event model* is a structure $\mathbf{E} = (E, \{\sim_i\}_{i \in G}, \{Pre_e\}_{e \in E}, e)$ with a set of *events* E, epistemic uncertainty relations $\sim_i$ for each agent,[4] a map assigning *preconditions* Pre_e to events e, stating just when these are executable, and finally, an *actual event* e.[5] ■

Event models are like static epistemic ones in endowing the events affecting our current model *themselves* with epistemic structure. Agents' uncertainty relations encode which events they cannot distinguish between in the scenario, because of observational limitations. When I open my envelope, you know it must be either Lecture or Night-Out, but you cannot distinguish the two events of 'my reading L' and 'my reading N'. Thus, both are in the event model, though not events we all consider irrelevant, like Mount Etna erupting. Next, the event model carries no propositional valuation, but instead, events come with *preconditions* for their occurrence. A public announcement $!P$ of a fact P presupposes truth of P, reading Night-Out has a precondition that this is the real card in my envelope, my asking a genuine question meant I did not know the answer. Most events carry information about when and where they occur. That is why they are informative!

This feature is at the heart of the following update mechanism, that describes how new worlds, after update, become pairs of old worlds with an event that has taken place:

DEFINITION Product update.
For any epistemic model $\mathbf{M}, s$ and event model $\mathbf{E}, e$, the *product model* $\mathbf{M} \times \mathbf{E}, (s, e)$ is an epistemic model with the following main components. Its domain is the set $\{(s, e) \mid s$ a world in $\mathbf{M}$, e an event in $\mathbf{E}$ and $\mathbf{M}, s \models Pre_e\}$, and its accessibility relations satisfy

$$(s, e) \sim_i (t, f) \text{ iff both } s \sim_i t \text{ and } e \sim_i f$$

Finally, the valuation for atoms p at (s, e) is the same as that at s in $\mathbf{M}$. ■

Explanation The new model uses the Cartesian product of the old model $\mathbf{M}$ and the event model $\mathbf{E}$, and this explains the possible growth in size.

[4] Accessibilities will often be equivalence relations, but we emphatically include event models with directed minimal accessibility for later scenarios of 'misleading' where knowledge gets lost.

[5] As with actual worlds in epistemic models, we assume one event actually takes place.

But some pairs are filtered out by the preconditions, and this elimination makes information flow. The *product rule* for epistemic accessibility reflects a simple idea: we cannot distinguish two new worlds if we could not distinguish them before and the new events cannot distinguish them either.

The above definition is incomplete: we have not specified *what language the preconditions come from*! We will assume here that they are in our epistemic language, either as atomic propositions ('I have the red card'), or as more complex epistemic ones ('I think you may know whether *P*'). Later on, we will allow the precondition language to contain dynamic update modalities,[6] but this feature is not required in the examples to follow. Finally, the intuitive understanding is that the preconditions are common knowledge in the group. Many of these assumptions can be lifted eventually to make the framework more general, but such variations will be obvious to readers who have grasped the simpler base case.

This mechanism can model a wide range of scenarios, keeping track of surprising changes in truth value for propositions generalizing the 'self-refuting' announcements of Chapter 3. We give some simple examples.

Example Public announcement.
The event model for a public announcement *!P* has just one event with precondition *P*, and reflexive accessibility arrows for all agents. The product model $\boldsymbol{M} \times \boldsymbol{E}$ then just retains the pairs *(s, !P)* where $\boldsymbol{M}, s \models P$, thus producing an isomorphic copy of our earlier $\boldsymbol{M}|P$. ■

Example The two envelopes.
Our initial model had '*L*' for my having the lecture, '*N*' for NightOut:

you
M L ●————○ N
me

You could not distinguish between the two events of my reading 'Lecture' and my reading 'NightOut' – and this shows in the following event model, where the black event is actual:

[6] Technically, this requires a simultaneous recursion with the definitions to follow.

E *events* **'I-Read-Lecture'** —— *you* —— *'I-Read-NightOut'*
preconditions *I have Lecture card* *I have NightOut card*

We also assume all reflexive arrows for both agents. In the resulting product model **M** × *E*, of the four possible pairs, two drop out by the preconditions. But this time, we do not just copy the original model: the product rule performs link-cutting, yielding the right outcome:

M × E ***(have-lecture, read-lecture)*** —— *you* —— *(have-nightout, read-nightout)* ■

Example The doubtful signal.
Here is how model size can increase. Consider our earlier model

M *p* ● —— *you* / *me* —— ○

Now take an event model for the earlier scenario where I hear *!p*, but think that you may also have witnessed just the trivial identity event *Id* that can happen anywhere:

E *event* ***!p*** —— *you* —— *Id*
precondition *p* *True*

This time, the product model **M** × **E** has *3* worlds, arranged as follows:[7]

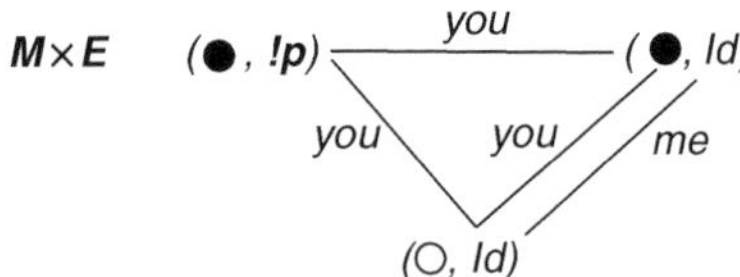

This indeed satisfies our intuitive expectations described earlier. ■

Product update in realistic scenarios can explode size. Ji (2004) analyses this for *bcc* in emails, whose complexity we feel when trying to track who knows what with this device. Van Ditmarsch (2000) analyses model changes in the

[7] The actual world in **M** × *E* is on the top left: the old actual world plus the actual event.

parlour game of *Clue*, that are much more delicate than our simple examples might suggest.[8] All this points the way to applications in the real-world setting of *security*: cf. van Eijck, Sietsma & Wang (2009).

4.3 Exploring the product update mechanism further

Product update provides something that is rare in logical semantics: a systematic mechanism for *constructing* and maintaining the right models. We will look at its theory later on. But descriptively, it is of interest in that it can be used to classify types of scenarios in communication. For instance, here is one important boundary:

Under-informed versus misinformed When you buy a board game, the rules printed on the cardboard cover may make things complicated, but they will not *necessarily* mislead you. Likewise, *bcc* in email will not mislead you if you are aware that it is available. There are degrees of public ignorance, and product update helps classify them. For instance, public resolution of questions does not increase domains, but only cuts links (cf. Chapter 6).

Arrows of belief In contrast with this, someone's illicit secret observation of a card will mislead you, inducing a false belief that nothing happened: it crosses a threshold toward *cheating*. Product update can also model this, using *pointed accessibility arrows* indicating what the current world could be like according to you. In particular, that world itself is not among these when you are mistaken. This issue will return in Chapter 7 on belief models and belief change, but for now, we just give one concrete illustration:

Example A secret peek.
We both do not know if p, but I secretly look and find out.

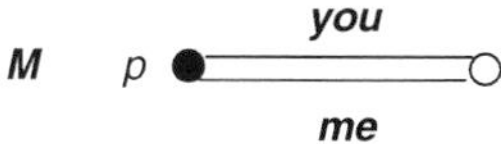

Here is the event model, with the pointed arrows as just indicated:

[8] In this setting, computer programs that perform update and related tasks are a useful tool, such as the *DEMO* system of Jan van Eijck: http://homepages.cwi.nl/~jve/demo/.

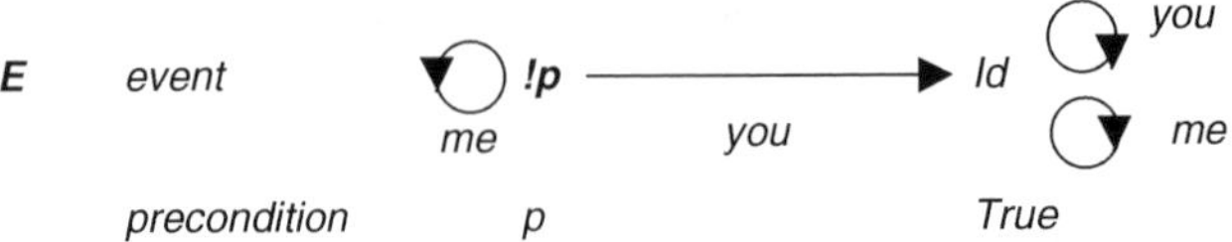

We get a product model **M** × **E** with three worlds, as in the following picture. For convenience, we leave out reflexive arrows (one for me in the real world, two for us both in the others):

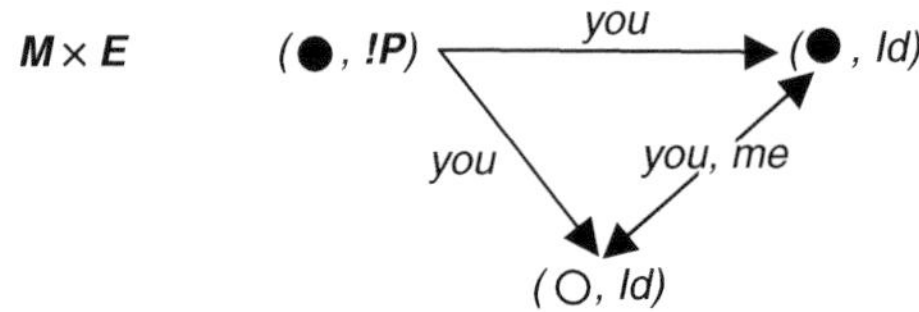

In the new actual world, I know exactly what has happened – while you still think, mistakenly, that we are in one of the other two worlds, where everything is just as before. ■

Thick events and agent-types So far, we had simple physical events, but the framework allows any type of event. This extends its power to much more complex cases, where we need to distinguish *agent types*, such as the beloved ones from logic puzzles, where one must separate Liars (who always lie) from Truth-Tellers (who always speak the truth).

Example Liars versus truth tellers.
You meet someone who tells you that *p*, but you do not know if she is a Liar, or a Truth-Teller. You hear her say that *p* is the case, which you know to be true. Here is how product update tells you that the person is a Truth-Teller. One forms new *pair events* of the form (agent type, proposition announced), with the following preconditions: $Pre_{\text{(Truth-Teller, !p)}} = p$, $Pre_{\text{(Liar, !p)}} = \neg p$. Only the first can happen in the actual world, and so you know that you are meeting a Truth-Teller. Of course, in general, more events will qualify, but the point is that, in scenarios like this, we can encode precisely what we know about the agents by using suitable preconditions. ■

Varying kinds of agents Which type of agent is hard-wired in product update? *DEL* says that uncertainties only arise from earlier uncertainties: this is perfect *memory*. It also says that old uncertainty can be removed by observation, giving agents powers of *observation* encoded in the event model. We will return to these features in Chapter 11. This does raise the interesting issue of how to modify *DEL* product update for other agents, say, with defective memory. Here is an illustration from van Benthem & Liu (2004):[9]

Example Update with memory-free agents.
Memory-free agents go by the last observation made. Thus, we can change their product update rule to read simply $(s, e) \sim (t, f)$ iff $e \sim f$. Liu (2008) has more discussion of this line, including modelling agents with k-cell bounded memories for any number k. ■

Real world change *DEL* so far assumed that atomic facts do not change in the process of getting information. This fits with its emphasis on *preconditions*. But actions can also *change* the world, as described by their *postconditions*. When the latter are simple, it is easy to incorporate change. In event models, now let each event e have a precondition Pre_e for its occurrence as before, but also an *atomic postcondition* $Post_e$ that assigns to each event e the set of atoms true after an occurrence of e. Equivalently, one can record this same information as a *substitution* σ_e^E from proposition letters p to either p, or $\neg p$.

DEFINITION Product update with change in ground facts.
The product model of an epistemic model **M** and an event model **E** is formed exactly as before, with just the following change in the propositional valuation: $(s, e) \in V_{\mathbf{M}x\mathbf{E}}(p)$ iff $p \in Post_e$.[10] ■

Why must we exercise care with the syntactic form of postconditions? The reason is that there is no canonical update method for making complex epistemic postconditions φ true. There may even be paradoxes in specifications, where the very act of bringing about φ blocks it.

4.4 Language, logic, and completeness

Given our extensive study of *PAL* in Chapter 3, the set-up for the dynamic logic *DEL* turns out familiar. The only novelty is the use of event models inside modalities, which takes some guts:[11]

[9] Van Ditmarsch & French (2009) have a more sophisticated account of *acts* of forgetting.

[10] Van Benthem, van Eijck & Kooi (2006) give a complete logic for this extended system.

[11] In practice, one often uses finite event models, though this is not strictly needed. We might then use suitable representations to make things look more linguistic.

Definition *DEL* language and semantics.

The *dynamic epistemic language* is defined by the following inductive syntax:

$$p \mid \neg\varphi \mid \varphi \wedge \psi \mid K_i\varphi \mid C_G\varphi \mid [\mathbf{E}, e]\varphi$$

where *(E, e)* is any event model with actual event *e*. There is a recursion here, since preconditions in event models come from the same language.[12] The semantic interpretation is standard, except for the dynamic modality:

$$\mathbf{M},\ s \models [\mathbf{E},\ e]\varphi \quad \text{iff} \quad \text{if } \mathbf{M},\ s \models Pre_e, \text{ then } \mathbf{M} \times \mathbf{E}, (s,\ e) \models \varphi$$ ■

This is a watershed. If you are down-to-earth, you will find a syntax putting models in a language weird, and your life will be full of fears of inconsistency. If you are born to be wild, however, you will see *DEL* as a welcome flight of the imagination.

Next, axiomatizing this language goes by the *PAL* methodology of recursion axioms (Baltag, Moss & Solecki 1998):

Theorem *DEL* is effectively axiomatizable and decidable.

Proof The recursion axioms are like those for *PAL*, with two key adjustments:[13,14]

$$[\mathbf{E},\ e]q \quad \leftrightarrow (Pre_e \rightarrow q)$$
$$[\mathbf{E},\ e]K_i\varphi \leftrightarrow (Pre_e \rightarrow \wedge_{e \sim i f \text{ in } \mathbf{E}} K_i[\mathbf{E},\ f]\varphi)$$

The latter axiom follows if we recall that the accessible worlds in **M** × **E** are those whose world-component is accessible in **M**, and whose event-component is accessible in **E**.[15] The overall completeness argument about reducing arbitrary dynamic-epistemic formulas to equivalent ones in the epistemic base language runs as before for *PAL*. ■

[12] *Caveat.* Preconditions with dynamic modalities referring to other event models might introduce circularities in the system. This is avoided by viewing the preconditions as literal *parts of* event models, so that the only event models referred to are 'smaller'.

[13] It is a useful exercise to check that the *PAL* axioms are special cases of what follows.

[14] We omit common knowledge here, for which we give a treatment later.

[15] *Caveat.* If the event model is infinite, this axiom may have an infinitary conjunction over all $f \sim_i e$. The axioms remain finite if the event models are only *finitely branching*.

One can adapt these axioms to deal with the two earlier extensions. With substitutions for world change, one amends the atomic case to

$$[\mathbf{E},\ e]q \leftrightarrow (Pre_e \rightarrow \sigma_e^E(q))$$

For memory-free agents, one amends the recursion axiom for knowledge to

$$[\mathbf{E}, e]K_i\varphi \leftrightarrow (Pre_e \rightarrow \wedge_{f \sim i\ e\ \text{in}\ E}\ U[\mathbf{E}, f]\varphi)$$

where U is the universal modality over all worlds in the model $\mathbf{M}$.

4.5 Conclusion

Event models are a truly new style of modelling, drastically increasing the analytical powers of dynamic-epistemic logic. One can now state systematically what agents know at successive stages of realistic scenarios like games, or any systems with restricted information flow, through a calculus of systematic model construction.

In principle, viewed as a logical system, *DEL* still behaves as with public announcements, but interested readers will see how, in the rest of this chapter, it raises new issues of its own. These include iterated update over longer-term temporal universes and extensions to group knowledge. We now continue with this, and at the end, we formulate a number of open problems linking *DEL* to other computational and mathematical frameworks in perhaps surprising ways.

4.6 Model theory of product update: bisimulation and update evolution

We study the semantic behaviour of product update a bit more, to get a better sense of what this mechanism of model change does. For a start, consider a traditional issue:

Preserving special logics Suppose that $\mathbf{M}$ belongs to a special model class for an epistemic logic defined by some frame condition on accessibility, and

E satisfies the same relational constraint. Will the product model **M** × **E** also fall in this class? This is definitely the case for the two extremes of *K* (no constraint) and *S5* (equivalence relations), but things are much less clear in between. In fact, things get worse than with public announcements. There, update always goes from models to submodels, and so we observed that all logics whose characteristic frame conditions are *universally definable* are preserved. But with product update, we only have the following observation from general model theory:

Fact All *universal Horn formulas* are preserved under product update, i.e., all first-order formulas of the special form $\forall x_1 \ldots \forall x_k\ (A \rightarrow B)$, where *A*, *B* are conjunctions of atoms.

This is easy to prove given the conjunctive nature of product update. There is a general result in first-order logic that a formula is preserved under submodels and direct products iff it is definable by a conjunction of universal Horn formulas. It is an open problem which precise syntactic class of first-order formulas is preserved under product update.

Example Properties non-preserved under product update.
The preceding fact excludes existential properties such as world *succession*: $\forall x\ \exists y\ Rxy$. Here is how this can fail. Take a two-world model **M** with $x \rightarrow y$, and *y* reflexive. Now take a two-event model **E** with $e \rightarrow f$ (this time, *f* is reflexive), where *x* satisfies Pre_f and *y* satisfies Pre_e. The pair (x, f) has no successor in **M** × **E**, as world and event arrows run in the wrong directions. But universal frame conditions with disjunction can also fail, such as *connectedness* of the ordering. In fact, the preceding counter-example also turns two connected orders into a non-connected one.[16] ■

Bisimulation invariance However this may be, at least, with product update, we have not left the world of basic epistemic logic, since the same semantic invariance still holds:

Fact All formulas of *DEL* are invariant for epistemic bisimulation.

[16] One might use the existence of such cases against product update. But one can also bite the bullet: the examples show how cherished properties for static logics no longer look plausible when we perform quite plausible dynamic model constructions. For instance, the fragility of *connectedness* has often been remarked upon in the theory of ordering – and in Chapter 9, losing it is just what we want.

Proof One proof is by the above reduction to basic *EL* formulas. More informative is a direct induction, where the case of the dynamic modalities uses an auxiliary result.

FACT For any two bisimilar epistemic models **M**, *s* and **N**, *t*, the product model *(**M**, s)* × *(**E**, e)* is bisimilar with the product model *(**N**, t)* × *(**E**, e)*. ■

This observation leads to a new notion and one more question. In Chapter 2, bisimulation was our structural answer to the question when two epistemic models represent the same information state. The present setting raises the issue when two event models are the same. One might say that two event models $(\mathbf{E}_1, e_1)$, $(\mathbf{E}_2, e_2)$ *emulate each other* when, for all epistemic models *(**M**, s)*, the product models $(\mathbf{M}, s) \times (\mathbf{E}_1, e_1)$ and $(\mathbf{M}, s) \times (\mathbf{E}_2, e_2)$ are bisimilar.[17] Van Eijck, Ruan & Sadzik (2006) provide a complete characterization.

Update evolution Like public announcement, product update becomes more spectacular when *repeated* to model successive assertions, observations, or other informational events. We discussed *PAL* with iterations already in Chapter 3, and it will return in Chapter 15. *DEL* scenarios with iteration generate *trees* or *forests* whose nodes are finite sequences of events, starting from worlds in the initial epistemic model. We will study *DEL*-over-time in Chapter 11, but here we look at one case.

Product update may blow up epistemic models starting from some initial **M**, creating ever larger models. But will it always do so? We discuss one concrete scenario with finite extensive *games of imperfect information* (cf. van Benthem 2001, and Chapter 10 below). First collect all moves with their preconditions (these restrict what is playable at each node of the game tree) into one event model **E**, that also records which players can observe which move. Now, rounds of the game correspond to updates with this game-event model:

DEFINITION Update evolution models.
Let an initial epistemic model **M** be given, plus an event model **E**. The *update evolution model Tree* *(**M**, **E**)* is the possibly infinite epistemic tree model whose

[17] In fact, this holds iff $\mathbf{E}_1$, $\mathbf{E}_2$ produce bisimilar results on all bisimilar inputs **M**, **N**.

successive layers are disjoint copies of all successive product update layers $M \times E$, $(M \times E) \times E, \ldots$[18] ■

The potential infinity in this sequence can be spurious – as in those games where the complexity of information first grows, but then decreases again toward the end game:

Example Stabilization under bisimulation.
Consider a model **M** with two worlds satisfying p, $\neg p$, between which agent *1* is uncertain, though *2* is not. The actual world on the left has p.

For convenience, we read the literals p, $\neg p$ as naming the worlds. Now an announcement of p takes place, and agent *1* hears this. But agent *2* thinks that the announcement might also just be a statement 'True' that could hold anywhere. The event model **E** for this scenario is one we had earlier, and we now draw it as follows:

!p (precondition: *p*) ——*2*—— *Id* (precondition: *T*)

E

The next two levels of *Tree* **(M, E)** then become as follows:

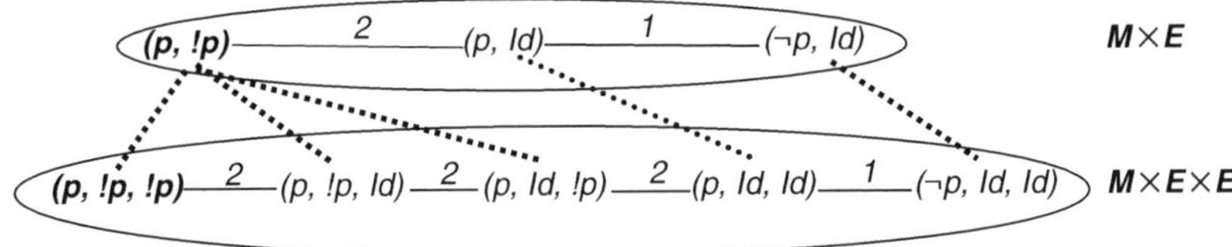

Now note that there is an epistemic *bisimulation* between these two levels, connecting the lower three worlds to the left with the single world ***(p, !p)*** in **M** × **E**. Thus, **M** × **E** × **E** is bisimilar with **M** × **E**, and the subsequent tree iteration is finite modulo bisimulation. ■

Update evolution also suggests that *DEL* event models may yield compact representations of complex processes. In Chapter 11, we look at the general

[18] This definition will be made more precise in the epistemic-temporal logics of Chapter 11.

background in branching *temporal logic*, and find precise conditions under which an extensive game with imperfect information can be represented as a tree induced by iterated product update.[19]

4.7 Language extensions: common knowledge

Is the dynamic-epistemic language strong enough for our models? Several new modalities make sense, as in Chapters 2 and 3. In particular, in communication scenarios, we need to deal with common knowledge – perhaps also implicit knowledge. Already the extension with common knowledge suggests further strengthenings to a propositional dynamic logic: van Benthem, van Eijck & Kooi (2006) supplies details for what follows.

Finding a *DEL*-style recursion axiom for common knowledge is not at all routine. Indeed, it is not clear intuitively what the right common knowledge axiom should be, even for simple scenarios. We will work with arbitrary models, not necessarily with equivalence relations. Here is about the simplest illustration of what we want to analyse.

Example Secret observation, or private subgroup announcement.
Consider the following two-event model involving two agents *1*, *2* – where *1* learns that the precondition of *e* holds, say *p*, while agent *2* thinks that nothing has happened (*T*):

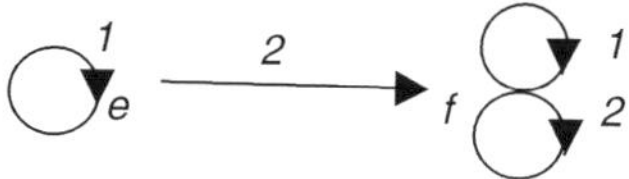

First, we want *2* to learn nothing significant here. This can be seen as follows. The second event with its accessibilities acts as a public announcement of its precondition $T =$ 'true'. Now here is a law of public announcement (announcing *T* does not change a model):

[19] Van Benthem (2001) stated the following *Finite Evolution Conjecture*: starting with finite **M** and *E*, the model *Tree* (**M**, *E*) remains finite modulo bisimulation. Hence some horizontal levels $\mathbf{M} \times E^k$ and $\mathbf{M} \times E^l$ in the tree must be bisimilar, with $k < l$. This was refuted in Sadzik (2005). Finite evolution up to bisimulation holds only in single-agent *S5*-models, but can fail with two *S5*-agents. (Sadzik uses *pebble games* over **M** and *E*, where *Tree* (**M**, *E*) is finite modulo bisimulation iff the response player has a winning strategy.) The Conjecture holds for special cases, like when the epistemic relations for all agents in the models *E* are linearly ordered by inclusion. This applies to some games.

Fact $<!T>\varphi \leftrightarrow \varphi$ is schematically valid in *PAL*.

Of course, agent *1* does learn that p is the case (note that only event e is accessible to e for *1*):

$$[\mathbf{E},\ e]K_1 p \leftrightarrow (p \rightarrow K_1[\mathbf{E},\ e]p) \leftrightarrow (p \rightarrow K_1(p \rightarrow p)) \leftrightarrow (p \rightarrow T) \leftrightarrow T$$

Now, try to write a recursion axiom for common knowledge formulas $[\mathbf{E}, e]$ $C_{\{1,\ 2\}}\varphi$ in this setting: it is not obvious![20] Here is an explicit solution, written for convenience with existential modalities, that may be computed using the techniques that prove our next theorem:

$$<\mathbf{E},\ f><C_G>\varphi \leftrightarrow <(1 \cup 2)^*>\varphi$$

$$<\mathbf{E},\ e><C_G>\varphi \leftrightarrow (p \wedge <(?p;\ 1)^*>(p \wedge \varphi)) \vee (p \wedge <(?p;\ 1)^*>(p \wedge <2><(1 \cup 2)^*>\varphi)) \quad \blacksquare$$

One solution here is to just add an ordinary common knowledge modality, give up on a recursion axiom, and prove completeness via special-purpose inference rules (as in Baltag, Moss & Solecki 1998). But if we want to stick with recursion axioms, the driving force in our principled information dynamics, we need another approach.

The only known solution right now is a drastic extension of our static language to an epistemic variant of *propositional dynamic logic*, a system we have encountered before in Chapter 3 as providing program structure on top of basic announcement actions:

Definition *EDL* language and semantics.
The *epistemic dynamic language EDL* is defined by the following syntax:

formulas	$p \mid \neg\varphi \mid \varphi \wedge \psi \mid [\pi]\varphi$
programs	$i \mid \pi \cup \pi \mid \pi;\pi \mid \pi^* \mid ?\varphi$

The semantics is over the standard epistemic models of Chapter 2, in a mutual recursion.[21] Formulas are interpreted as usual, where $\mathbf{M}, s \models [\pi]\varphi$ means that φ is true at all successors of s in $\mathbf{M}$ according to the relation

[20] Draw a picture of how $C_G\varphi$ is true in a product model $\mathbf{M} \times \mathbf{E}$, with finite accessibility sequences $(s_1, e_1), \ldots, (s_k, e_k)$, and now try to write the same information at the level of $\mathbf{M}$ itself, with only s-sequences. You need to keep track of all the different events attached, and it is unclear how to do this.

[21] For details, see Blackburn, de Rijke & Venema (2000), Harel, Kozen & Tiuryn (2000), as well as van Benthem (to appearB).

denoted by the program π. Simultaneously, program expressions π denote binary accessibility relations between worlds defined inductively along with the given operations: atomic i are agents' epistemic uncertainty relations, $\cup$ stands for *union* ('choice'), ; for *sequential composition*, and * for *reflexive-transitive closure* (Kleene star). Finally, $?\varphi$ is the *test program* $\{(s, s) \mid \boldsymbol{M}, s \models \varphi\}$. ■

Caveat: EDL versus DEL This language is a dynamic logic in the sense of logics of computation, but it should not be confused with the dynamic-epistemic logic *DEL* of this chapter that changes models. The preceding semantics stays inside one fixed model.[22]

Complex program expressions now define epistemic accessibilities for complex agents. An example is common knowledge: $C_G\varphi$ can be defined as $[(\cup_{i\in G}\ i)^*]\varphi$, with a matching agent $(\cup_{i\in G}\ i)^*$. Conditional common knowledge $C_G{}^{\psi}\varphi$ as in Chapter 3 involves a complex agent defined using tests: $(?\psi\ ;\ \cup_{i\in G}\ i\ ;\ ?\psi)^*$. Of course, not every program corresponds to a natural epistemic agent: we leave the study of natural fragments open here. Still, *EDL* makes an interesting general point about group agency (cf. Chapter 12): one can use program logics to give an explicit account of epistemic group agents.

Back to our original topic, the language *EDL* does provide recursion axioms for common knowledge, and much more. The following result states this abstractly. Consider a version of *DEL* defined as earlier in this chapter, but where the static language is now *EDL*. The crucial recursion axiom will now be of the form $<E, e><\pi>\varphi$, for any program expression π of *EDL*, where we use existential versions for convenience in what follows.

Definition Product closure.

We call a language *L* *product closed* if for every formula φ in *L*, the expression $<E, e>\varphi$, interpreted using product update as before, is equivalent to a formula already inside *L*. We speak of *effective* product closure if there is an algorithm producing the equivalents. ■

Our earlier completeness result showed that standard epistemic logic *EL* is product-closed for *finite* event models – and indeed the recursion axioms provide an effective algorithm. We will restrict ourselves to finite event

[22] Even so, a standard propositional dynamic perspective makes sense for *DEL*. In Chapter 3, we briefly viewed the universe of all epistemic models as a big state space (the 'Super-model'), with transition relations matching model updates. See Chapter 11 for more.

models henceforth. Now product closure fails for the epistemic language *EL-C* that just adds plain common knowledge (Baltag, Moss, & Solecki 1998; van Benthem, van Eijck & Kooi 2006). But here is a case of harmony:

THEOREM The logic *E-PDL* is effectively product-closed.[23]

Proof The argument involves two simultaneous inductions. Fix some finite pointed event model *(E, e)* where, in what follows, we will assume that the domain of E is enumerated as a finite set of numbers $\{1, \ldots, n\}$. Now, we show by induction on *EDL*-formulas that

LEMMA A For φ in *EDL*, $\langle E, e\rangle\varphi$ is in *EDL*.

The inductive steps for atomic formulas and Booleans are just as for the completeness of *DEL*. The essential step is the case $\langle E, e\rangle\langle\pi\rangle\varphi$ with program π, that we analyse as follows:

DEFINITION Path expressions.
Given any *EDL* program π, and any two events *e*, *f* in the given event model E, we define the *EDL* program $T(e, f, \pi)$ by the following induction:

$T(e, f, i)$	$=$	$?Pre_e\ ;\ i\ ;\ ?Pre_f$	if $e \sim_i f$ in E
		$?\bot$	otherwise[24]
$T(e, f, \pi_1 \cup \pi_2)$	$=$	$T(e, f, \pi_1) \cup T(e, f, \pi_2)$	
$T(e, f, \pi_1\ ;\ \pi_2)$	$=$	$\bigcup_{g\in E} (T(e, g, \pi_1)\ ;\ T(g, f, \pi_2))$	
$T(e, f, ?\varphi)$	$=$	$?\langle E, e\rangle\varphi$	if $e = f$
		$?\bot$	otherwise
$T(e, f, \pi^*)$	$=$	$P(e, f, n, \pi)$	n is largest in E

Here the auxiliary program $P(e, f, i, \pi)$ is itself defined by induction on the natural number *i*:

$P(e, f, 0, \pi)$	$=$	$T(e, f, \pi) \cup ?T$	if $e = f$
		$T(e, f, \pi)$	if $e \neq f$
$P(e, f, i{+}1, \pi)$	$=$	$P(e, f, i, \pi) \cup$	
		$P(e, i, i, \pi); P(i, i, i, \pi)^*; P(i, f, i, \pi)$	if $i \neq e, i \neq f$
		$P(e, i, i, \pi); P(i, i, i, \pi)^*$	if $i \neq e, i = f$
		$P(i, i, i, \pi)^*; P(i, f, i, \pi)$	if $i = e$ ■

[23] The first proof of this result was in van Benthem & Kooi (2004) using finite automata.
[24] Here, the 'falsum' $\bot$ is a formula that is always false.

All these programs are in the language *EDL*. Their meanings are clarified in the following assertion that is proven simultaneously with all other assertions in the main proof:

Claim (a) $(w, v) \in [[T(e, f, \pi)]]^M$ iff $((w, e), (v, f)) \in [[\pi]]^{M \times E}$

(b) $(w, v) \in [[P(e, f, i, \pi)]]^M$ iff there is a finite sequence of transitions in $\mathbf{M} \times \mathbf{E}$ of the form $(w, e)\ [[\pi]]\ x_1 \ldots x_k\ [[\pi]]\ (v, f)$ such that none of the stages $x_j = (w_j, e_j)$ has an event e_j with index $j \geq i$.

We will not spell out the inductive steps in the proof of the Claim. The most complex case in the clauses for $P(e, f, i+1, \pi)$ expresses that a finite path from e to f can at worst pass through the event with index i some finite number of times – and marking these, it can then be decomposed as an initial part starting from e with indices lower than i for the intermediate stages, an iteration of cases from i to i with lower indices, and a final part from i to f with lower indices:

$$e \quad \ldots < i \ldots \quad i \ \ldots.. < i \ \ldots.. \quad i \ \ldots < i \ldots \quad f$$

With this explanation, here is our desired equivalence:

LEMMA

$$<E, e><\pi>\varphi \leftrightarrow \bigvee_{i \in E} <T(e, i, \pi)><E, i>\varphi$$

The proof follows easily from the explanation for the relations $T(e, i, \pi)$. ■

The preceding argument is constructive, and it yields an effective algorithm for writing correct recursion axioms for common knowledge and, in fact, all other *EDL*-definable notions.[25] Van Benthem, van Eijck & Kooi (2006) use this to provide axioms for common knowledge in subgroups after complex scenarios such as uses of *bcc* in email.[26]

25 *Open Problem*: Does *EDL* have natural smaller fragments satisfying product closure?

26 Van Benthem & Ikegami (2008) prove product closure by analysing the recursion in an *epistemic μ-calculus* (cf. Chapter 15) extending *EDL* with fixed-point operators over positive formulas. The key is the equivalence $<E, e><\pi^*>\psi \leftrightarrow <E, e>\psi \vee <E, e><\pi><\pi^*>\psi$. One views formulas $<E, e><\pi^*>\psi$ as propositional variables p_e for all events e, that can be solved in one simultaneous smallest fixed-point equation, with clauses matching the above transition predicates $T(i, j, \pi)$ for the program π. To show that the solution is in *EDL*, the authors use the fact that propositional dynamic logic is a μ-calculus fragment whose recursions involve only modal diamonds and disjunctions. The same analysis shows that the epistemic μ-calculus is effectively product-closed.

4.8 Further issues and open problems

We conclude with a list of further developments in the *DEL* framework, actual or potential.

Agents and channels *PAL* and *DEL* have no explicit agency, as they just analyse events. How can agents be added? And how to model their communication channels, that restrict the sort of knowledge achievable in a group? Van Benthem (2006b) discusses agents with one-to-one telephone lines, Pacuit & Parikh (2007) discuss realistic general social networks. Roelofsen (2006) translates communication channels in terms of event models. A general *DEL* theory of channels (cf. Apt, Witzel & Zvesper 2009; van Eijck, Sietsma & Wang 2009 for recent progress) meets with the analysis of security and cryptography in computer science (Dechesne & Wang 2007; Kramer 2007).

Computational complexity The complexity of *DEL* core tasks is like for *PAL*. Model checking is in **P**, model comparison as bisimulation checking is in **P**, and satisfiability is probably *Pspace*-complete, though this has not yet been proved. But other questions seem of a different nature, having to do with difficulty for agents, not complexity of our logic. There are thresholds in model size in going from public to private communication, or from telling the truth to lying and cheating. Likewise, there is complexity in update evolution, when predicting maximal size of intermediate models between the opening of a game and its end game. What is the right notion of agent-oriented complexity here? Chapter 11 discusses this same issue in somewhat more detail – but we offer no solution.

Abstract update postulates The update rule of *DEL* seems just one option. Chapter 3 had a *postulational* correspondence analysis of *PAL*, showing how world elimination is the only operation satisfying intuitive postulates on update. Can one extend this, showing how, in a space of model constructions, the *DEL* axiom $[\mathbf{E}, e]K_i\varphi \leftrightarrow (Pre_e \rightarrow \wedge\{K_i[\mathbf{E}, f]\varphi \mid f \sim_i e \text{ in } \mathbf{E}\})$ fixes product update? But there are also other ways of capturing update rules. Given a space of models and an ordering or distance function, one can look for models closest to the current one satisfying some condition. Thus, a hard information update *!P* on a current model **M** might be a model *closest to* **M** that has undergone some minimal change to make *P* common knowledge.[27]

[27] More abstractly, updates have been studied as minimal jumps along inclusion in models that contain all relevant information stages: cf. van Benthem (1989), Jaspars (1994).

Sorting out connections between these approaches seems another desideratum to better understand the mechanics of *DEL* information update.

Backward or forward-looking? In line with the preceding point, *DEL* is *past-oriented*: its preconditions tell us what was the case when the informative event occurred. In contrast with this, many temporal logics of agency (cf. Chapter 11) are *future-oriented*, talking about what will be the case with postconditions of 'coming to know' or 'seeing to it that φ' (Belnap, Perloff & Xu 2001). What is the connection between these approaches?

Generic event languages *DEL* has a striking asymmetry. Epistemic models **M** interpret a language, but event models **E** do not. In particular, preconditions Pre_e are not *true* at event *e* in **E**: they state what must be true *in* **M** for *e* to happen. But one might define a second epistemic language for event models, describing generic properties of events, with atoms, Booleans, and modalities over indistinguishable events. Van Benthem (1999a) has an 'epistemic arrow logic' for this purpose, that describes product update in a joint modal language for epistemic and event models. This fits the remark in Gerbrandy (1999a) that event models are often too specific. We want to make *generic assertions* like 'for every epistemic event of type *X*, result *Y* will obtain'. Sadrzadeh & Cirstea (2006) show how algebraic formalisms can abstract away from specifics of *DEL*. What would be an optimal framework? See Aucher (2009) for a recent proposal that stays closer to logic.

Links with other disciplines: events, information, and computation Events are important in philosophy, linguistics, and artificial intelligence (McCarthy 1963; Sergot 2008). It would be of interest to connect our logics with these other traditions where events are first-class citizens.[28] *DEL* also meets computer science in the study of specific computational and informational events. We saw how it relates to fixed-point logics of computation, and to logics of agents and security – and these initial links should be strengthened.

Logic of model constructions From a technical point of view, perhaps the most striking feature of *DEL* is its explicit logic of *model constructions*, such as *PAL*-style definable submodels or *DEL*-style products. Van Benthem (1999b) links the latter to a basic notion in logic: *relative interpretations* between theories

[28] Cf. the 2010 issue of the *Journal of Logic, Language and Information* on relating different paradigms in the temporal logic of agency, edited by van Benthem & Pacuit.

using pair formation, predicate substitution, and domain relativization. Formalizing properties of such constructions seems a natural continuation of standard model theory. In this line, van Benthem & Ikegami (2008) use product closure as a criterion for expressiveness of logical languages, extending the usual closure under translation or relativization. But there are also other technical perspectives on the product update mechanism of *DEL*. For instance, the discussion in Chapter 3 of making iterated announcements $[!P][!Q]\varphi$ suggests a more general axiom

$$[\mathbf{E}_1,\ e_1][\mathbf{E}_2,\ e_2]\varphi \leftrightarrow [\mathbf{E}_1\ o\ \mathbf{E}_2,\ (e_1,\ e_2)]\varphi,$$

where o is some easily definable operation of *composition* for event models. Another such operation would be *sum* in the form of disjoint union. What is a natural *complete algebra of event models*? Here, we meet computer science once more. Like *DEL*, Process Algebra (cf. Bergstra, Ponse & Smolka 2001) is all about definable bisimulation-invariant model constructions.[29] Can we connect the two approaches in a deeper manner?

From discrete dynamic logics to continuous dynamical systems We conclude with what we see as a major challenge. Van Benthem (2001, 2006c) pointed out how update evolution suggests a long-term perspective that is like the evolutionary dynamics found in *dynamical systems*. Sarenac (2009) makes a strong plea for carrying Logical Dynamics into the latter realm of emergent system behaviour, pointing at the continuous nature of real events in time – and also in space. Technically, this would turn the *DEL* recursion axioms into differential equations, linking the discrete mathematics of *DEL* and epistemic-temporal logic with classical continuous mathematics. Interfacing current dynamic and temporal logics with the continuous realm is a major issue, also for logic in general.

4.9 Literature

A significant first step extending *PAL* to private announcements in subgroups was made in Gerbrandy & Groeneveld (1997). Gerbrandy (1999a) is a

[29] In fact, speaking simultaneously and matching public announcements $!(P \wedge Q)$ are already a rudimentary form of concurrency, obeying laws very different from those for sequential composition. Also relevant is van Eijck, Ruan & Sadzik (2006) on bisimulation-invariant operations on event models.

trail-blazer covering many more examples, raising many conceptual issues concerning epistemic dynamics for the first time, and technically defining a broad class of 'epistemic updates' over non-wellfounded sets. Van Ditmarsch (2000) studies the logic of a family of partly private learning events that suffice for modelling the moves in some significant parlour games. Extending Gerbrandy's work decisively, it was Baltag, Moss & Solecki (1998) who added the key idea of product update via event models, giving *DEL* its characteristic flavour and modelling power, and clarifying the mathematical theory. Baltag & Moss (2004) is a mature later version of what is now known as the '*BMS* approach'. Van Benthem (2001) is a systematic description of update evolution in *DEL* linked with temporal logic and games of imperfect information. Van Benthem, van Eijck & Kooi (2006) is a version of *DEL* with factual change and recursion axioms for common knowledge. These are just a few basic sources. Further contributions to the area will be referenced in the chapters where they make their appearance.

5 Dynamics of inference and awareness

As we have seen in our Introduction, agents have many ways of getting information for their purposes. Observation, inference, and communication are all respectable means, and logical dynamics should incorporate all on a par. Some of this was done in the logics of public and private observation of Chapters 3, 4. But *PAL* and *DEL* still fail to do justice to even the simple story of The Restaurant in Chapter 1, where the waiter mixes observation and inference in one task. The proof system of *PAL* is a classical static description of properties of agency, not a dynamic account of events of inferential information flow. This mismatch reflects a conceptual problem of long standing, that was briefly raised in our introductory Chapter 1: the variety of notions of information in logic, and the matching variety of acts that handle information. In this chapter, we will explore a dynamic-epistemic approach, showing how these issues can be dealt with in the general framework of this book. The approach comes in two flavours: one based on access to worlds, and one on awareness. Our proposals are quite recent, and they are mainly meant as a point of departure. In Chapter 13, we will discuss their philosophical significance.

5.1 Information diversity and missing actions

The problem of inferential information Of our three sources of information, observation and communication seem similar. An observation is an answer to a question to Nature, hearing an answer to a question is like observing a fact. Our logics *PAL* and *DEL* handled information flow of this sort. But though entangled with observation, the third fundamental process generating information poses a problem. There is no consensus on what information flows in inference. Here is the difficulty (cf. Chapter 1):

Example Deduction adds no semantic information.
Consider the valid inference *from* $A \vee B$ *and* $\neg A$ *to* B. The premises may be viewed as updating some initial information state with four worlds: the first rules out *1*, the second *2*:

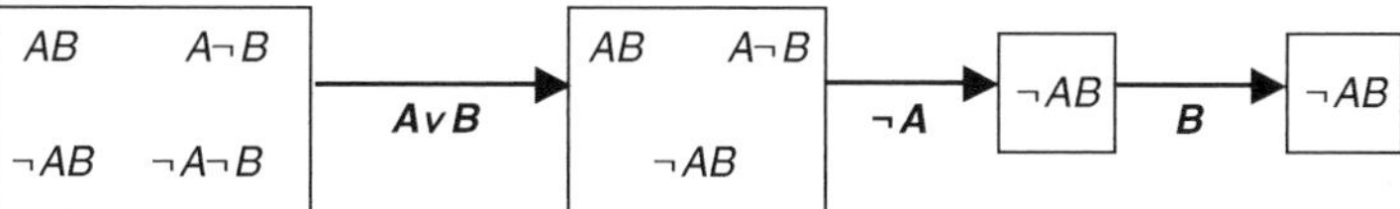

The final inference step does not do anything useful: we are stuck.

But clearly, inferences *are* useful. All of science is a grand co-existence of experiment and deduction.[1] Van Benthem & Martinez (2008) discusses attempted solutions to this 'scandal of deduction', that also occurs in the philosophy of science.[2] We need a fine-grained account of what changes when we make valid inferences, and this must work together with successful semantic elimination accounts of observation-based update.

Information dynamics on syntax There are many approaches to inferential dynamics.[3] In this chapter, we mainly follow a simple syntactic sense of information: inferences add new formulas to a current set Σ of formulas that we have already seen to be true. Thus, an act of inference from $A \vee B$ and $\neg A$ to B adds B to any set of formulas Σ containing $A \vee B, \neg A$. We will give this viewpoint a general twist, making inference a special case of actions that elucidate the syntactic information given to us by different sources.

Logical omniscience, and what is missing in action But before starting with our systems, we restate the issue as it arises in an epistemological setting, in the so-called problem of 'logical omniscience'. Recall the distribution axiom in epistemic logic (Chapter 2):

$$K(\varphi \to \psi) \to (K\varphi \to K\psi)$$

1 Van Benthem (1996) points out how puzzles like Master Mind, that can be described with semantic update only, in practice involve crucial syntactic inferences.

2 A related problem is explaining how *computation* is informative (Abramsky 2008).

3 Some say that single steps are trivial, but information emerges by *accumulation* of many inferences. Single inferences as real information-changing events occur in logics of proofs (Artemov 1994), the clausal approach of d'Agostino & Floridi (2007), or the modal logic of proof of Jago (2006).

This seems to make knowledge closed under Modus Ponens and, repeating single steps, under logical inference in general. The usual objection is that this idealizes agents too much. But this is misguided on two counts. First, as we have seen already, the $K\varphi$ modality really describes implicit semantic information of the agent, which does have the stated closure property. The point is rather that closure need not hold for a related but different intuitive notion, viz. explicit 'aware-that' knowledge $Ex\varphi$, in a sense yet to be defined. So, what we need is not epistemic logic bashing, but a richer account of agents' attitudes. And there is more. The real issue is not if explicit knowledge *has* deductive closure. It is rather: 'What do agents *have to do* to make implicit knowledge explicit?' Consider the two assumptions $Ex(\varphi \rightarrow \psi)$, $Ex\varphi$, saying that the agent explicitly knows both $\varphi \rightarrow \psi$ and φ. These imply $K\psi$, that is, the agent knows ψ implicitly in our semantic sense. But to make this information explicit, the agent has to do work, by an act of awareness raising that leads to $Ex\psi$. Stated syntactically,

the implication $Ex(\varphi \rightarrow \psi) \rightarrow (Ex\varphi \rightarrow Ex\psi)$ *contains a gap* $[\,]$:

having the real logical form $Ex(\varphi \rightarrow \psi) \rightarrow (Ex\varphi \rightarrow [\,]\ Ex\psi)$

and in that gap, we should place an informational *action*. Thus, the agent is no longer omniscient: postulating a closure property just means a refusal to do honest work. But the agent is not defective either: with the right actions available, she can raise awareness as needed. This chapter explores these ideas in the dynamic-epistemic methodology.

5.2 Worlds with syntactic access

Creating access to worlds Here is a first approach (van Benthem 2008e, f). Adding syntax to a semantic information range, we attach sets of formulas to worlds w as an agent's *access* to w. This access is the code the agent has available as explicit representation of relevant features of w, and for cognitive tasks manipulating this, such as inference and computation. For instance, the inference from $A \vee B$ and $\neg A$ to B, though adding no semantic information, did increase access to the single final world w by adding an explicit property B:

from $w, \{A \vee B, \neg A\}$, we went to $w, \{A \vee B, \neg A,\ B\}$

In standard epistemic models, agents implicitly know things encoded in the accessibility structure. But they need not have conscious access to this knowledge via linguistic code. But code is essential to many logical activities, such as inference or model checking.

As for dynamic events, the observations $!P$ of Chapter 3 increased implicit knowledge by eliminating worlds, while acts of inference increase internal access to worlds, upgrading implicit to explicit knowledge. We will model this in a two-level semantic-syntactic format with pairs (w, Σ) of worlds w plus a set of formulas Σ partially describing w, merging dynamic-epistemic logic with the awareness logics of Fagin, Halpern & Vardi (1990).[4]

External versus internal acts of 'seeing' Two-level models allow for information update at both levels: worlds may be removed, or sentences may be added. In this setting, the basic distinction is between *implicit observation* of true facts and acts of *explicit realization* or 'promotion' making implicit knowledge explicit. Likewise, in colloquial terms, the phrase 'I see' can mean that I am observing some fact, or that something becomes clear to me. We will treat both types of act, returning to inferential dynamics later.

5.3 Dynamic-epistemic logic with informational access

Epistemic logic with informational access to worlds A slight change in the standard epistemic models of Chapter 2 suffices to model the desired notion:

DEFINITION Static epistemic access language.
The *epistemic access language* extends a propositional base with an epistemic modality, plus one of 'explicit information' to be explained below. Its syntax rules are these:[5]

$$p \mid \neg\varphi \mid \varphi \wedge \psi \mid K\varphi \mid I\varphi$$

In this language, the *factual propositions* are those constructed without using the operators K or I. ■

[4] Related ideas occur with belief bases in belief revision theory (Gärdenfors & Rott 2008).

[5] For simplicity, we will work with single-agent versions in this chapter – though both observation and inference can be social: think of experiments performed by groups, or multi-agent argumentation.

Later on we will have modalities for dynamic events. For a start, here is our static semantics:

DEFINITION Models and truth conditions.
Epistemic access models $\mathbf{M} = (W, W^{acc}, \sim, V)$ are epistemic models with an additional set W^{acc} of *access worlds (w, X)* consisting of a standard world *w* and a set of *factual* formulas X. In these pairs, we require that all formulas in X be true at *(w, X)* according to the truth conditions below, while epistemically indistinguishable worlds have the same set.[6] The key truth conditions are as follows (with Booleans interpreted as usual):

$\mathbf{M}, (w, X) \vDash p$	iff	$w \in V(p)$
$\mathbf{M}, (w, X) \vDash K\varphi$	iff	*for all* $v \sim w : \mathbf{M}, (v, X) \vDash \varphi$
$\mathbf{M}, w, X \vDash I\varphi$	iff	$\varphi \in X$ *and* $\mathbf{M},\ w,\ X \vDash \varphi$ ■

Intuitively, $I\varphi$ says that the agent is explicitly *informed about* the truth of φ. The logic is easy. We get the usual principles of epistemic logic, while the *I*-operator has a few of its own:

FACT $I\varphi \to \varphi$ and $I\varphi \to KI\varphi$ are valid principles of our static logic.

Proof Veridicality $I\varphi \to \varphi$ was in the above definition, and so is Introspection $I\varphi \to KI\varphi$ by the uniformity in assigning access sets. The two principles together imply $I\varphi \to K\varphi$. ■

Two limitations We have only put restricted factual formulas in access sets. Our reason is maintaining Veridicality. Unlike factual formulas, complex epistemic ones can change truth values in update: e.g., true Moore sentences $\neg Kq \wedge q$ became false when announced. Thus, an access set may become obsolete. We also assumed that sets are the same across epistemic ranges: only modelling access to implicit knowledge, not to individual worlds.

Two ways of changing access models Different events can change these models. Indeed, *PAL* announcement splits into two versions. One changes the range of worlds but not the access, an implicit seeing that something is the case.

[6] One can also have a *function* from worlds to access sets respecting ~.

The other is a conscious act where the observed proposition also moves into the agent's access set.[7]

Definition Implicit and explicit observation.
Let *M, (w, X)* be an epistemic access model whose actual world *(w, X)* satisfies the formula φ. An *implicit public observation* $!\varphi$ transforms this into a new access model $\mathbf{M}|\varphi, (w, X)$ with $\mathbf{M}|\varphi = (W, \{(w, X) \in W^{acc} \mid \mathbf{M}, (w, X) \models \varphi\}, \sim, V)$. An *explicit public observation* $+\varphi$ yields the model $\mathbf{M}|^{+}\varphi, (w, X \cup \{\varphi\})$ with $\mathbf{M}|^{+}\varphi = (W, \{(w, X \cup \{\varphi\}) \in W^{acc} \mid \mathbf{M}, (w, X) \models \varphi\}, \sim, V)$.

This implements what we have discussed. But still, we want a cleaner distinction:

5.4 Separating world dynamics and access dynamics

The above explicit and implicit seeing overlap in their effects on a model. We would like orthogonal acts of bare observation $!\varphi$ on worlds, and enrichment of access sets:

Definition Acts of realization.
Over epistemic access models, *realizing that* φ (written as $\#\varphi$) means adding a formula φ to all awareness sets in our current range, provided that $K\varphi$ holds in the actual world. ■

This is a special case of explicit public observation, when φ is already true in all worlds. We can also see it as a proto-typical act of promoting implicit to explicit knowledge. Taking the latter as primitive, with a model operation $(\mathbf{M}|^{\#}\varphi, (w, X))$, we can factor explicit seeing into consecutive actions of implicit public observation, plus an act of realizing:

Fact Explicit seeing is definable using implicit observation and realization.[8]

But realization is more general. One can realize any φ, factual or not, using many means: perception, inference, reflection, memory, and so on. We return to these points later.

7 The implicit form occurs in the sentence 'Isolde saw her lover's ship land', that does not imply the lady was aware of the nature of the ship. Explicit seeing entails awareness, reflected by linguistic *that*-clauses as in 'Isolde saw that her lover's ship was landing' (cf. Barwise & Perry 1983).

8 This only works for factual assertions. For non-factual assertions, more care is needed.

Realization and inference We can now fill the gap of Section 5.1, modelling inference acts on factual formulas by realization. Say, we have implicit knowledge $K\varphi$ and $K(\varphi \to \psi)$.[9] The axiom $K(\varphi{\to}\psi) \to (K\varphi \to K\psi)$ tells us that $K\psi$, too, is implicit knowledge. Thus we have a basis for an act of realization *licensed* by the epistemic axiom:

$$K(\varphi \to \psi) \to (K\varphi \to [\#\psi]I\psi)$$

But our logic can also describe different scenarios where observations yielded the premises:

$[!p][!(p \to q)] < \# q > Iq$ observation enables realization

Any sound inference in our logic is available for an event of realization.

5.5 A complete dynamic-epistemic logic of realization

Next, we introduce new dynamic modalities for implicit observation and explicit realization:

$$\varphi ::= p \mid \neg\varphi \mid \varphi \wedge \psi \mid K\varphi \mid I\varphi \mid [!\varphi]\psi \mid [\#\varphi]\psi$$

The semantics of these operators is straightforward. For instance,

$$\mathbf{M}, (w, X) \models [\#\varphi]\psi \;\text{ iff }\; \text{if } \mathbf{M}, w \models K\varphi, \text{then } \mathbf{M}|^{\#}\varphi, (w, X) \models \psi$$

It is easy to write a complete set of recursion axioms for these actions:

THEOREM The dynamic logic of informational access is axiomatized by

(a) the static logic of the epistemic access language with K and I,
(b) the *PAL* recursion axioms augmented with one for the I modality:

$$[!\varphi]I\psi \leftrightarrow (\varphi \to I\psi)$$

(c) the following recursion axioms for the dynamic modality of realizing:

[9] In *PAL*, $[!\varphi]K\varphi$ is valid for factual φ. Of course, $[!\varphi]I\varphi$ is *not* a valid law in our semantics.

$[\#\varphi]q \quad \leftrightarrow \quad K\varphi \rightarrow q$ for atomic propositions q

$[\#\varphi]\neg\psi \quad \leftrightarrow \quad K\varphi \rightarrow \neg[\#\varphi]\psi$

$[\#\varphi](\psi \wedge \chi) \quad \leftrightarrow \quad [\#\varphi]\psi \wedge [\#\varphi]\chi$

$[\#\varphi]K\psi \quad \leftrightarrow \quad K\varphi \rightarrow K[\#\varphi]\psi$

$[\#\varphi]I\psi \quad \leftrightarrow \quad K\varphi \rightarrow (I\psi \vee \psi = \varphi)$[10]

Proof Soundness. The axiom $[!\varphi]I\psi \leftrightarrow (\varphi \rightarrow I\psi)$ says that implicit observation does not change access, and the first axiom under (c) says that realization does not change ground facts. The next two axioms hold for a partial function. The recursion axiom for knowledge states the precondition for an act of realization, and then says that epistemic accessibility has not changed. The final axiom says that only φ has been added to the access set.

Completeness. Working inside out on innermost occurrences of the two dynamic modalities $[!\varphi]$, $[\#\varphi]$, the axioms reduce every formula to an equivalent static one. ■

Illustration Explicit knowledge from acts of realization.

$$<+\varphi> I\varphi \leftrightarrow \neg[+\varphi]\neg I\varphi \leftrightarrow \neg(K\varphi \rightarrow \neg[+\varphi]I\varphi) \leftrightarrow (K\varphi \wedge [+\varphi]I\varphi) \leftrightarrow$$

$$(K\varphi \wedge (K\varphi \rightarrow (I\varphi \vee \varphi = \varphi))) \leftrightarrow (K\varphi \wedge (K\varphi \rightarrow T)) \leftrightarrow K\varphi$$

A similar calculation proves the earlier validity $[!p][!(p \rightarrow q)]<+q>Iq$. ■

Thus, a dynamic-epistemic logic with observational access works with the same simple methodology as Chapter 3, while extending the scope to different kinds of logical information, and events of observation, inference, and general awareness raising.

5.6 Lifting the limitations: an entertainment approach

Beyond factual propositions: introspection The *PAL* dynamics of $!\varphi$ can announce complex epistemic formulas φ. Likewise, acts of realization may concern any φ that the agent knows implicitly. A clear instance are acts of *introspection*. Let an agent know implicitly that φ. This may become explicit knowledge $I\varphi$ through realization. But also, thanks to the *K4* axioms in the

[10] Here, we indulge in shaky notation. The symbol '=' stands for *syntactic identity* of formulas.

epistemic structure of the model, the introspective act of realizing $K\varphi$ *itself* should be available, leading to the truth of $IK\varphi$.[11]

Indeed, our dynamic axioms are valid on a broader reading with formulas of arbitrary complexity. But the issue is how to deal then with Veridicality when a model is updated, since, as we noted, formulas may get 'out of synch'. There are various options, including drastic promotion acts that delete all old non-factual formulas from access sets. Or one may accept that there are *two primitive notions*: implicit and explicit knowledge, that are only weakly related, and give dynamics for both. The latter may be inescapable eventually. Here, however, we will discuss a simpler proposal from van Benthem & Velázquez-Quesada (2009). It moves closer to the static awareness logics of Fagin, Halpern & Vardi (1990) – that we now dynamify in a slightly different manner.

Entertaining a formula This time we take a modal operator $E\varphi$ ('the agent *entertains* φ') saying the agent is actively aware of φ, for any reason: approval, doubt, or rejection. The syntax of the static language is now completely recursive: φ can be of any complexity. It may make things a bit simpler if we assume weak implicit introspection: $E\varphi \rightarrow KE\varphi$. Our models are as before now, with a set of formulas Σ_w assigned to each epistemic world w.

DEFINITION Explicit knowledge.
We define *explicit knowledge* $Ex\varphi$ as $K(\varphi \wedge E\varphi)$: the agent knows implicitly that φ is true and also that she is entertaining φ.[12,13]

Dynamic actions Actions are now two natural operations on access sets:

$+\varphi$	adding the formula φ	('pay attention')
$-\varphi$	deleting the formula φ	('neglect', 'ignore')

The new set-up is close to what we had before: an act of realization is an addition, the only difference is that we have dropped the precondition of implicit knowledge, since it is up to an agent to entertain formulas for whatever reason. For factual formulas φ, an earlier successful act $\#\varphi$ of

[11] Again, the relevant *K4*-axiom would not *be* introspection, but *license* an act of introspection.

[12] There are variants like $K\varphi \wedge E\varphi$, that may be equivalent depending on one's base logic.

[13] One can criticize $K(\varphi \wedge E\varphi)$ for being too weak. I might implicitly know that φ and also entertain φ, while failing to make the connection, not realizing that I am entertaining a true statement. Thus, a primitive of explicit knowledge and its dynamics may be good things to have around.

promotion from implicit to explicit knowledge might now be defined as a sequential composition $!K\varphi\ ;\ +\varphi$ where we first announce the precondition.

Here is a major virtue of the new set-up. Suppose φ is explicit knowledge at some stage. Now we make a *PAL*-style implicit observation of some new fact ψ. This may change the truth value of φ, and so we are not sure that φ remains explicit knowledge. But thanks to the decoupling, the fact that φ remains in the syntactic awareness set just expresses its still being entertained. Whether we have $[!\psi]Ex\varphi$, i.e., $[!\psi]K(\varphi \wedge E\varphi)$, totally depends on the circumstances of the update, and the new truth value of $Ex\varphi$ will adjust automatically.

The new setting is also richer, in having acts of dropping formulas. From an epistemic point of view, these make obvious sense: gaining knowledge and losing knowledge are natural duals, and indeed, forgetting is crucial to one's sanity.

Complete dynamic logic As always, the dynamic actions have corresponding modalities. These obey obvious recursion axioms that can be written down easily by a reader who has understood the preceding system. Here are some laws for knowledge and entertaining:

$$[!\psi]E\varphi \quad\leftrightarrow\quad \psi \rightarrow E\varphi$$

$$[+\psi]K\varphi \quad\leftrightarrow\quad K[+\psi]\varphi$$

$$[+\psi]E\varphi \quad\leftrightarrow\quad E\varphi \vee \varphi = \psi$$

$$[-\psi]E\varphi \quad\leftrightarrow\quad \psi \neq \varphi$$

THEOREM The dynamic-epistemic logic of entertainment is axiomatizable.

Derivable formulas of this logic encode interesting facts of entertaining. Some are obvious counterparts for our earlier dynamic laws of realization. New laws include

$$[-\psi]Ex\varphi \leftrightarrow K([-\psi]\varphi \wedge E\varphi) \quad \text{for } \varphi \neq \psi,$$

$$\neg[-\psi]Ex\psi$$

that state what explicit knowledge remains after dropping a formula from consideration. Finally, we state a perhaps less obvious feature of the system:

FACT The implication $\varphi \rightarrow [+\varphi]\varphi$ is valid for all formulas φ.

The reason is that an act $+\varphi$ of entertaining can only change truth values for $E\varphi$ and formulas containing it. Hence φ itself is too short to be affected by this. This shows how a dynamic logic like this mixes semantic and syntactic features in its valid principles.

Finally, let us return to the earlier 'action gap' in epistemic distribution for explicit knowledge: $Ex(\varphi \rightarrow \psi) \rightarrow (Ex\varphi \rightarrow [\ ]\ Ex\psi)$. Indeed, the following principle is valid in our logic:

$$K(\varphi \rightarrow \psi) \rightarrow (K\varphi \rightarrow [+\psi]Ex\psi)$$

This says that even implicit knowledge of the premises licenses an awareness raising act $+\psi$ that makes the conclusion explicit knowledge. This solves the earlier problem of making the epistemic effort visible that is needed for explicit knowledge.[14]

5.7 Conclusion

We have shown how a fine-grained syntactic notion of information can be treated in the same dynamic style as semantic information, using awareness as access or entertainment. This led to complete dynamic logics in the same style as before, resulting in a joint account of observation, inference, and other actions for agents. We do not claim to have solved all problems of inferential information and omniscience in this simple way, and indeed, we have pointed at several debatable design choices along the way. Even so, we feel that some of the usual mystery surrounding these issues gets dispelled by placing them in a simple concrete framework that has served us for other aspects of agent dynamics.

5.8 Further directions and open problems

There are many further issues once we have this platform.

Inferential dynamics once more We have not modelled specifics of inference rules, or knowing one's inferential apparatus. Following Jago (2006), Velázquez Quesada (2008) places explicit rules in models as objects of implicit or explicit knowledge, bringing in specifics of proof rules. One concrete

[14] Just $+\psi$ may seem a deus ex machina. Tying $Ex\psi$ to a conscious inferential act of *drawing this conclusion* would require more formal machinery than we have available at present.

format for testing intuitions are Horn-clause rules $p_1 \wedge \ldots \wedge p_n \rightarrow q$ that update syntactic Herbrand sets, as in the following sequence:

$$\boxed{p, \neg q, r} \xrightarrow{(p \wedge r) \rightarrow s} \boxed{p, \neg q, r, s} \xrightarrow{(t \wedge r) \rightarrow q} \boxed{p, \neg q, r, s, \neg t}$$

Many puzzles involve reasoning of this special kind, with *Sudoku* puzzles as a prime example. Our logics can then express dynamic inferential facts like $Exp \rightarrow [+q]\langle(p \wedge q) \rightarrow r\rangle Exr$: if the agent realizes that p in her current information state, and then explicitly observes that q, she can come to realize that r by performing an inference step invoking the rule $(p \wedge q) \rightarrow r$.[15]

Structured acts of inference More generally, a striking limitation of our approach so far is the bleak nature of our technical acts of promotion, that seem far removed from real acts of inference. Concrete reasoning has a good deal of further structure, including different roles for various kinds of assertions and rules. A more satisfactory dynamic account of inference in our style might have to perform a dynamification of the richer syntactic formats found in Argumentation Theory (Toulmin 1958; Gabbay & Woods 2004). But even pure mathematical proof has many different dynamic acts of 'supposing', 'inferring', 'withdrawing', and others, that should be brought within the scope of our analysis.

Forgetting Our actions of awareness management included both adding and dropping formulas. They also suggest a treatment of a phenomenon outside of *DEL* so far: losing facts from memory. Van Ditmarsch & French (2009) propose a more sophisticated account.

Algebra of actions and schematic validity With this richer algebra, one technical issue that came up with *PAL* returns. It is easy to see that the syntactic operations $\#\varphi$, $+\varphi$, $-\varphi$ validate interesting equations under composition, and also jointly with announcements $!\varphi$. But the *schematic validities* of this algebra are not directly encoded in our dynamic logics. As before, since action modalities are removed inside out in the completeness argument applied to *concrete formulas*, the logic needed no explicit schematic recursion axioms for combinations like $[+\varphi][-\psi]\alpha$. But there are interesting valid principles

[15] Rules will also have *refutational* effects, as inference works both ways. If $Ex\neg q$: the agent knows that q fails, consciously applying the rule $p \rightarrow q$ will make her know that the premise fails, resulting in $Ex\neg p$.

that may remain hidden then, witness the above valid law $\varphi \rightarrow [+\varphi]\varphi$. Can we axiomatize them?

From private to public Access or entertainment is typically a private matter, and hence a more proper setting for this chapter is eventually multi-agent *DEL*, allowing for differences in access for different agents, and private acts of inference or realization. Van Benthem & Velázquez-Quesada (2009) give a product update logic for this generalization that has one interesting twist. If you come to entertain φ in private, your awareness set should now get φ, while mine stays the same. The general mechanism is events with both preconditions and *postconditions*, letting product update change awareness sets in new worlds (w, Σ_w) like in the treatment of world-changing factual update with *DEL* in Chapter 4.

Multi-agent perspectives, argumentation, and observation Most prominently, in the public sphere, inference has a multi-agent version in the form of *argumentation*. How can we extend the above systems to deal with argumentation and dialogue between agents? Grossi & Velázquez-Quesada (2009) present a multi-agent logic that can handle inferential information flow between participants, intertwined with observation updates. It would be of interest to extend this to merges between *PAL* and existing logic games for proof and consistency maintenance.[16] On a related note, in Chapter 13, we take a look at intuitionistic logic as a proof-oriented perspective on mathematics, and show how its semantic tree models of information involve a mixture of public announcements and steps of syntactic realization making propositions explicit at certain stages of the process.

Belief and default inference In addition to classical inference and knowledge, an equally important feature of agency is management of beliefs and default inferences. Velázquez-Quesada (2010) studies this generalization along the above lines, where awareness raising acts can now also affect beliefs, and interact with *soft announcements* (Chapter 7), and non-monotonic inference rules like abduction or circumscription. In particular, the later rules for default reasoning may now license 'soft updates' of beliefs (cf. Chapter 13 for more on this theme).

Impossible worlds and paraconsistency One can recast the systems in this chapter closer to the elimination dynamics of *PAL* by using *impossible worlds*

[16] Grossi (2009) is a first systematic study of argumentation in dynamic logic.

from paraconsistent logic, i.e., arbitrary maps from formulas to truth values (cf. Priest 1997). Initially, we allow all maps, not just those obeying our semantic truth definition. Acts of realization make us see that some formula must get value *true*, pruning the set of valuations to leave more standard ones. The impossible worlds approach is richer than ours, as it suggests other dynamic acts that change valuations, say, actions of *consistency management* for agents whose beliefs are inconsistent. A merge between dynamic-epistemic logic and logics tolerating contradictions seems natural as a step toward a realistic account of fallible agents.

Syntax, semantics, and in between Here is a perennial problem in logic. Semantic entities are too poor to deal with intensional notions like 'proposition' or 'information', but brute syntax is too detailed. Intermediate levels have been proposed, but there is no consensus. Can we formulate our logics at some good intermediate abstraction level? One candidate is the modal 'neighbourhood models' that occur at various places in this book. These support static modalities of evidence, as well as belief referring to the intersection of all evidence sets (cf. Parikh 1991). Natural dynamic actions here include hard update by world elimination, 'adding evidence' in the form of a new neighbourhood, and internal 'evidence combination' by taking intersections of neighbourhoods. Van Benthem & Pacuit (2011) is a first exploration of the resulting dynamic-epistemic and doxastic logics. More generally, studying inference steps and related fine-grained acts may throw new light on appropriate representation levels for information – an approach advocated in a philosophical setting in Sections 13.2, 13.3.

5.9 Literature

Fagin, Halpern & Vardi (1990) is a classic source on syntactic awareness logic. Explicit dynamic logics of inference acts were proposed in van Benthem (1991) and Jago (2006), while the classic paper Dung (1995) is also still relevant. Different logical theories of information are surveyed in depth in van Benthem & Martinez (2008). The dynamic-epistemic logics in this chapter are based on van Benthem (2008e), Velázquez-Quesada (2008), and van Benthem & Velázquez-Quesada (2009).

6 Questions and issue management

By now, our logical dynamics of agents can deal with their observations and inferences. But the examples in our Introduction crucially involved one more informational act. The waiter in The Restaurant asked *questions* to get the relevant information, the father in the Muddy Children asked questions to direct the conversation, and questions were behind many other scenarios that we discussed. So far, we just looked at the information provided by answers. But questions themselves are just as important as statements in driving reasoning, communication, and investigation: they set up an *agenda of issues* that directs information flow, and makes sense of what takes place. It has even been said that all of science is best viewed as a history of questions, rather than of answers. In this chapter, we will see how this agenda setting fits into our approach. We follow our usual set-up with static logics, dynamic actions and logics, and an 'opt-out point' for the reader who has seen enough – listing some further directions afterwards.

6.1 Logics of questions

The idea that questions are basic logical acts goes back to Ajdukiewicz in the 1930s, sparking work in Poland in the 1960s (Wisniewski 1995). A defender of this view since the 1970s has been Hintikka, who shows how any form of inquiry depends on an interplay of inference and answers to questions, resulting in the interrogative logic of Hintikka *et al.* (2002).[1] These logics are mainly about general inquiry and learning, with relations to epistemology and philosophy of science. But there is also a stream of work on questions in natural language, as basic speech acts with a systematic vocabulary

[1] Another well-known source on logic and questions is Belnap & Steele (1976).

(cf. Ginzburg 2009; Groenendijk & Stokhof 1997, and the 'inquisitive semantics' of Groenendijk 2008). Logic of inquiry and logic of questions are related, but there are differences in thrust. A study of issue management in inquiry is not necessarily one of speech acts that must make do with the expressions that natural language provides.

In this chapter, we do not choose between these streams, but study the role of questions in dynamic-epistemic logic as events that affect information flow. This seems an obvious task given earlier scenarios, but so far, we have just focused on public announcement of answers. Baltag (2001) is a first account of questions in *DEL*, and we take this further, proposing various acts of what we call *issue management* plus complete dynamic logics for them, once the right recursion axioms are identified. One advantage of this approach, as we shall see, is that multi-agent and temporal perspectives find a natural place from the start.

6.2 A static system of information and issues

Epistemic issue models We first need a convenient static semantics to dynamify. The relevant idea occurs from linguistics to learning theory: a current issue is a *partition* of a set of options, with cells for areas where we want to arrive. The partition can arise from a conversation, a game, a research programme, and so on. The alternative worlds may range from finite deals in a card game to infinite histories representing a total life experience. We start without indicating agents: multi-agent models for questions will come later.

DEFINITION Epistemic issue model.
An *epistemic issue model* $\mathbf{M} = (W, \sim, \approx, V)$ has a set W of worlds (epistemic alternatives), $\sim$ is an equivalence relation on W (epistemic indistinguishability), $\approx$ is another equivalence relation on W (the abstract *issue relation*), and V is a valuation for atomic p. ■

Models with more general relations account for different epistemic logics, or belief instead of knowledge. While this is important, equivalence relations will suffice for our points.

Example Information and issues.
In the model pictured here, dotted black lines are epistemic links, while cells of the issue relation are marked by rectangles. There is no knowledge of the

real situation with p, q, though the agent knows whether p. Moreover, the issue whether q has been raised:

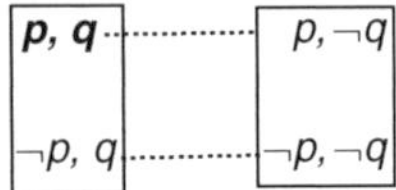

■

A suitable language Now we want to describe these situations. Our language has a universal modality U, a knowledge modality K, a modality Q ('question') describing the current issue, and also a more technical modality R ('resolution') talking about what happens when the issue is resolved by combining with the available information:

DEFINITION Epistemic issue language.
The static *epistemic issue language* has a set of proposition letters p, over which formulas are constructed by the following inductive syntax rule:

$$p \mid \neg\varphi \mid (\varphi\wedge\psi) \mid U\varphi \mid K\varphi \mid Q\varphi \mid R\varphi$$

Other Boolean operations and existential modalities $<U>$, $<K>$, $<Q>$, $<R>$ are defined as usual. ■

DEFINITION Relational semantics.
Interpretation is as usual, and we only highlight three clauses:

$\mathbf{M}, s \models K\varphi$	iff	for all t with $s \sim t$: $\mathbf{M}, t \models \varphi$
$\mathbf{M}, s \models Q\varphi$	iff	for all t with $s \approx t$: $\mathbf{M}, t \models \varphi$
$\mathbf{M}, s \models R\varphi$	iff	for all t with $s\ (\sim \cap \approx)\ t$: $\mathbf{M}, t \models \varphi$

■

With this language we can say things about information and issues like

$U(Q\varphi \vee Q\neg\varphi)$	the current issue settles fact φ[2]
$<K>(\varphi \wedge <Q>\neg\varphi)$	the agent considers it possible that φ is true but not at issue
$\neg K\varphi \wedge \neg Q\varphi \wedge R\varphi$	φ is not known and not at issue, but true upon resolution.

[2] Here 'settling' means that the issue answers (explicitly or implicitly) the question whether φ holds. 'Settling an issue', however, is an informational event of issue management: our later *resolution*.

In this way, we can define many notions from the literature on questions.

Static logic of information and issues As for reasoning with this language, we have valid implications like $U\varphi \rightarrow Q\varphi$, $(K\varphi \vee Q\varphi) \rightarrow R\varphi$, etcetera.[3] A proof system for the logic has the laws of S5 for U, K, Q, R separately, plus obvious connection principles. These laws derive many intuitive principles, such as introspection about the current issue:

$$U(Qp \vee Q\neg p) \rightarrow UU(Qp \vee Q\neg p) \rightarrow KU(Qp \vee Q\neg p)$$

A completeness result can be proved by standard modal techniques (cf. Blackburn, de Rijke & Venema 2000), but as usual, details of a static base system are not our concern:

THEOREM The static epistemic logic of information, questions, and resolution is axiomatizable.[4]

6.3 Actions of issue management

In Chapters 3, 4, 5, the next step was to identify basic events of information flow, and find their dynamic logic. This is similar with questions and events of issue management.

Basic actions of issue management Basic actions in our setting start with update of information, and for this we will just take *announcements* $!\varphi$ as in Chapter 3, though in a special *link-cutting version* where no worlds are eliminated from the model, but all links get cut between φ-worlds and $\neg\varphi$-worlds.[5] More interesting is an action $?\varphi$ that refines the current issue by also forcing it to distinguish between φ-worlds and $\neg\varphi$-worlds:

Example Information and issue management.
Here is a video sequence that shows what can happen in this setting. First there is no genuine issue and no information, then q becomes an issue, and next, we get the information that p:

[3] Typically non-valid implications are $K\varphi \rightarrow Q\varphi$, $R\varphi \rightarrow (K\varphi \vee Q\varphi)$.

[4] The intersection modality R can be axiomatized if we add *nominals* to the language, denoting single worlds. Cf. van Benthem & Minica (2009) for the resulting hybrid logic.

[5] Cf. van Benthem & Liu (2007). This change is for technical reasons that are irrelevant here.

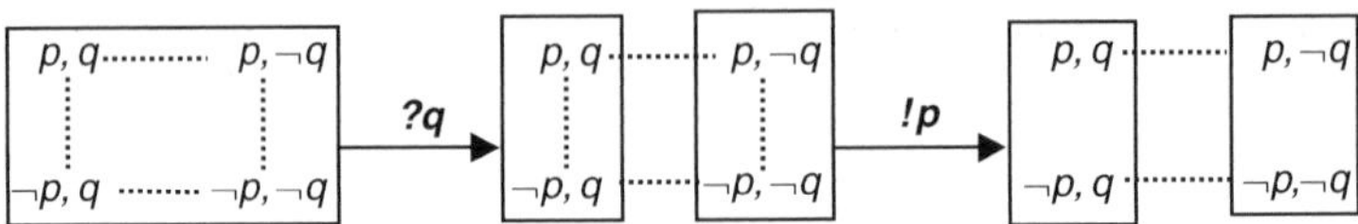

In the final situation, the issue still has not been resolved by the information. To at least align the two, one could form a refinement of the issue to what we now know:

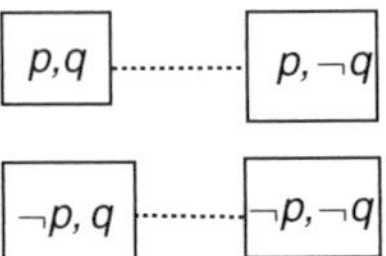

■

Definition Announcements and questions as model transformers.
Let $\mathbf{M} = (W, \sim, \approx, V)$ be an epistemic issue model. Let the relation $s =_\varphi t$ hold between worlds s, t in W if these give the same truth value to the formula φ. The *information-updated model* $\mathbf{M}^{!\varphi}$ is the structure $(W, \sim^{!\varphi}, \approx, V)$ with $s \sim^{!\varphi} t$ iff $s \sim t$ *and* $s =_\varphi t$. The *issue-updated model* $\mathbf{M}^{?\varphi}$ is the structure $(W, \sim, \approx^{?\varphi}, V)$ with $s \approx^{?\varphi} t$ iff $s \approx t$ *and* $s =_\varphi t$.[6] ■

Note the symmetry between the two operations. We do not want to make too much of this, but one way of looking at the situation is to view the issue itself as an epistemic agent guiding the process, who already knows all the answers.[7]

Resolving and refining actions But further management actions make sense, not driven by specific formulas, but by the structure of the models themselves. Both amount to taking a common refinement, as suggested in an earlier example:

Definition Resolution and refinement.
An execution of the *resolve action* ! in a model $\mathbf{M}$ yields a model $\mathbf{M}^!$ that is the same except that the new epistemic relation $\sim^!$ is the intersection $\sim \cap \approx$. An execution of the *refine action* ? in a model $\mathbf{M}$ yields a model $\mathbf{M}^?$ that is the same except that the new issue relation $\approx^?$ is the intersection $\sim \cap \approx$. Performing both is the *resolve-refine action* #. ■

Here is a summary of our repertoire of actions: $!\varphi$, $?\varphi$, !, ?, #.

[6] Please note: the question mark here is *not the test operation* of propositional dynamic logic *PDL*!

[7] The symmetry is lost if we let $!\varphi$ be executable only if φ is *true*, as in *PAL*. For, $?\varphi$ is always executable in every world in a model. Our later results hold for both actions.

Semantic properties of issue management Our basic actions satisfy several intuitive laws. In particular, !, ?, # form an algebra under composition, witness the following table:

;	!	?	#
!	!	#	#
?	#	?	#
#	#	#	#

The total picture is more diverse. Composing operations is complex, and many prima facie laws do not hold when you think a bit more. For instance, $?\varphi$; $!\psi$ is the same operation as $!\psi$; $?\varphi$ for factual formulas φ, ψ, performing two independent refinements – but not in general, since effects of the first on the model can influence the action of the second.

It is of interest to compare the situation with that in *PAL* (Chapter 3). *Repeating* the same assertion $!\varphi$ has no new effects when its content φ is factual. But as the Muddy Children puzzle showed, repeating the same epistemic assertion can be informative, and lead to new effects. Likewise, asking a factual question twice has the same effect as asking it once. But with Q modalities that are sensitive to the effects of questions, things change:

FACT The identity $\varphi? \,;\, \varphi? = \varphi?$ does not hold generally for questions.

This is not hard to see, but we refer to van Benthem & Minica (2009) for a counter-example. Next comes a difference with *PAL*. Public announcement satisfied a valid *composition principle* that gives the effects of two consecutive announcements with just a single one:

$$!\varphi \,;\, !\psi \;=\; !(\varphi \wedge [!\varphi]\psi)$$

FACT There is no valid principle for composing questions.

The reason is simply that two consecutive questions can turn one equivalence class into four by two consecutive cuts, which is more than a single question can do.[8]

[8] This expressive failure can easily be remedied by introducing *list-questions* into our language of binary Yes/No questions. It should be obvious how to do this.

6.4 Dynamic logic of issue management

Dynamic language of information and issues Here is how one can talk explicitly about the above model changes and their laws:

DEFINITION Dynamic language and semantics.
The *dynamic language of information and issues* is defined by adding the following clauses to the static base language:

$$[!\varphi]\psi \mid [?\varphi]\psi \mid [?]\varphi \mid [!]\varphi$$

These are interpreted by adding the following truth conditions:

$$\mathbf{M}, s \vDash [?\varphi]\psi \quad \text{iff} \quad \mathbf{M}^{?\varphi}, s \vDash \psi$$
$$\mathbf{M}, s \vDash [?]\psi \quad \text{iff} \quad \mathbf{M}^{?}, s \vDash \psi$$

The clauses for the information modalities are completely analogous. ■

Insights relating knowledge, questions, and answers can now be expressed in our language. We can, for instance, express that *question* $?\psi$ *is entailed by question* $?\varphi$ in a model:

$$[?\varphi]U(Q\psi \vee Q\neg\psi)$$

Intuitively this says that the question $?\psi$ does not change the issue structure after $?\varphi$.[9] The dynamic language also encodes basic laws of interrogative logic. For instance, one knows in advance that a question followed by resolution leads to knowledge of the issue:

$$K[?\varphi][!]U(K\varphi \vee K\neg\varphi)$$

A complete dynamic logic To find a complete logic, we need the right recursion axioms for new information and issues after basic acts of information and issue management. Here is a sufficient set for the reduction argument. We start with the laws for *asking*:

1. $[?\varphi]p \leftrightarrow p$
2. $[?\varphi]\neg\psi \leftrightarrow \neg[?\varphi]\psi$
3. $[?\varphi](\psi \wedge \chi) \leftrightarrow [?\varphi]\psi \wedge [?\varphi]\chi$

[9] One can also express *compliance of answers* (Groenendijk 2008) in these terms.

4. $[?\varphi]U\psi \leftrightarrow U[?\varphi]\psi$
5. $[?\varphi]K\psi \leftrightarrow K[?\varphi]\psi$

These express features that are close to those we saw for *PAL*: no factual change, update is a function – while simple operator reversals for modalities indicate that the action did not change the corresponding component in the model. In what follows, we will not repeat such standard clauses for the other actions, as the reader can easily supply them by herself. This leaves the more interesting cases where something really happens:

6. $[?\varphi]Q\psi \leftrightarrow (\varphi \wedge Q(\varphi \to [?\varphi]\psi)) \vee (\neg\varphi \wedge Q(\neg\varphi \to [?\varphi]\psi))$
7. $[?\varphi]R\psi \leftrightarrow (\varphi \wedge R(\varphi \to [?\varphi]\psi)) \vee (\neg\varphi \wedge R(\neg\varphi \to [?\varphi]\psi))$

These principles make a case distinction matching the action. For instance, if φ is true right now, the new issue relation only looks at formerly accessible worlds that satisfy φ. Thus, we have to conditionalize the old Q-modality to the φ-zone – and likewise with $\neg\varphi$.

Next we consider the *global refinement action* ? with all current information. Its atomic and Boolean clauses are obvious as we said. It commutes with all modalities not affected by it, as in $[?]K\varphi \leftrightarrow K[?]\varphi$. Here is the only case where something interesting happens:

8. $[?]Q\varphi \leftrightarrow R[?]\varphi$

The new issue relation is the intersection of the old issue relation with the epistemic one, and this is what the resolution modality R describes. Finally, there are recursion axioms for soft announcement $!\varphi$ and global resolution $!$. Most are again standard, and we only list those with some interest (by an earlier-noted analogy, even these reflect earlier ones):

9. $[!\varphi]K\psi \leftrightarrow (\varphi \wedge K(\varphi \to [!\varphi]\psi)) \vee (\neg\varphi \wedge K(\neg\varphi \to [!\varphi]\psi))$
10. $[!\varphi]R\psi \leftrightarrow (\varphi \wedge R(\varphi \to [!\varphi]\psi)) \vee (\neg\varphi \wedge R(\neg\varphi \to [!\varphi]\psi))$
11. $[!]K\varphi \leftrightarrow R[!]\varphi$

It is easy to see that this list suffices for completeness. Working inside out, these principles remove innermost occurrences of dynamic modalities:

THEOREM The dynamic logic of information and issues is completely axiomatized by our static base logic plus the stated recursion axioms.

Deductive power This calculus can derive many further principles, for instance:

FACT The formula $\varphi \rightarrow [?\varphi][!]K\varphi$ is derivable for all factual formulas φ.

Proof First observe that for factual φ, $[a]\varphi \leftrightarrow \varphi$ for all the above actions a: management does not change ground facts.[10] Now a derivation goes like this:

(a) $\varphi \rightarrow (\varphi \wedge R(\varphi \rightarrow \varphi)) \vee (\neg\varphi \wedge R(\neg\varphi \rightarrow \varphi))$	in propositional logic
(b) $\varphi \rightarrow (\varphi \wedge R(\varphi \rightarrow [?\varphi]\varphi)) \vee (\neg\varphi \wedge R(\neg\varphi \rightarrow [?\varphi]\varphi))$	by our observation
(c) $\varphi \rightarrow [?\varphi]R\varphi$	recursion axiom
(d) $\varphi \rightarrow [?\varphi]R[!]\varphi$	by our observation
(e) $\varphi \rightarrow [?\varphi][!]K\varphi$	recursion axiom ■

The subtlety of our setting shows in how this obvious-looking principle can fail with complex φ:

Example Side effects of raising issues.
In the following model **M**, a Moore-type scenario unfolds. In the actual world x, $\varphi = p \wedge \neg Kp$ is true. Now we raise φ as an issue, resulting in the second model. Then we resolve to get the third model. But there, φ is false at **x**, and we have not achieved knowledge:

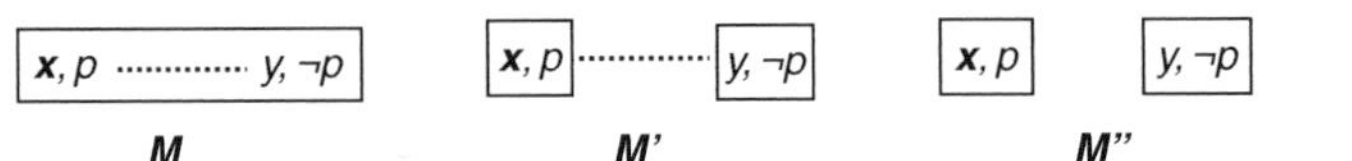

■

Hidden validities While this looks good, as with *PAL* and the dynamic logic of inference in Chapter 4, there is unfinished business with hidden validities. Recursion axioms do not describe all substitution-closed *schematic laws* of the system. This showed with the earlier algebra of management actions, that requires recursions with *stacked dynamic modalities* $[?\varphi][!\psi]\chi$. It is an open problem if the schematic validities can be axiomatized, but doing it would be important to understanding the total mechanics of our system.

6.5 Multi-agent question scenarios

Questions are typical multi-agent events, even though many logics of questions ignore this feature. In our setting, it should be easy to add more agents,

[10] This observation can be proved by a straightforward induction on φ.

and it is. We briefly explain the issues that arise, and solutions that are possible when using *DEL* (cf. Chapter 4).

Static multi-agent logic of information and issues The first step is routine. A static language, semantics, and logic are as before, now adding agent indices where needed. This language can make crucial distinctions like something being an issue for one agent but not (yet) for another, or some agent knowing what the issue is for another. Perhaps less routine would be adding *group knowledge* to our static base, since, as we saw in Chapter 3, an answer to a question often makes a fact common knowledge in the group {Questioner, Answerer}. Also, groups might have collective issues not owned by individual members, and this might be related to the earlier global actions of refinement and resolution. Even so, we leave extensions with common or distributed knowledge as an open problem.

Agent-specific preconditions So far, we only had impersonal public questions raising issues. Now consider real questions asked by one agent to another:

i asks j: 'Is φ the case?'

As usual, even such simple Yes/No questions show what is going on. In Chapters 1, 3, we noted that the preconditions for normal episodes of this kind are at least the following:

(a) Agent i does not know whether φ is the case,
(b) Agent i thinks it possible that j knows whether φ is the case.

Of course, we are not saying that all questions have these preconditions. For instance, rhetorical or Socratic questions do not. Typology of questions is an empirical matter, beyond the scope of our logic. We will just assume that some preconditions are given for a question/answer episode, say $pre(?\varphi, i, j)$. Then we can proceed with our analysis.

Which question actions? But do we really need new actions in this setting? One might just split the event into two: publicly announcing the precondition,[11] and then raising the issue in the earlier pure non-informative sense:

$!pre(?\varphi, i, j); ?\varphi$

[11] Recall the analogy with *i tells j that* φ (cf. Chapter 3) where one often absorbs preconditions like 'i knows that φ', 'i thinks that j does not know that φ' into the assertion.

This may have side-effects, however, since announcing the precondition may change the model and hence truth values of φ, making the effect of $?\varphi$ different from what it would have been before the announcement. One option (cf. van Benthem & Minica 2009) is then to change both the epistemic relation and the issue relation of the model in *parallel*. Still, things works fine for factual φ, and we do not pursue this complication here.[12]

Multi-agent scenarios with privacy The real power of dynamic-epistemic logic only unfolds with informational actions that involve privacy of information and issues, and differences in observation. Say, someone asks a question of the teacher, but you do not hear what has been asked. Private questions can be followed by public answers, and vice versa. Typical scenarios in this setting would be these: (a) One of two agents raises an issue, but the other only sees that it is $?p$ or $?q$. What will happen? (b) One agent raises an issue $?p$ privately, while the other does not notice. What will happen?

These are the things that *product update* was designed to solve (Chapter 4). We will not do a *DEL* version of our logic in detail, but only show the idea. First, we need *issue event models*, where events can have epistemic links, but also an issue relation saying intuitively which events matter to which agent. Here is the essence of the update mechanism:

DEFINITION Product update for questions.
Given an epistemic issue model **M** and issue event model **E**, the *product model* **M** × **E** is the *DEL* product of Chapter 4 with the new issue relation $(s, e) \approx (t, f)$ iff $s \approx t$ *and* $e \approx f$. ■

As a special case, think of our earlier actions $?p$ for a single agent. Say, **M** has just a p-world and a $\neg p$-world that the agent considers possible, while she does not have an issue yet ($\approx$ is universal). Now we take an event model **E** with two *signals* $!p$, $!\neg p$ that the agent cannot epistemically distinguish, though she cares about their outcomes (the issue relation $\approx$ in **E** separates them). Preconditions are as usual. The model **M** × **E** is a copy of the original epistemic model, but the issue relation has split into cells {*p-world*}, {*¬p-world*}. This is what we had before. Now we do a more complex example:

[12] Adding a *parallel operator* $!\varphi \mid\mid ?\psi$ to *DEL* raises interesting logical issues of its own.

Example Uncertainty about the question.
Let the above scenario (a) start with the following model *M*:

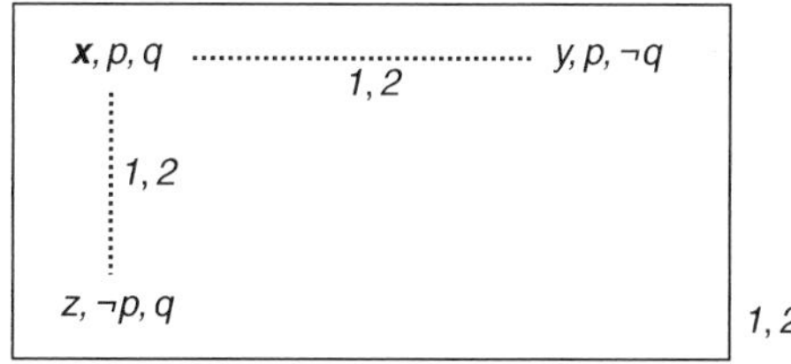

The issue event model **E** has four events, two for each question. $!p$, $!\neg p$ mark answers to the first, and $!q$, $!\neg q$ mark answers to the second, all with obvious factual preconditions. Epistemically, neither agent can distinguish between the answers $!p$, $!\neg p$ and $!q$, $!\neg q$, but agent *1* (though not *2*) can distinguish between these two groups. Next, the issue relation for *1* cares about all answers, and has four singletons $\{!p\}$, $\{!\neg p\}$, $\{!q\}$, $\{!\neg q\}$, while *2* does not care about the issue, and lumps them all as $\{!p, !\neg p, !q, !\neg q\}$. Now the resulting product model $M \times E$ is this (omitting lines for *2*):

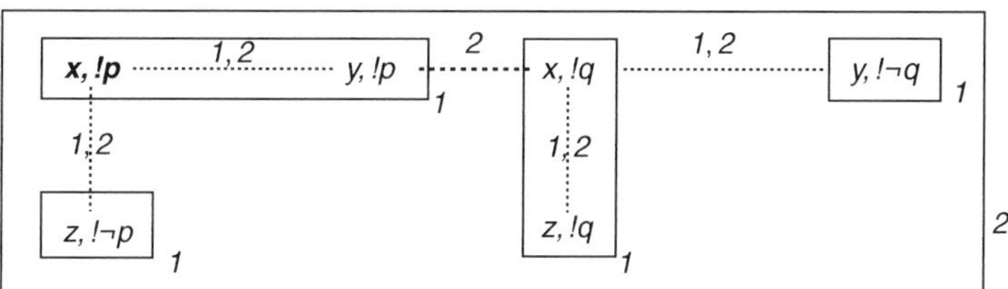

Here the actual issue has been raised for agent *1*, while *2* has only learnt that *p* or *q* has been raised. Neither agent has learnt the answer (this would need a further resolution step). The reader may want to check this outcome against her intuitions about this scenario. ■

This mechanism can also deal with private questions where other agents are misled about what has taken place. This is similar to product update for *belief* in Chapter 7, where equivalence relations give way to arbitrary directed relations (not necessarily reflexive) indicating what worlds an agent considers possible from the current one. Scenario (b) above will work with an event model with $!p$, $!\neg p$, *id* (the identity event that can happen anywhere), where agent *2* has a directed arrow pointing from the real event, say signal $!p$ in a world satisfying *p*, to *id*. We do not pursue these issues here, but refer to

Baltag (2001), Minica (2010) for more sophisticated scenarios, including informative preconditions of questions that we have omitted here. Minica (2010) also has a complete *DEL*-style logic for question product update that generalizes the ones in this chapter.[13]

6.6 Conclusion

Our logical calculus of questions is just a pilot system. But it does show how one can incorporate issue management into the logical dynamics of information. The system also seems a natural complement to existing logics of questions. Moreover, studies of questions can now tap into everything else in this book: many agents, beliefs, preferences, games, and temporal perspective. In what follows, we make this slightly more concrete.

6.7 Further directions and open problems

Desirable extensions Some challenges have been mentioned already. We need extensions of our logics to multiple questions and general *Wh*-questions, and also to group knowledge and group issues. We also found technical problems like axiomatizing the schematic validities of issue management, or adding a parallel composition of actions of information and issue management. Here are a few more general issues on our agenda:

Temporal protocols for inquiry One crucial theme has emerged in earlier chapters: *long-term temporal procedure* in agency. Single information updates usually make sense only in an ongoing process of conversation, experiment, study, and so on. This is especially true for questions. Not everything can be asked, due to social convention, limited measuring apparatus, and so on. Thus, it makes sense to incorporate procedure into our dynamic logics, towards a more realistic theory of inquiry. Like the earlier multi-agent perspective, this essential temporal structure seems largely absent from existing logics of questions.

Here are a few illustrations. There is a hierarchy of factual questions ('*p*?'), epistemic ('Did you know that *p*?') and procedural ones ('What would your

[13] The intuitive interpretation of issue event models is non-trivial. Van Benthem & Minica (2009) raise worries about actual events, and the complex entanglement of issue changes and preconditions.

brother say if I asked him *p*?') that determines how informational processes perform. In general, atomic Yes/No questions take more time to reach a propositional goal than complex ones. Or beyond that, consider the waiter in The Restaurant of Chapter 1, who must reduce six options at the table to one. This requires more than two bits of information, i.e., more than two Yes/No questions of any kind, but the waiter uses two *Wh*-questions instead. Finally, an informational process may have procedural goals. Consider a standard Weighing Problem: '*You have nine pearls, one lighter than the others that are of equal weight. You can weigh two times with a standard balance. Find the lighter pearl.*' The process is a tree of histories of weighing acts (cf. the *DEL* update evolution of Chapter 4) with both solutions and failures.

To deal with such structure, Chapter 11 adds temporal 'protocols' to *PAL* and *DEL* (cf. van Benthem *et al.* 2009). In the same line, van Benthem & Minica (2009) define temporal *question protocols* that constrain histories of investigation. These support an epistemic temporal logic of issues that can be axiomatized. It differs from our dynamic logic of questions in that there is no more reduction to a static base language. For instance, in the crucial recursive step for issue management, questions now obey the following modified equivalence (here with an existential action modality):

$$<?\varphi>Q\psi \leftrightarrow <?\varphi>T \wedge ((\varphi \wedge Q(\varphi \rightarrow <?\varphi>\psi)) \vee (\neg\varphi \wedge Q(\neg\varphi \rightarrow <?\varphi>\psi)))$$

$<?\varphi>T$ says the question is allowed by the protocol. This procedural information does not reduce to ground facts.

Questioning and knowledge games Further procedural restrictions arise in learning, where asking questions is a natural addition to inputting streams of mere answers (cf. Kelly 1996). A concrete way of linking questions, announcements and temporal procedure are *knowledge games* for public announcements, where players must reach perhaps conflicting epistemic goals. Ågotnes and van Ditmarsch (2009) give a precise meaning to the *value* of a question.[14,15] The subject is new to logic: but Feinberg (2007) has a game-theoretic treatment.

Beliefs and preferences Agents' *beliefs* are just as important as their knowledge, and we can merge the question logics of this chapter with the logics of soft information and belief change in Chapter 7. In fact, asking a question can be

[14] For game-theoretic values of questions in concrete signalling games, cf. van Rooij (2005).

[15] Ågnotes *et al.* (2010) merge this approach with the question logic of this chapter.

a subtle way of influencing beliefs. But questions also affect other attitudes. They can give us information about agents' goals and *preferences*. Indeed, 'Why' questions typically concern reasons for behaviour.

Questions, structured issues, and syntactic dynamics Next comes a major desideratum from our point of view. Our approach so far has been semantic, and this may be too coarse. Chapter 5 extended *DEL* to include fine-grained information produced by inference or introspection. Now asking a question is a typical way of making agents aware that something is an issue. Thus we also need a finer *syntactic dynamics* of questions, where they modify a current set of relevant propositions: the agenda of inquiry. Such a take fits well with the combination of questions and deduction that drives inquiry. More concretely, consider the realization acts $\#\varphi$ of Chapter 5. In that setting, we can have both explicit raising of issues, and implicit raising of issues ('derived questions') that only become explicit by acts of inference.[16] More generally, *structured issues* have been studied in Girard (2008), Liu (2008), where 'agendas' are priority graphs of assertions: surely a great improvement from the usual semantic approaches where one cannot distinguish between more and less urgent questions. Chapters 9, 12 have more details on working with priority graphs, including a richer repertoire of inserting and deleting questions from an agenda.

Connections to other disciplines Dynamic logic of inquiry has natural connections with other disciplines. Linking with interrogative logic in Hintikka's style would bring us to epistemology and philosophy of science.[17] The latter connection gets even stronger when we view agendas as research programmes, noting that asking questions is as much of a talent in science as giving answers. Van Benthem (2010d) develops the idea that theories in science consist not just of propositions (answers to past questions), but also a dynamically evolving set of live questions. Next, linking with the inquisitive semantics of Groenendijk (2008) would get us closer to linguistics, where further ideas live of interest to our logics, such as replacing partitions by mere 'covers' in the basic models. We also mentioned links with game theory and decision theory: cf. van Rooij (2003, 2005) and again Feinberg (2007).

[16] As in Chapter 11, the epistemic temporal question protocols of van Benthem & Minica (2009) have a strong syntactic aspect that fits well with a richer awareness perspective.

[17] The thesis Hamami (2010) provides a first formal exploration.

6.8 Literature

Pioneering works on logics of questions are Wisniewski (1995), Hintikka (1973), Hintikka, Halonen & Mutanen (2002), Belnap & Steele (1976), Groenendijk & Stokhof (1985). Baltag (2001) is a first treatment in *DEL* style. Groenendijk (2008) is a recent line in dynamic semantics and pragmatics of natural language that is related to our approach, though it favours intuitionistic rather than epistemic logic. This chapter is based on van Benthem & Minica (2009).

7 Soft information, correction, and belief change

So far, we have developed dynamic logics that deal with knowledge, inference, and questions, all based on information and truth. Now we want to look at another pervasive attitude that agents have toward information, namely, their *beliefs*. This chapter will show how belief change fits well with our dynamic framework, and develop some of its logical theory.[1] This puts one more crucial aspect of rational agents in place: not their being right about everything, but their being wrong, and acts of self-correction.

7.1 From knowledge to belief as a trigger for actions

While knowledge is important to agency, our actions are often driven by fallible beliefs. I am riding my bicycle this evening because I believe it will get me home, even though my epistemic range includes worlds where the San Andreas Earthquake strikes. Decision theory is about choice and action on the basis of beliefs, as knowledge may not be available. Thus, our next step in the logical dynamics of rational agency is the study of beliefs, viewed as concretely as possible. Think of our scenarios in Chapter 3. The cards have been dealt. I know that there are 52 of them, and I know their colours. But I have more fleeting beliefs about who holds which card, or about how the other agents will play.[2]

Hard versus soft information With this distinction in attitude comes a richer dynamics. A public announcement $!P$ of a fact P was an event of *hard*

[1] Later on, we discuss how this relates to the alternative *AGM* style (Gärdenfors & Rott 1995).

[2] I could even be wrong about the cards (perhaps the Devil added his visiting card) – but this worry seems morbid, and not useful in investigating normal information flow.

information that changes irrevocably what I know. When I see the Ace of Spades played, I come to know that no one has it any more. This is the trigger that drove our dynamic-epistemic logics in Chapters 3 and 4. Such events of hard information may also change our beliefs – and we will find a complete logical system for this. But there are also events of *soft information*, affecting my beliefs without affecting my knowledge about the cards. I see you smile. This makes it more likely that you hold a trump card, but it does not rule out that you do not. To describe this, we will use worlds with plausibility orderings supporting dynamic updates.

The tandem of jumping ahead and self-correction Here is what is most important to me in this chapter from the standpoint of rational agency. As acting agents, we are bound to form beliefs that go beyond the hard information that we have. And this is not a concession to human frailty or to our mercurial nature. It is rather the essence of creativity, jumping ahead to conclusions we are not really entitled to, and basing our beliefs and actions on them. But there is another side to this coin, that I would dub our capacity for *self-correction*, or if you wish, for *learning*. We have an amazing capacity for standing up after we have fallen informationally, and to me, rationality is displayed at its best in intelligent responses to new evidence that contradicts what we thought so far. What new beliefs do we form, and what amended actions result? Chapter 1 saw this as a necessary *pair of skills*: jumping to conclusions (i.e., beliefs) and correcting ourselves in times of trouble. And the hallmark of a rational agent is to be good at both: it is easy to prove one theorem after another, it is hard to revise your theory when it has come crashing down. So, in pursuing the dynamic logics of this chapter, I am trying to chart this second skill.

7.2 Static logic of knowledge and belief

Knowledge and belief have been studied together ever since Plato proposed his equation of knowledge with 'justified true belief', and much of epistemology is still about finding an ingredient that would turn true belief into knowledge. Without attempting this here (see Chapter 13 for our thoughts), how can we put knowledge and belief side by side?

Reinterpreting PAL One easy route reinterprets dynamic-epistemic logic so far. We read the earlier *K*-operators as beliefs, again as universal quantifiers over the accessible range, placing no constraints on the accessibility relation: just pointed arrows. One test for such an approach is that it must be possible for beliefs to be *wrong*:

Example A mistaken belief.
Consider the following model with two worlds that are epistemically accessible to each other, but the pointed arrow is the only belief relation. Here, in the actual black world to the left, the proposition p is true, but the agent mistakenly believes that $\neg p$:

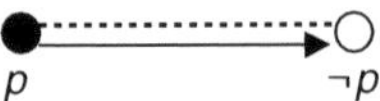

■

With this view of doxastic modalities (cf. Hintikka 1962), the machinery of *DEL* works exactly as before. But there is a problem:

Example, continued
Consider a public announcement $!p$ of the true fact p. The *PAL* result is the one-world model where p holds, with the inherited *empty* doxastic accessibility relation. But on the universal quantifier reading of belief, this means the following: the agent believes that p, but also that $\neg p$, in fact $B\bot$ is true at such an end-point. ■

In this way, agents who have their beliefs contradicted are shattered and start believing anything. Such a collapse is unworthy of a rational agent in the sense of Chapter 1, and hence we will change the semantics to allow for more intelligent responses.

World comparison by plausibility A richer view of belief follows the intuition that an agent believes the things that are true, not in all her epistemically accessible worlds, but only in those that are 'best' or most relevant to her. I believe that my bicycle will get me home on time, even though I do not know that it will not suddenly disappear in a seismic chasm. But the worlds where it stays on the road are more plausible than those where it drops down, and among the former, those where it arrives on time are more plausible than those where it does not. Static models for this setting are easily defined:

DEFINITION Epistemic-doxastic models.
Epistemic-doxastic models are structures $\mathbf{M} = (W, \{\sim i\}_{i \in I}, \{\leq_{i,s}\}_{i \in I}, V)$ where the relations $\sim_i$ stand for epistemic accessibility, and the $\leq_{i,s}$ are ternary comparison relations for agents read as follows, $x \leq_{i,s} y$ if, in world s, agent i considers world y at least as plausible as world x. ■

Now epistemic accessibility can be an equivalence relation again. Models like this occur in conditional logic, Shoham (1988) on preference relations in AI, and the 'graded models' of Spohn (1988). One can impose several conditions on the plausibility relations, depending on their intuitive reading. Burgess (1981) has *reflexivity* and *transitivity*, Lewis (1973) also imposes *connectedness*: for all worlds s, t, either $s \leq t$ or $t \leq s$. The latter yields the well-known geometrical nested spheres for conditional logic.[3] As with epistemic models, our dynamic analysis works largely independently from such design decisions, important though they may be. In particular, connected orders yield nice pictures of a line of equi-plausibility clusters, in which there are only three options for worlds s, t:

either *strict precedence* $s < t$ *or* $t < s$, *or equiplausibility* $s \leq t \wedge t \leq s$.

But there are also settings that need a fourth option of *incomparability* of worlds: $\neg\, s \leq t \wedge \neg\, t \leq s$. This happens when comparing worlds with conflicting criteria, as with some preference logics in Chapter 9. Sometimes also, partially ordered graphs are just the mathematically more elegant approach (Andréka, Ryan & Schobbens 2002; cf. Chapter 12).

Languages and logics One can interpret many logical operators in this richer semantic structure. In what follows, we choose intuitive maximality formulations for belief $B_i\varphi$.[4] First of all, there is plain belief, whose modality is interpreted as follows:[5]

DEFINITION Belief as truth in the most plausible worlds.
In epistemic-doxastic models, knowledge is interpreted as usual, while we put $\boldsymbol{M}, s \models B_i\varphi$ iff $\boldsymbol{M}, t \models \varphi$ for all worlds t that are maximal in the ordering $\lambda xy.\ x \leq_{i,\,s} y$.[6] ■

But absolute belief does not suffice. Reasoning about information flow and action involves *conditional belief*. We write this as follows: $B^{\psi}\varphi$, with the intuitive reading that, conditional on ψ, the agent believes that φ. This is close to conditional logic:

3 The natural *strict variant* of these orderings is defined as follows: $s < t$ iff $s \leq t \wedge \neg\, t \leq s$.

4 These must be modified in non-wellfounded models allowing *infinite descent* in the ordering. This issue is orthogonal to the main thrust of this chapter.

5 For convenience, henceforth we drop subscripts where they do not add insight.

6 Here we used lambda notation to denote relations, but plain '$x \leq_{i,\,s} y$' would serve, too.

Definition Conditional beliefs as plausibility conditionals.
In epistemic-doxastic models, $\mathbf{M}, s \models B^{\psi}\varphi$ iff $\mathbf{M}, t \models \varphi$ for all worlds t that are maximal for the ordering $\lambda xy.\ x \leq_{i,\ s} y$ in the set $\{u \mid \mathbf{M}, u \models \psi\}$. ■

Absolute belief $B\varphi$ is the special case $B^{T}\varphi$. It can be shown that conditional belief is not definable in terms of absolute belief, so we have a genuine language extension.[7,8]

Digression on conditionals As with epistemic notions in Chapters 2, 3, conditional beliefs *pre-encode* absolute beliefs that we would have if we were to learn certain things.[9] A formal analogy is this. A conditional $\psi \Rightarrow \varphi$ says that φ is true in the closest worlds where ψ is true, along some comparison order on worlds. This is exactly the above clause. Thus, results from conditional logic apply. For instance, on reflexive transitive plausibility models, we have this completeness theorem (Burgess 1981, Veltman 1985):

Theorem The logic of $B^{\psi}\varphi$ is axiomatized by the laws of propositional logic plus obvious transcriptions of the basic laws of conditional logic:

$$(a)\ \varphi \Rightarrow \varphi,\ (b)\ \varphi \Rightarrow \psi \textit{ implies } \varphi \Rightarrow \psi \vee \chi,\ (c)\ \varphi \Rightarrow \psi, \varphi \Rightarrow \chi \textit{ imply } \varphi \Rightarrow \psi \wedge \chi,$$
$$(d)\ \varphi \Rightarrow \psi, \chi \Rightarrow \psi \textit{ imply } (\varphi \vee \chi) \Rightarrow \psi,\ (e)\ \varphi \Rightarrow \psi, \varphi \Rightarrow \chi \textit{ imply } (\varphi \wedge \psi) \Rightarrow \chi.$$

Epistemic-doxastic logics In line with the general approach in this book, we do not pursue completeness theorems for static logics of knowledge and belief. But for greater ease, this chapter makes one simplification that

7 Likewise, the binary quantifier *Most A are B* is not definable in first-order logic extended with just a unary quantifier 'Most objects in the universe are *B*' (cf. Peters & Westerståhl 2006).

8 One can also interpret richer languages on epistemic-doxastic models. E.g., maximality suggests a binary relation *best* defined as '*t* is maximal in $\lambda xy.\ \leq_s xy$'. One can then introduce a modality for this, defining conditional belief as $[best\psi]\varphi$. Dynamic extensions of our language will come below.

9 *Static pre-encoding versus dynamics.* A conditional belief $B^{\psi}\varphi$ does not quite say that we would believe φ if we learnt that ψ. For an act of learning ψ *changes the current model* **M**, and hence the truth value of φ might change, as modalities in φ now range over fewer worlds in $\mathbf{M}|\psi$. Similar things happened with epistemic statements after communication in Chapter 3 – and they also occur in logic in general. The relativized quantifier in 'All mothers have daughters' does not say that, if we relativize to the subset of mothers, all of them have daughters who are mothers themselves.

reflects in the logic. Epistemic accessibility will be an *equivalence relation*, and plausibility a *pre-order over the equivalence classes*, the same as viewed from any world inside the class. This makes the following axiom valid:

$B\varphi \rightarrow KB\varphi$ *Epistemic-Doxastic Introspection*

While this is a debatable simplification, it helps focus on the core ideas of the belief dynamics.

7.3 Belief change under hard information

Our first dynamic logic of belief revision puts together the logic *PAL* with our static models for conditional belief, following the same methodology as earlier chapters. We will move fast, as the general points of Chapters 2, 3 apply, and indeed thrive here.

A complete axiomatic system For a start, we must locate the key recursion axiom for the new beliefs, something that can be done easily, using update pictures as before:

FACT The following formula is valid for beliefs after hard information:

$[!P]B\varphi \leftrightarrow (P \rightarrow B^{P}[!P]\varphi)$

This is like the *PAL* recursion axiom for knowledge under announcement. But note the conditional belief in the consequent, that does not reduce to an absolute belief $B(P \rightarrow \ldots$ Still, to keep the language in harmony, this is not enough. We need to know, not just which beliefs are formed after new information, but which conditional beliefs.[10] What is the recursion law for change in conditional beliefs under hard information? There might be a vicious regress here toward conditional conditional beliefs, but in fact, we have:

THEOREM The logic of conditional belief under public announcements is axiomatized completely by (a) any complete static logic for the model class

[10] This is overlooked in classical belief revision theory, which says only how new absolute beliefs are formed. One gets stuck in one round, as the new state does not pre-encode what happens in the next round. This so-called *Iteration Problem* cannot arise in our systematic logical set-up.

chosen, (b) the *PAL* recursion axioms for atomic facts and Boolean operations, (c) the following new recursion axiom for conditional beliefs:

$$[!P]B^{\psi}\varphi \leftrightarrow (P \rightarrow B^{P\wedge[!P]\psi}\ [!P]\varphi).$$

Proof First we check the soundness of the new axiom. On the left-hand side, it says that in the new model $(\boldsymbol{M}|P, s)$, φ is true in the best ψ-worlds. With the usual precondition for true announcement, on the right-hand side, it says that in $\boldsymbol{M}$, s, the best worlds that are P now and will become ψ after announcing that P, will also become φ after announcing P. This is indeed equivalent. The remainder of the proof is our earlier stepwise reduction analysis, noting that the above axiom pushes announcement modalities inside. ■

For a joint version with knowledge, just combine with the *PAL* axioms.

Clarifying the Ramsey Test Our dynamic logic sharpens up the *Ramsey Test* that says: 'A conditional proposition $A \Rightarrow B$ is true, if, after adding A to your current stock of beliefs, the minimal consistent revision implies B.' In our perspective, this is ambiguous, as B need no longer say the same thing after the revision. That is why our recursion axiom carefully distinguishes between formulas φ before update and what happens to them after: $[!P]\varphi$. Even so, there is an interesting special case of *factual propositions* φ without modal operators (cf. Chapter 3), that do not change their truth value under announcement. In that case, with Q, R factual propositions, the above recursion axioms read as follows:[11]

$$[!P]BQ \leftrightarrow (P \rightarrow B^{P}Q), \quad [!P]B^{R}Q \leftrightarrow (P \rightarrow B^{P\wedge R}Q)$$ ■

Belief change under hard update is not yet revision in the usual sense, that can be triggered by weaker information (see below). Nevertheless, we pursue it a bit further, as it links to important themes in rational agency: variety of attitudes, and of consequence relations. The first will be considered right now, the second a bit later in this chapter.

[11] For some paradoxes in combining the Ramsey test with belief revision, cf. Gärdenfors (1988). The nice thing about a logic approach is that every law is automatically sound.

7.4 Exploring the framework: safe belief and richer attitudes

The above setting may seem simple, but it contains some tricky scenarios:

Example Misleading with the truth.
Consider a model where an agent believes that p, which is indeed true in the actual world to the far left, but for the wrong reason: she finds the most plausible world the one to the far right. For convenience, assume each world verifies a unique proposition letter q_i:

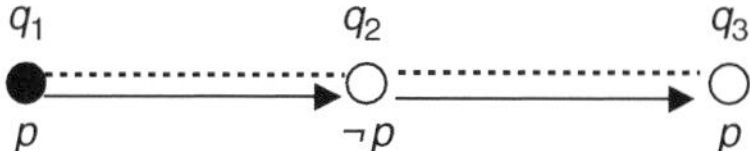

Now giving the true information that we are not in the final world ('$\neg q_3$') updates to a model in which the agent believes mistakenly that $\neg p$.[12]

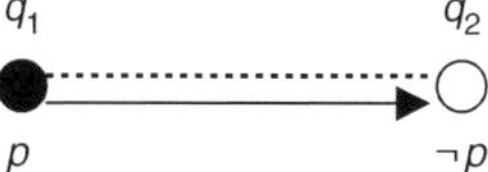

■

Agents have a rich repertoire of attitudes In response, an alternative view of our task in this chapter makes sense. So far, we have assumed that knowledge and belief are the only relevant attitudes. But in reality, agents have a rich repertoire of attitudes concerning information and action, witness the many terms in natural language with an epistemic or doxastic ring: being certain, being convinced, assuming, etcetera.[13]

Language extension: safe belief Among all possible options in this plethora of epistemic-doxastic attitudes, the following new notion makes particular sense, intermediate between knowledge and belief. It has stability under new true information:

DEFINITION Safe belief.
The modality of *safe belief* $B^+\varphi$ is defined as follows: $M, s \models B^+\varphi$ iff for all worlds t in the epistemic range of s with $t \geq s$, $M, t \models \varphi$. In words, this says

[12] Observations like this have been made in philosophy, computer science, and game theory.

[13] Cf. Lenzen (1980) for similar views. Krista Lawlor has pointed me also at the richer repertoire of epistemic attitudes found in pre-modern epistemology.

that φ is true in all epistemically accessible worlds that are at least as plausible as the current one.[14] ■

The modality $B^+\varphi$ is stable under hard information, at least for factual assertions φ that do not change their truth value as the model changes.[15] And it makes a lot of technical sense, as it is the *universal base modality* $[\leq]\varphi$ for the plausibility ordering.[16] This idea occurs in Boutilier (1994), Halpern (1997) (cf. Shoham & Leyton-Brown 2008), Baltag & Smets (2006, 2007a) (following Stalnaker), and independently in our later Chapter 9 on preference logic. In what follows, we make safe belief part of the static doxastic language – as a pilot for a richer theory of attitudes in the background. Pictorially, one can think of this as follows:

Example Three degrees of doxastic strength.
Consider this picture, now with the actual world in the middle:

$K\varphi$ describes what we know: φ must be true in all worlds in the epistemic range, less or more plausible than the current one. $B^+\varphi$ describes our safe beliefs in further investigation: φ is true in all worlds from the middle toward the right. Finally, $B\varphi$ describes the most fragile thing: our beliefs as true in all worlds in the current rightmost position. ■

In addition, safe belief simplifies things, if only as a technical device:

Fact On finite epistemic connected plausibility models,
(a) safe belief can define its own conditional variant,
(b) with a knowledge modality, safe belief can define conditional belief.

Proof (a) is obvious, since we can conditionalize to $B^+(A \to \varphi)$ like a standard modality. (b) uses a well-known fact about finite connected plausibility models, involving the existential dual modality $<B^+>$ of safe belief (cf. van Benthem & Liu 2007, Baltag & Smets 2006):

[14] Safe belief in this style uses an *intersection* of epistemic accessibility and plausibility. We could also decouple the two, and introduce a modality for plausibility alone.

[15] Note that new true information will never remove the actual world, our vantage point.

[16] Van Benthem & Pacuit (2011) point out that, on *pre-orders*, safe belief in its dynamic sense might rather have to be a modality ranging over all worlds that are not proper predecessors of the current world. This includes incomparable worlds.

Claim Conditional belief $B^{\psi}\varphi$ is equivalent to the iterated modal statement

$$K((\psi \wedge \varphi) \rightarrow <B^{+}> (\psi \wedge \varphi \wedge B^{+}(\psi \rightarrow \varphi)))$$

This claim can be proved with a little puzzling.[17] ■

Safe belief also has some less obvious features. For instance, since its accessibility relation is transitive, it satisfies Positive Introspection, but since that relation is not Euclidean, it fails to satisfy Negative Introspection. The reason is that safe belief mixes purely epistemic information with *procedural information* (cf. Chapters 3, 11). Once we see that agents have a richer repertoire of doxastic-epistemic attitudes than K and B, old intuitions about epistemic axioms need not be very helpful in understanding the full picture.

Finally, we turn to belief dynamics under hard information:

THEOREM The complete logic of belief change under hard information is the one whose principles were stated before, plus the following recursion axiom for safe belief:

$$[!P]\ B^{+}\varphi \leftrightarrow (P \rightarrow B^{+}(P \rightarrow [!P]\varphi))$$

This axiom for safe belief under hard information in fact implies the earlier one for conditional belief, by unpacking the above modal definition.

7.5 Belief change under soft information: radical upgrade

It is time to move to a much more general and flexible view of our subject.

Soft information and plausibility change Our story so far is a hybrid: we saw how a soft attitude changes under hard information. The more general scenario has an agent aware of being subject to continuous belief changes,

[17] The result generalizes to other models, and this modal translation is *itself* a good candidate for lifting the maximality account of conditional belief to infinite models, as well as non-connected ones. Alternative versions would use modalities for the strict ordering corresponding to reflexive plausibility $\leq$ to define maximal ψ-worlds directly, in the format $\psi \wedge \neg<<>\psi$: cf. Girard (2008).

and taking incoming signals in a softer manner, without throwing away options forever. But then, public announcement is too strong:

Example No way back.
Consider the earlier model where the agent believed that $\neg p$, though p was in fact the case:

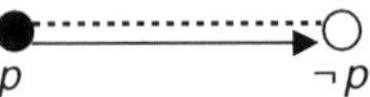

Announcing p removes the $\neg p$-world, making belief revision impossible. ■

What we need is a mechanism that just makes incoming information P more plausible, without burning our boats behind us. An example are the *default rules* $A \Rightarrow B$ in Veltman (1996). Accepting a default conditional does not say that all A-worlds must now be B-worlds. It makes the counter-examples (the $A \wedge \neg B$-worlds) less plausible until further notice. This soft information does not eliminate worlds, it just changes their ordering. More precisely, a triggering event that makes us believe P need only *rearrange worlds* making the most plausible ones P: by 'promotion' or 'upgrade' rather than elimination. Thus, in our doxastic models $\mathbf{M} = (W, \sim_i, \leq_i, V)$, we will change the relations $\leq_i$, rather than the world domain W or the epistemic accessibilities $\sim_i$. Rules for plausibility change exist in models of belief revision (Grove 1988; Rott 2006) as different *policies* that agents can adopt toward new information. We now show how our dynamic logics deal with them.[18]

Radical revision One very strong policy is like a radical social revolution where some underclass P now becomes the upper class. In a picture, we get the following reversal:

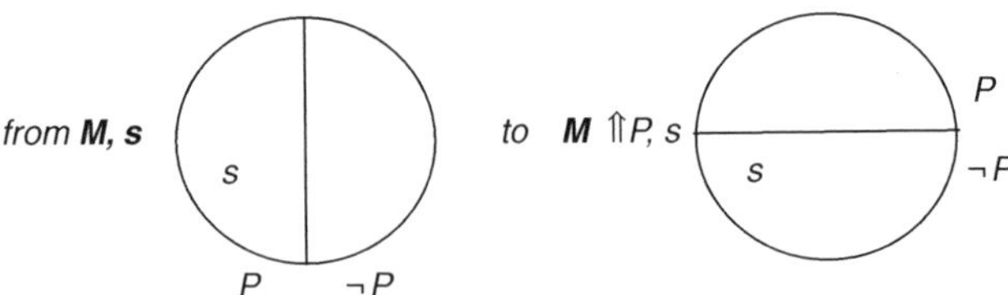

[18] In formal learning theory (Kelly 1996), these are different learning strategies.

Definition Radical, or lexicographic upgrade.
A *lexicographic upgrade* $\Uparrow P$ changes the current ordering $\leq$ between worlds in $\boldsymbol{M}$, s to a new model $\boldsymbol{M}\Uparrow P$, s as follows: all P-worlds in the current model become better than all $\neg P$-worlds, while, within those two zones, the old plausibility ordering remains.[19] ■

With this definition in place, our earlier methodology applies. As we did with public announcement, we introduce a corresponding upgrade modality in our dynamic doxastic language:

$$\boldsymbol{M}, s \models [\Uparrow P]\varphi \quad \text{iff } \boldsymbol{M} \Uparrow P, s \models \varphi$$

Here is a complete account of how agents' beliefs change under soft information, in terms of the key recursion axiom for changes in conditional belief under radical revision:

Theorem The dynamic doxastic logic of lexicographic upgrade is axiomatized completely by the following principles:
(a) any complete axiom system for conditional belief on the static models,
(b) the following recursion axioms:[20]

$$\begin{array}{lll}
[\Uparrow P]q & \leftrightarrow & q \quad \text{for all atomic proposition letters } q \\
[\Uparrow P]\neg\varphi & \leftrightarrow & \neg[\Uparrow P]\varphi \\
[\Uparrow P](\varphi \wedge \psi) & \leftrightarrow & [\Uparrow P]\varphi \wedge [\Uparrow P]\psi \\
[\Uparrow P]K\varphi & \leftrightarrow & K[\Uparrow P]\varphi \\
[\Uparrow P]B^{\psi}\varphi & \leftrightarrow & (\Diamond(P \wedge [\Uparrow P]\psi) \wedge B^{P \wedge [\Uparrow P]\psi}\, [\Uparrow P]\varphi) \\
 & & (\neg\Diamond(P \wedge [\Uparrow P]\psi) \wedge B^{[\Uparrow P]\psi}\, [\Uparrow P]\varphi)
\end{array}$$

Proof The first four axioms are simpler than those for *PAL*, since there is no precondition for $\Uparrow P$ as there was for $!P$. The first axiom says that upgrade does not change truth values of atomic facts. The second says that upgrade is a function on models, the third is a general law of modality, and the fourth that no change takes place in epistemic accessibility.

The fifth axiom is the locus where we see the specific change in the plausibility ordering. The left-hand side says that, after the P-upgrade, all best ψ-worlds satisfy φ. On the right-hand side, there is a case distinction.

[19] This is known as the 'lexicographic policy' for relational belief revision.

[20] Here, as in Chapter 2, $\Diamond$ is the dual *existential epistemic modality* $\neg K \neg$.

Case (1): there are accessible P-worlds in the original model $\mathbf{M}$ that become ψ after the upgrade. Lexicographic reordering $\Uparrow P$ makes the best of these worlds the best ones over-all in $\mathbf{M}\Uparrow P$ to satisfy ψ. In the original $\mathbf{M}$ – viz. its epistemic component visible from the current world s – the worlds of Case 1 are just those satisfying the formula $P \wedge [\Uparrow P]\psi$. Therefore, the conditional formula $B^{P \wedge [\Uparrow P]\psi}[\Uparrow P]\varphi$ says that the best among these in $\mathbf{M}$ will indeed satisfy φ after the upgrade. These best worlds are the same as those described earlier, as lexicographic reordering does not change order of worlds inside the P-area. Case (2): no P-worlds in the original model $\mathbf{M}$ become ψ after upgrade. Then lexicographic reordering $\Uparrow P$ makes the best worlds satisfying ψ after the upgrade just the same best worlds over-all as before that satisfied $[\Uparrow P]\psi$. Accordingly, the formula $B^{[\Uparrow P]\psi}[\Uparrow P]\varphi$ in the reduction axiom says that the best worlds become φ after upgrade.

The rest of the proof is the reduction argument of Chapter 3.[21] ■

The final equivalence describes which conditional beliefs agents have after soft upgrade. This may look daunting, but try to read the principles of some default logics existing today! Also, recall the earlier point that we need to describe how conditional beliefs change, not just absolute ones, to avoid getting trapped in the Iteration Problem.

Special cases Looking at special cases may help. First, consider absolute beliefs $B\varphi$. Conditioning on 'True', the key recursion axiom simplifies to:

$$([\Uparrow P]B\varphi \leftrightarrow (\Diamond P \wedge B^{P}[\Uparrow P]\varphi) \vee (\neg\Diamond P \wedge B[\Uparrow P]\varphi)$$

And here is the simplified recursion axiom for *factual propositions* that did not change their truth values under update or upgrades:[22]

$$[\Uparrow P]B^{\psi}\varphi \leftrightarrow (\Diamond(P \wedge \psi) \wedge B^{P\wedge\psi}\varphi) \vee (\neg\Diamond(P \wedge \psi) \wedge B^{\psi}\varphi)$$ ■

Safe belief once more As a final simplification, recall the earlier notion of safe belief, that defined conditional belief using the modality K. We can also derive the above recursion axiom from the following:

FACT The following axiom is valid for safe belief under radical revision:

$$[\Uparrow P]B^{+}\varphi \leftrightarrow (P \wedge B^{+}(P \rightarrow [\Uparrow P]\varphi)) \vee (\neg P \wedge B^{+}(\neg P \rightarrow [\Uparrow P]\varphi) \wedge K(P \rightarrow [\Uparrow P]\varphi))$$

21 Details on this result and the next are in van Benthem (2007b), van Benthem & Liu (2007).

22 To us, this is the paradox-free sense in which a Ramsey Test holds for our logic.

Proof For any world s, the two disjuncts describe its more plausible worlds after upgrade. If $\mathbf{M}, s \models P$ in the initial model $\mathbf{M}$, these are all former P-worlds that were more plausible than s. If not $\mathbf{M}, s \models P$, these are the old more plausible $\neg P$-worlds plus *all* P-worlds. ■

Static pre-encoding Our compositional analysis says that any statement about effects of hard or soft information is encoded in the initial model: the epistemic present contains the epistemic future. We have used this line to design the right static languages, with a crucial role for conditional belief. As in earlier chapters, we may want to drop this reduction when considering global informational procedures: Chapter 11 shows how to do this.[23]

Radical upgrade will be used at various places in this book, especially in our study of game solution in Chapters 10, 15. This will throw further light on its semantic features.

7.6 Conservative upgrade and general revision policies

Radical revision was our pilot, but its specific plausibility change is just one way of taking soft information. A more conservative policy for believing a new proposition puts not all P-worlds on top qua plausibility, but just *the most plausible P-worlds*. After the revolution, this policy co-opts just the leaders of the underclass – the sage advice that Machiavelli gave to rulers pondering what to do with the mob outside of their palace.

DEFINITION Conservative plausibility change.
Conservative upgrade $\uparrow P$ replaces the current plausibility ordering $\leq$ in a model $\mathbf{M}$ by the following: the best P-worlds come on top, but apart from that, the old ordering remains. ■

Technically, $\uparrow P$ is a special case of radical revision: $\Uparrow(best(P))$, if we have the latter in our static language. But it seems of interest per se. Our earlier methods produce its logic:

THEOREM The dynamic doxastic logic of conservative upgrade is axiomatized completely by the following principles:

[23] Technically, this design involves a form of closure beyond syntactic relativization (Chapter 3). We now also need closure under syntactic *substitutions* of defined predicates for old ones.

(a) a complete axiom system for conditional belief on the static models,

(b) the following reduction axioms:

$$
\begin{array}{lll}
[\uparrow P]q & \leftrightarrow & q \quad \text{for all atomic proposition letters } q \\
[\uparrow P]\neg\varphi & \leftrightarrow & \neg[\uparrow P]\varphi \\
[\uparrow P](\varphi \wedge \psi) & \leftrightarrow & [\uparrow P]\varphi \wedge [\uparrow P]\psi \\
[\uparrow P]K\varphi & \leftrightarrow & K[\uparrow P]\varphi \\
[\uparrow P]B^{\psi}\varphi & \leftrightarrow & (B^{P}\neg[\uparrow P]\psi \wedge B^{[\uparrow P]\psi}[\uparrow P]\varphi) \vee (\neg B^{P}\neg[\uparrow P]\psi \wedge B^{P\wedge[\uparrow P]\psi}[\uparrow P]\varphi)
\end{array}
$$

We leave a proof to the reader. Of course, one can also combine this logic with the earlier one, to combine different sorts of revising behaviour, as in mixed formulas $[\Uparrow][\uparrow]\varphi$.

Policies Many further changes in a plausibility order can be responses to an incoming signal. This reflects the host of belief revision policies in the literature: Rott (2006) lists twenty-seven. General relation transformers were proposed in van Benthem, van Eijck & Frolova (1993), calling for a dynamification of preference logic. The same is true for defaults, commands (Yamada 2006), and other areas where plausibility or preference can change (cf. Chapter 9). Our approach suggests that one can take any definition of change, write a matching recursion axiom, and then a complete dynamic logic. But how far does this go?[24]

Relation transformers in dynamic logic One general method works by inspection of the format of definition in the above examples. For instance, it is easy to see the following:

FACT Radical upgrade $\Uparrow P$ is definable as a program expression in propositional dynamic logic.

Proof The format is this, with 'T' the universal relation between worlds:

$$\Uparrow P(R) := (?P; T; ?\neg P) \cup (?P; R; ?P) \cup (?\neg P; R; ?\neg P)$$ ■

[24] Maybe 'policy' is the wrong term, as it suggests a persistent habit over time, like being stubborn. But our events describe local responses to particular inputs. Speech act theory has a nice distinction between information per se (what is said) and the *uptake*, how a recipient reacts. In that sense, the softness of our scenarios is in the uptake, rather than in the signal itself.

Van Benthem & Liu (2007) then introduce the following format:

DEFINITION *PDL*-format for relation transformers.
A definition for a new relation R on models is in *PDL-format* if it can be stated in terms of the old relation, *union*, *composition*, and *tests*. ■

A further example is an act of 'suggestion' $\#P$ (cf. the preference logics of Chapter 9) that merely takes out R-pairs with '$\neg P$ over P':

$$\#P(R) = (?P; R) \cup (R; ?\neg P)$$

This format generalizes our earlier procedure with recursion axioms:

THEOREM For each relation change defined in *PDL*-format, there is a complete set of recursion axioms that can be derived via an effective procedure.

Proof Here are two examples of computing modalities for the new relation after the model change, using the recursive program axioms of *PDL*. Note how the second calculation uses the existential epistemic modality $<E>$ for the occurrence of the universal relation:

(a) $<\#P(R)><R>\varphi \leftrightarrow <(?P; R) \cup (R; ?\neg P)>\varphi \leftrightarrow <(?P; R)>\varphi \vee <(R; ?\neg P)>\varphi$
$\leftrightarrow <?P><R>\varphi \vee <R><?\neg P>\varphi \leftrightarrow (P \wedge <R>\varphi) \vee <R>(\neg P \wedge \varphi)$

(b) $<\Uparrow P(R)>\varphi \leftrightarrow <(?P; T; ?\neg P) \cup (?P; R; ?P) \cup (?\neg P; R; ?\neg P)>\varphi$
$\leftrightarrow <(?P; T; ?\neg P)>\varphi \vee <(?P; R; ?P)>\varphi \vee <(?\neg P; R; ?\neg P)>\varphi$
$\leftrightarrow <?P><T><?\neg P>\varphi \vee <?P><R><?P>\varphi \vee <?\neg P><R><?\neg P>\varphi$
$\leftrightarrow (P \wedge E(\neg P \wedge \varphi)) \vee (P \wedge <R>(P \wedge \varphi)) \vee (\neg P \wedge <R>(\neg P \wedge \varphi))$[25]

This gives uniformity behind earlier results. For instance, Example (b) easily transforms into an axiom for safe belief after radical upgrade $\Uparrow P$, equivalent to the one we gave before. ■

7.7 Conclusion

This chapter has realized the second stage of our logical analysis of agency, extending the dynamic approach for knowledge to belief. The result is one merged theory of information update and belief revision, using standard modal techniques instead of ad-hoc formalisms. We found many new topics,

[25] Here, 'E' stands for the *existential modality* 'in some world'.

like using the dynamics to suggest new epistemic-doxastic modalities. After this opt-out point for the chapter, we will pursue some more specialized themes for the interested reader, including transfer of insights between *DEL* and frameworks such as *AGM*.

7.8 Further themes: belief revision with *DEL*

DEL formats, event models as triggers Another approach uses *DEL* (Chapter 4) rather than *PAL* as a role model. Event models for information can be much more subtle than announcements, or the few specific policies we have discussed. While its motivation came from partial observation, *DEL* also applies to receiving signals with different strengths. Here is a powerful idea from Baltag & Smets (2006) (with a precursor in Aucher 2004):

DEFINITION Plausibility event models.
Plausibility event models are event models just as in Chapter 4, but now expanded with an additional plausibility relation over their epistemic equivalence classes. ■

In this setting, radical upgrade $\Uparrow P$ can be implemented in an event model as follows: we do not throw away worlds, so we need two 'signals' $!P$ and $!\neg P$ with obvious preconditions P, $\neg P$ that will copy the old model. But we now say that signal $!P$ is more plausible than signal $!\neg P$, relocating the revision policy in the nature of the input:

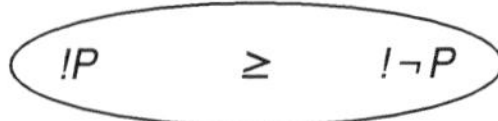

Different event models will then represent a great variety of update rules. But we still need to state the update mechanism more precisely, since it is not quite that of Chapter 4:

'One Rule To Rule Them All' The relevant product update rule radically places the emphasis on the last event observed, but it is conservative with respect to everything else:

DEFINITION Priority Update.
Consider an epistemic plausibility model **M**, *s* and a plausibility event model **E**, *e*. The *product model* **M** × **E**, *(s, e)* is defined entirely as in Chapter 4, with the

following new rule for the plausibility relation, with $<$ the earlier-mentioned strict version of the relation:

$$(s, e) < (t, f) \text{ iff } (s \leq t \wedge e \leq f) \vee e < f \qquad \blacksquare$$

Thus, if the new P induces a preference between worlds, that takes precedence: otherwise, we go by the old plausibility order. This rule places great weight on the last observation or signal received. This is like belief revision theory, where receiving just one signal $*P$ leads me to believe that P, even if all of my life I had been receiving evidence against P. It is also in line with 'Jeffrey Update' in probability (Chapter 8) that imposes a new probability for some proposition, while adjusting all other probabilities proportionally.[26,27]

THEOREM The dynamic logic of priority update is axiomatizable completely.

Proof As before, it suffices to state the crucial recursion axioms reflecting the above rule. We display just one case, for the modality of safe belief, stated in existential format:

$$<E, e><\leq>\varphi \leftrightarrow (Pre_e \wedge (\bigvee_{e \leq f \text{ in E}} <\leq><E,f>\varphi \vee \bigvee_{e < f \text{ in E}} \Diamond <E,f>\varphi))$$

where $\Diamond$ is again the existential epistemic modality. $\blacksquare$

This *shifts the locus of description*. Instead of many policies for processing a signal, each with its own logic, we now put the policy in the input E. This has some artificial features: the new event models are much more abstract than those in Chapter 4. Also, even to describe simple policies like conservative upgrade, the language of event models must be extended to event preconditions of the form *most-plausible(P)*. But the benefit is clear: infinitely many policies can be encoded in event models, while belief change now works with just one update rule, and the common objection that belief revision theory is non-logical and messy for its proliferation of policies evaporates.[28]

Digression: abrupt revision and slow learning An update rule with so much emphasis on the last signal is special. Chapter 12 brings out how, using social

[26] There may be a worry that this shifts from *DEL's precondition* analysis to a forward style of thinking in terms of *postconditions*: cf. Chapter 3, but we will not pursue this objection.

[27] As in Chapter 4, product update with event models generalizes easily to real world change, taking on board the well-known Katsuno–Mendelzon sense of temporal 'update'.

[28] A comparison between the earlier *PDL*-style and *DEL*-style formats remains to be made.

choice between old and new signals. Learning theory also has gentler ways of merging new with old information, without overriding past experience. This theme will return with probabilistic update rules in Chapter 8, and with score-based rules for preference dynamics in Chapter 9.

7.9 Further themes: belief revision versus generalized consequence

Update of beliefs under hard or soft information is also an alternative to the current proliferation of non-standard notions of consequence. Here is a brief illustration (van Benthem 2008d): Chapter 13 has a more extensive discussion of the issues.

Update versus inference: non-monotonic logic Classical consequence says that all models of premises P are models for the conclusion C. McCarthy (1980) pointed out how problem solving goes beyond this. A *circumscriptive* consequence from P to C says that

C is true in all the *minimal* models for P

Here, minimality refers to a relevant comparison order $\leq$ for models: inclusion of object domains, extensions of predicates, and so on. The general idea is minimization over any order (Shoham 1988), supporting non-monotonic consequence relations that are like the earlier conditional logics. This is reminiscent of our plausibility models for belief, and indeed, one can question the original view of problem solving. We are given initial information and need to find the goal situation, as new information comes in. The crucial process here is our responses: solving puzzles and playing games is all about information update and belief change. Non-monotonic logics have such processes in the background, but leave them *implicit*. But making them *explicit* is the point of our dynamic logics.

Dynamic consequence on a classical base Our logic suggests two kinds of dynamic consequence. First, *(common) knowledge* may result, and we get classical consequence for factual assertions (cf. the dynamic inference of Chapter 3).[29] Or *belief* may result, referring to the minimal worlds. Thus, what is usually cast as a notion of consequence

[29] Factual assertions drive most non-standard consequence relations. But as we saw in Chapter 3, structural rules get dynamic twists when we consider the full language.

$P_1, \ldots, P_k \Rightarrow \varphi$

gets several dynamic variants definable in our language:

either $[!P_1] \ldots [!P_k]K\varphi$ or $[!P_1] \ldots [!P_k]B\varphi$

whose behaviour is captured by our earlier complete logics. This suggests a truly radical point of view. Once the relevant informational events have been made explicit, there is no need for 'non-standard logics'. The dynamic logic just works with a classical static notion of consequence. *Non-monotonic logic is monotonic dynamic logic of belief change.*

New styles Our dynamic logics for soft information even suggest *new* consequence relations, such as:

$P_1, \ldots, P_k \Rightarrow_{circ-soft} \varphi$ iff $[\Uparrow P_1] \ldots [\Uparrow P_k]B\varphi$

FACT For factual assertions P, Q, (i) $P, Q \Rightarrow_{circ\text{-}hard} P$, (ii) not $P, Q \Rightarrow_{circ\text{-}soft} P$.

Proof (i) New hard updates yield subsets of the P-worlds. (ii) The last upgrade with Q may have demoted all P-worlds from their former top positions. ■

Thus, we have an interesting two-way interplay between logical dynamics of belief change and the design of new non-monotonic consequence relations.[30]

7.10 Further themes: postulates and correspondence results

A brief comparison with AGM The best-known account of belief revision is *AGM theory* (Gärdenfors 1988; Gärdenfors & Rott 1995; Rott 2007) that deals with three fundamental abstract operations of $+A$ ('update'), $*A$ ('revision'), and $-A$ ('contraction') for factual information of a single agent. In contrast with the *DEL* mechanism for transforming plausibility models and generating complete logics, *AGM* analyses belief change without proposing a specific construction, placing abstract algebraic postulates on the above operations instead.[31] This is the contrast between constructive and postulational approaches that we have seen in Chapter 3.

[30] Technically, this suggests new open problems about complete sets of structural rules.

[31] We refer to the cited literature for a precise statement of the postulates.

The *AGM* postulates claim to constrain all reasonable revision rules. Do they apply to ours? Here is a simple test. The 'Success Postulate' says that all new information comes to be believed in the theory revised with this information: $A \in T^*A$. But even *PAL* fails this test, and the difference is instructive. Success follows from our axioms for *factual* propositions, but it fails for complex epistemic or doxastic ones. For instance, true Moore-sentences cannot be believed after announcement. The main intuitions of belief revision theory (or generalized consequence relations) apply to factual assertions only, whereas a logic approach like ours insists on dealing with all types of proposition at once.[32]

Here are two more differences between the frameworks. *AGM* deals with single agents only, while *DEL* is essentially multi-agent, including higher-order information about what others believe. And also tellingly, *DEL* analyses not three, but an *infinity* of triggers for belief change, from public announcements to complex informational events.

Revision postulates as frame correspondences Despite the difference in thrust, the postulational approach to belief revision makes good sense. A matching modal framework at the right level of generality is Dynamic Doxastic Logic (*DDL*, Segerberg 1995, 1999; Leitgeb & Segerberg 2007). This merely assumes some relation change on the current model, functional or relational, without specifying it further. The main operator then becomes:

DEFINITION Abstract modal logic of model change.
Let **M** be any doxastic-epistemic model, $[[P]]$ the set of worlds in **M** satisfying P, and $\mathbf{M}^*[[P]]$ just some new model. For the matching dynamic modal operator, we set $\mathbf{M}, s \models [^*P]\varphi$ iff $\mathbf{M}^*[[P]], s \models \varphi$. ■

[32] The same point about complex propositions and order dependence returns with other *AGM* postulates. For instance, the 'Conjunction Postulate' that compresses two updates into one mixes events that we distinguish: processing a conjunction of propositions, and processing two new propositions successively. For factual propositions in *PAL*, this amounts to the same thing, but not for complex epistemic ones. And in our dynamic logic of belief, even the factual case is problematic. Here is why. In *PAL*, successive public announcements could still be compressed by the law $[!P][!Q]\varphi \leftrightarrow [!(P \wedge [!P]Q)]\varphi$. But two successive upgrades $\Uparrow P; \Uparrow Q$ rearrange a model as follows. First, P-worlds come on top of $\neg P$-ones, then the same happens with Q. The result is the order pattern $PQ \geq \neg PQ \geq P\neg Q \geq \neg P\neg Q$. No single upgrade does this, and no iteration law compresses the effect of two revision steps to just one with the same effects on conditional belief.

DDL models resemble Lewis spheres for conditional logic, or their neighbourhood versions (Girard 2008). The minimal modal logic *K* is valid, and on top, further axioms constrain relation changes for bona fide revision policies. In the limit, imposing a particular set of axioms might even determine one particular policy.

We show how this ties up with our approach in terms of *frame correspondence*, just as we did for *PAL* update in Chapter 3 by postulating key recursion axioms, and then seeing which update operations qualified on an abstract universe of models and transitions.

Usually, frame correspondences analyse semantic content of given axioms in one model for a static modal language. But one can just as well take the above setting of relation changing operations ♥*P* over a family of models (with worlds and a ternary comparison relation $x \leq_s y$). ♥*P* takes any model **M** and set of worlds *P* in it,[33] and yields a new model **M**♥*P* with the same worlds but a new relation $\leq_s$. Further axioms may then constrain this.[34]

Analysing a few AGM postulates For a start, the Success Postulate says something weak, that holds for both earlier operations $\Uparrow P$ and $\uparrow P$:[35]

Fact The formula *[*♥*p]Bp* says that the best worlds in **M**♥*p* are all in *p*.

But we can also demand something stronger, that the best worlds in **M**♥*p* are precisely the best *p*-worlds in **M** (the upper class law '*UC*'). This, too, can be expressed. But we need a stronger dynamic formula, involving two different proposition letters *p* and *q*:

Fact The formula $B^p q \leftrightarrow$ *[*♥*p]Bq* expresses *UC*.

But this preoccupation with the upper classes still fails to constrain the total relation change. For that, we must look at the new social order in all classes

33 Here we have dropped the above double denotation brackets *[[P]]* for convenience.

34 Even more abstract spaces of models can be used here as general background for analysing the content of dynamic axioms, but our setting suffices to make our main points.

35 Frame correspondence has a format like this (cf. van Benthem to appearB). The modal axiom $\Box p \rightarrow \Box\Box p$ is true at world *s* in frame $\mathbf{F} = (W, R)$ iff *R* is transitive at *s*: i.e., $\mathbf{F}, s \models \forall y (Rxy \rightarrow \forall z(Ryz \rightarrow Rxz))$. Frame truth is truth under all valuations on frame **F** for its proposition letters. Thus, it does not matter whether we use formula $\Box p \rightarrow \Box\Box p$ or the schema $\Box\varphi \rightarrow \Box\Box\varphi$. Not so for *PAL* and *DEL*, where plain and *schematic* validity differ. In the following proofs, we use proposition letters for sets of worlds, by-passing issues of changes in truth value across updates.

after the Revolution, i.e., at conditional beliefs following relation upgrade. As an illustration, we consider the key reduction axiom for $\Uparrow P$, using proposition letters instead of schematic variables.[36] The following shows how this determines lexicographic reordering of models completely (again we use the earlier auxiliary existential modality E):

THEOREM The axiom $[\heartsuit p] B^r q \leftrightarrow (E(p \wedge r) \wedge B^{p \wedge r} q) \vee (\neg E(p \wedge r) \wedge B^r q)$ holds in a universe of frames iff the operation $\heartsuit p$ is lexicographic upgrade.

Proof That the principle holds for lexicographic upgrade was our earlier soundness. Next, let the principle hold for all set values of q and r (but p is kept fixed). First, we show that, if $x \leq_s y$ in $\mathbf{M}\heartsuit p$, then this pair was produced by lexicographic upgrade. Let r be the set $\{x, y\}$ and $q = \{y\}$. Then the left-hand side of our axiom is true. Hence the right-hand side is true as well, and there are two cases. *Case 1*, $E(p \wedge r)$: one of x, y is in p, and so $p \wedge r = \{x, y\}$ (1.1) or $\{x\}$ (1.2) or $\{y\}$ (1.3). Moreover, $B^{p \wedge r} q$ holds in $\mathbf{M}$ at s. If (1.1), we have $x \leq_s y$ in $\mathbf{M}$, with both x, y in p. If (1.2), we have $y = x$, and by reflexivity, again $x \leq_s y$ in $\mathbf{M}$. Case (1.3) can only occur when $y \in p$ and $x \notin p$: which is the typical case for upgrade. *Case 2*, $\neg E(p \wedge r)$: x, y are not in p. The true disjunct $B^r q$ then says that $x \leq_s y$ in $\mathbf{M}$.

Conversely, we show that all pairs satisfying the definition of lexicographic upgrade in $\mathbf{M}$ make it into the new order in $\mathbf{M}\heartsuit p$. Here is one example: the other case is similar. Suppose that $y \in p$ while $x \notin p$. Set $r = \{x, y\}$ and $q = \{y\}$, whence $p \wedge r = \{y\}$. This makes $(E(p \wedge r) \wedge B^{p \wedge r} q)$ true at world s in $\mathbf{M}$, and hence also the whole disjunction to the right. By our axiom, the left-hand formula $[\heartsuit p]\ B^r q$ then also holds at s in $\mathbf{M}$. But this tells us that *in the new model* $\mathbf{M}\heartsuit p$, $B^r q$ holds at s. Thus, the best worlds in $\{x, y\}$ are in $\{y\}$: i.e., $x \leq_s y$ in $\mathbf{M}\heartsuit p$. ■

This can be generalized to abstract universes of transitions, quantifying over sets of worlds inside and across plausibility models.[37,38] But even our simple setting shows how frame correspondence for languages with

[36] Thus we suppress general modalities $[\Uparrow P]\psi$ that were sensitive to transfer effects.

[37] The above arguments then work uniformly by standard modal substitution techniques.

[38] Further *AGM*-postulates mix two operations that change models: *update* $!P$ and *upgrade* $\heartsuit P$, with laws like (a) $[\heartsuit(p \wedge q)]Br \rightarrow [!q][\heartsuit p]Br$, (b) $([\heartsuit p]Eq \wedge [!q][\heartsuit p]Br) \rightarrow [\heartsuit(p \wedge q)]Br$. These constrain simultaneous choice of two abstract model changing operations.

model-changing modalities is an illuminating way of doing abstract postulational analysis of update and revision.[39]

7.11 Still more issues, and open problems

Variations on the static models We assumed agents with epistemic introspection of their plausibility order. Without this, we would need *ternary* world-dependent plausibility relations, as in conditional logic. What do our $\{K, B\}$-based systems look like then?

Also, safe belief suggests having just one primitive plausibility pre-order $\leq$, defining knowledge as truth in all worlds, whether *less* or more plausible (cf. van Eijck & Sietsma 2009 on *PDL* over such models). What happens to our themes in the latter setting?

Finally, many authors have proposed basing doxastic logic on more general *neighbourhood models* (cf. Girard 2008; Zvesper 2010). How should we lift our theory to that setting?

Common belief and social merge We have not analysed common belief, satisfying the fixed-point equation $CB_G\varphi \leftrightarrow \bigwedge_{i \in G} B_i(\varphi \wedge CB_G\varphi)$. Technically, a complete dynamic logic seems to call for a combination of relation change with the *E-PDL* program techniques of Chapter 4. More generally, belief revision policies describe what a single agent does when confronted with surprising facts. But beliefs often change because *other agents* contradict us, and we need interactive settings where agents create one new plausibility ordering. These include events of *belief merge* (Maynard-Reid & Shoham 1998) and *judgment aggregation* (List & Pettit 2004). Construed either way, we need to look at *groups*: Chapter 12 has more on this.

Model theory of plausibility models and upgrade Our static logics for belief raise standard modal issues of appropriate notions of bisimulation and frame correspondence techniques for non-standard modalities that maximize over orderings. Like for conditional logic, these model-theoretic issues seem largely unexplored. In the dynamic setting, relation-changing modalities also raise additional issues, such as respect for plausibility bisimulation.

[39] Still, this is not the only abstract perspective on our logics. In Chapter 12, we do a postulational analysis of the Priority Update rule in terms of social choice.

Many of the themes in Chapters 2, 3 remain to be investigated for our systems here. Demey (2010b) is a technical exploration of bisimulation.

Syntactic belief revision There is also belief revision in a more fine-grained inferential sense, where agents may have fallible beliefs about what follows from their data, or about the consistency of their views. Such beliefs can be refuted by events like unexpected turns in argumentation or discussion. These scenarios are not captured by the semantic models of this chapter. How to do a syntactic dynamic logic in the spirit of Chapter 5?

Proof theory of logics for revision Our logics had complete Hilbert-style axiomatizations. Still, we have not looked at more detailed proof formats dealing with dynamic modalities, say, in natural deduction or semantic tableau style. What would these look like? The same question makes sense, of course, for *PAL* and *DEL* in Chapters 3, 4.

Backward versus forward once more Recall a basic contrast from earlier chapters. *AGM* is *forward-looking*, with postconditions of coming to believe.[40] But like *DEL* (backward-looking in its version without factual change), our logics compute what agents will believe only via preconditions. More generally, we want to merge local dynamics of belief change with a temporal Grand Stage where an instruction **P* wants a *minimal* move to some future state where one believes that *P*. We will do this in Chapter 11.[41]

7.12 Literature

A key source for belief revision theory in the postulational style is Gärdenfors (1988). Segerberg (1995) connects this to a 'dynamic doxastic logic' on neighbourhood models. Aucher (2004) treats quantitative belief revision rules in a *DEL* format, mixing event models with Spohn-style graded models. Van Benthem (2007b) is the main source for this chapter, axiomatizing dynamic logics for qualitative transformations of plausibility orders in a style similar to van Benthem & Liu (2007). Baltag & Smets (2006) gave the first general

[40] This gets harder with complex instructions like 'make agent i not believe that φ'.

[41] My first analysis of *AGM* in van Benthem (1989) was in this style, with modalities $[+P]$, $[-P]$, and $[{}^{*}P]$ for update, contraction, and revision on a universe of information stages.

qualitative *DEL* version of belief change, with priority product update using plausibility event models. Baltag & Smets (2008a) provide a mature version. Girard (2008) connects *DDL* with *DEL*, while Baltag, van Ditmarsch & Moss (2008) give broader framework comparisons and more history.

8 An encounter with probability

Our dynamic logics deal with agents' knowledge and beliefs under information update. But update has long been the engine of *probability theory*, a major lifestyle in science and philosophy. This chapter is an intermezzo linking the two perspectives, without any pretence at completeness, and assuming the basics of probability theory without further ado. We will show how probability theory fits well with dynamic-epistemic logics, leading to a new update mechanism that merges three aspects: prior world probability, occurrence probability of events, and observation probability. The resulting logic can be axiomatized in our standard style, leading to interesting comparisons with probabilistic methods.

8.1 Probabilistic update

The absolute basics A *probability space* $\mathbf{M} = (W, \mathbf{X}, P)$ is a set of worlds W with a family $\mathbf{X}$ of propositions that can be true or false at worlds, plus a probability measure P on propositions. This is like single-agent epistemic models, but with refined information on agent's views of propositions. In what follows, we use high-school basics: no σ–algebras, measurable sets, and all that. Intuitions come from finite probability spaces, putting each subset in $\mathbf{X}$. Under this huge simplification, the function P assigns values to single worlds, while its values for larger sets are automatic by addition, subject to the usual laws:

$$P(\neg\varphi) = 1 - P(\varphi), \quad P(\varphi \vee \psi) = P(\varphi) + P(\psi), \text{if } \varphi, \psi \text{ are disjoint}$$

Of crucial relevance to update are *conditional probabilities* $P(\varphi|A)$ giving the probability for proposition φ given that A is the case, using P rescaled to the set of worlds satisfying A:[1]

[1] We now use '*A*' for propositions, since the earlier '*P*' is taken for probability.

$$P(\varphi | A) = P(\varphi \wedge A)/P(A)$$

Bayes' Rule then computes conditional probabilities through the equation

$$P(\varphi \mid A) = P(A \mid \varphi) \bullet P(\varphi)/P(A)$$

and more elaborate versions. This describes probability change as new factual information A comes in. Other basic mechanisms update with non-factual information. E.g., the *Jeffrey Rule* updates with probabilistic information $P_i(A) = x$ by setting the new probability for A to x, and apart from that, redistributing probabilities among worlds inside the A and $\neg A$ zones proportionally to the old probabilities. And there are even further update rules.

Probabilistic and logical models Conditional probability resembles the conditional belief of Chapter 7, while the informal explanations surrounding it also have a whiff of *PAL* (Chapter 3). One zooms in on the worlds where the new A holds, and re-computes probabilities. This is like eliminating all $\neg A$-worlds, and re-evaluating epistemic or doxastic modalities. And Jeffrey Update is like the radical revision policy $\Uparrow A$ in Chapter 7 that fixes a belief to be achieved, but only minimally changes the plausibility order otherwise.

There are also differences. Crucially, dynamic logics make a difference between truth values of propositions φ before and after update, and hence a conditional belief $B^{\psi}\varphi$ was not quite the same as a *belief after update*: $[!\psi]B\varphi$. This made no difference for factual φ, ψ, but it did when formulas are more complex, as our logics want to have them. Also, our logics are about many agents, with syntactic *iterations* like 'I know that your probability is high', or 'My probability for your knowing φ equals y', that are scarce in probability theory. Finally, *DEL* update (cf. Chapter 4) did not just select subsets of a model: it *transforms* current models $\mathbf{M}$ into perhaps complex new ones. Thus, dynamic formulas $[E, e]\varphi$ conditionalize over drastic model-transforming actions. It is time to start:

Model constructions are not alien to a probabilistic perspective. Competent practitioners make such moves implicitly. Consider the following famous puzzle:

Example Monty Hall dilemma.
There is a car behind one of three doors: the quizmaster knows which one, you do not. You choose a door, say *1*, and then the quizmaster opens another door that has no car behind it. Say, he opens Door *3*. Now you are offered the chance to switch doors. Should you? If you conditionalize on the new information that the car is not behind Door 2, Bayes' Rule tells you it does not matter: Doors *1* and *3*

have equal probability now. But the real information is more complex: if the car is behind *1*, the quizmaster can open either Door *2* or Door *3* with probability *1/2*. But if it is behind Door *3*, then he must open Door *2*. This drives a construction of the probability space, as pictured in the following tree, where probabilities of branches arise by multiplying those of their successive transitions:

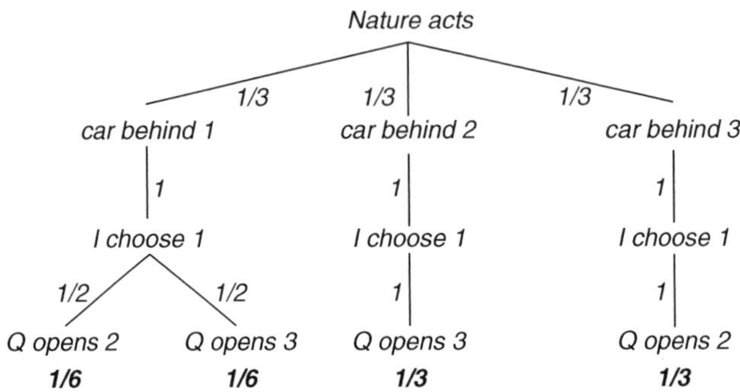

I chose Door *1*, Q opened Door *3*. We want the conditional probability that the car is behind Door *1*, given what we saw. The tree suggests A = *Q opens 3*, leaving two branches. Analysing that subspace, we find that *P(Car behind 1*$|$*A)* = *1/3* – and hence we should switch. ■

Now we look at this scenario using dynamic-epistemic product update.

Example Monty Hall meets *DEL*.
Nature's actions are indistinguishable for me (*I*), but not for the quizmaster Q. The result is the epistemic model at the second level. Now I choose Door *1*, which copies the same uncertainties to the third level. Then come public events of Q's opening a door, each with preconditions (a) I did not choose that door, and (b) Q *knows that* the car is not behind it. In principle this could generate $3 \times 3 = 9$ worlds, but the preconditions leave only *4*:

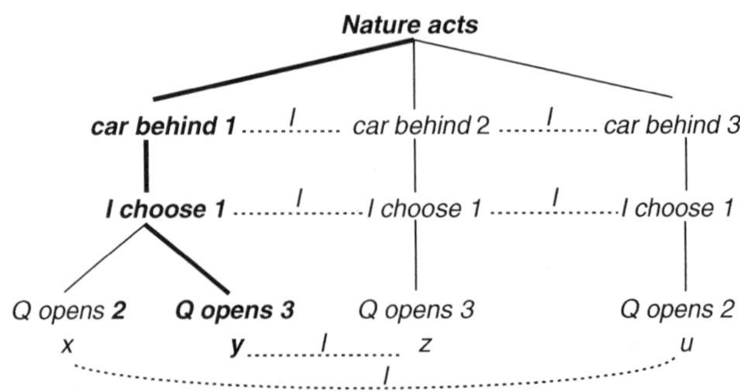

The actual world is y on the bold-face branch. In the bottom-most epistemic model, I know the world is either y or z. Throughout the tree, Quizmaster knows exactly where he is. ■

All this suggests finding a richer dynamic-epistemic logic on top of this that can accommodate probabilistic fine-structure. This requires the usual steps from earlier chapters: a suitably expressive static language, a product update rule, and a complete dynamic logic. This is not totally routine, and we will be forced to think about which probabilities play a role in update.

8.2 Static epistemic probabilistic logic

Epistemic probabilistic languages describing what agents know plus the probabilities they assign were introduced by Halpern and Tuttle (1993), Fagin and Halpern (1993):

Definition Epistemic probability models.
An *epistemic probability model* $\mathbf{M} = (W, \sim, \mathbf{P}, V)$ has a set of worlds W, a family $\sim$ of equivalence relations $\sim_i$ on W for each agent i, a set $\mathbf{P}$ of probability functions P_i that assign probability distributions for agents i at each world $w \in W$, and a valuation V assigning sets of worlds to proposition letters.[2] ■

These combined models represent both non-probabilistic and probabilistic information of agents.

Definition Static epistemic probabilistic language.
The *static epistemic probabilistic language* has the following inductive syntax:

$\varphi ::= p \mid \neg\varphi \mid (\varphi \wedge \psi) \mid K_i\varphi \mid P_i(\varphi) = q$, where q is a rational number,[3] plus linear inequalities $\alpha_1 \bullet P_i(\varphi_1) + \ldots + \alpha_n \bullet P_i(\varphi_n) \geq \beta$, with $\alpha_1, \ldots, \alpha_n, \beta$ rational numbers. ■

This syntax allows mixed formulas like $K_iP_j(\varphi) = k$, or $P_i(K_j\varphi) = k$ talking about agents' knowledge of the others' probabilities, or probabilities they give to someone knowing some fact.[4] Formulas $P_i(\varphi) = q$ are evaluated by summing over the worlds where φ holds:

[2] An important and intuitive special case is when the $P_i(w)$ are probability distributions defined only on the equivalence class of epistemically accessible worlds $\{v \mid v \sim_i w\}$.

[3] Another widely used notation has superscripts for agents: $P^i(\varphi) = q$.

[4] For convenience, we will drop agent indices whenever they do not help the presentation.

DEFINITION Semantics for epistemic probabilistic logic.
The clauses for proposition letters, Boolean operations, and epistemic modalities are as in Chapter 2. Here is the key clause for the probabilistic modality:

$$\mathbf{M}, s \models P_i(\varphi) = q \text{ iff } \Sigma_{t \text{ with } \mathbf{M},\, t \models \varphi} P_i(s)(t) = q$$

The semantics for the linear inequalities then follows immediately. ■

Epistemic probability models suggest constraints linking probability with knowledge (cf. Halpern 2003). Say, one can let $P_i(s)$ assign positive probabilities only to worlds $\sim_i$-linked to s. Such constraints define models with special logics. In particular, epistemically indistinguishable worlds often get the same probability distribution. Thus, agents will know the probabilities they assign to propositions, and hence we have a valid principle

$$P_i(\varphi) = q \rightarrow K_i\, P_i(\varphi) = q \qquad \textit{Probabilistic Introspection}$$

This may be compared with our introspective treatment of beliefs in Chapter 7.

The reader should feel free to assume this special setting in what follows, if it helps for concreteness – but as always, our analysis of probabilistic update mostly does not hinge on such options.[5] What matters is expressive harmony between the static and dynamic languages. This *pre-encoding* uses the above probabilistic linear inequalities, whose purpose will become clear later.

8.3 Three roles of probability in update scenarios

Earlier update rules Merges of *PAL* with probabilistic update occur in Kooi (2003). In line with Chapter 3, there are prior world probabilities in an initial model **M**, and one then conditionalizes to get the new probabilities in the model **M**|*A* after a public announcement *!A*.[6] This validates the following key *recursion axiom*:

$$[!A]P_i(\varphi) = q \leftrightarrow P_i([!A]\varphi \,|\, A) = q$$

reducing a probability after update to a conditional probability before.[7]

[5] There are more general probabilistic models in the literature, such as Dempster–Shaefer theory, but we feel confident that our dynamic style of analysis will work there, too.

[6] This update rule, as well as our later ones, works as long as the new proposition does not have *probability zero*. For more on the latter case, see the final section of this chapter.

[7] Kooi's paper also has a notion of *probabilistic bisimulation* that fits the language.

This cannot yet deal with Monty Hall, as the Quizmaster's actions had different probabilities, depending on the world where they occur. To deal with this new feature, van Benthem (2003) introduced 'occurrence probabilities' for publicly observable events in event models, assigning the following probabilities to new worlds *(s, e)*:

$$P^{\mathbf{M}\times \mathbf{E}}(s,e) = P^{\mathbf{M}}(s) \bullet P^{\mathrm{E}}{}_{s}(e), \quad \text{followed by a normalization step.}$$

More precisely, when all events are public, and their occurrence probabilities are common knowledge, the product rule reads as follows:[8]

$$P_{i,\,(s,\,e)}(t,e) = \frac{P_{i,\,s}(t) \bullet P_t(e)}{\Sigma_{\;u\sim i\ s\ in\ \mathbf{M}}\, P_{i,s}(u) \bullet P_u(e)}$$

This subsumes Kooi's rule for the special event of public announcement:

$$P_{i,(s,\,!A)}(\varphi) = \frac{\Sigma\{P_{i,s}(u)|s \sim_i u \;\&\; \mathbf{M}, u \models A \wedge [!A]\varphi\}}{\Sigma\{P_{i,s}(u)|s \sim_i u \;\&\; M, u \models A\}}$$

Example Monty Hall via product update.
Here is how one computes the earlier tree probabilities for Monty Hall:

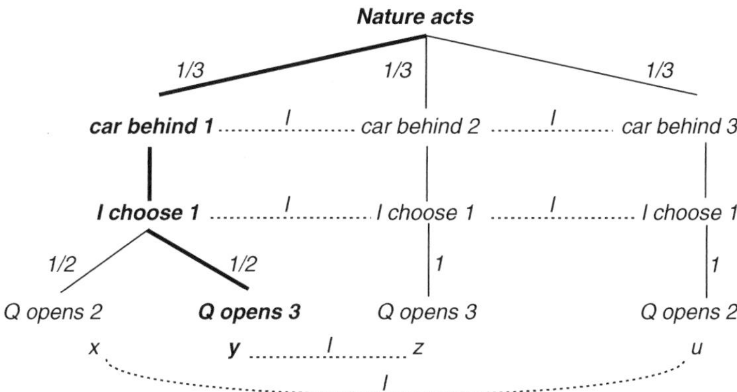

[8] In general encoding information about occurrence probabilities was done via 'generalized preconditions', that map worlds in arbitrary models to probabilities. This is a huge leap away from *DEL*, but we can often make do with finite definable versions.

It is easy to check that the probabilities in the final set $\{x, y\}$ work out to

for y:	$(1/3 \bullet 1/2) / (1/3 \bullet 1/2 + 1/3 \bullet 1)$	$=$	$1/3$
for z:	$(1/3 \bullet 1) / (1/3 \bullet 1/2 + 1/3 \bullet 1)$	$=$	$2/3$

■

Repercussions: Bayes' Rule This dynamic perspective has some features that may be controversial. One difference with probability theory is that we conditionalize on *events*. We observe events, and hence we have probabilities with heterogeneous arguments

$P(\varphi | e)$ where φ is a proposition, and e is an event.[9]

Now consider Bayes' Law. Is this principle plausible in a dynamic reading? Obviously, its static form $P(\varphi | A) = P(A | \varphi) \bullet P(\varphi) / P(A)$ holds for probabilities in a fixed model. It is also valid if we restrict attention to update with purely factual assertions. But the Rule turns out problematic as a general update principle relating a new probability model to an old one, since formulas can change truth values as new information comes in:

Proposition Bayes' Rule fails as a law of epistemic public announcement.

Proof Consider the following epistemic model **M** with two agents:

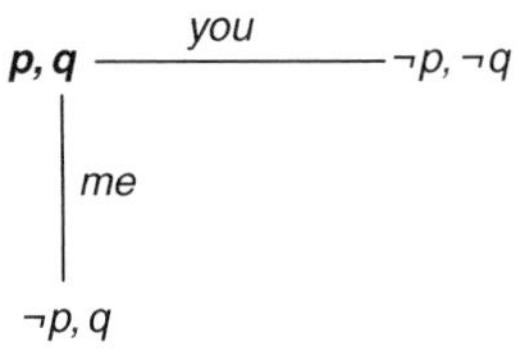

$\boldsymbol{p, q}$ —— *you* —— $\neg p, \neg q$
| *me*
$\neg p, q$

The actual world on the top left has p, q true. Consider the assertion

A You do not know whether p is the case

This is true in the two uppermost worlds, but not at the bottom. Next, take

φ I know whether p is the case

[9] Events of public announcement *!A* matched a propositional *precondition* *A*, and Q's opening Door 3 a *postcondition* 'Q opened Door 3'. But there need not be a general reduction.

that only holds in the world to the right. Let each world have probability *1/3*.[10] In the initial model, $P_{\mathbf{M}}(A) = 2/3$, while $P_{\mathbf{M}}(\varphi) = 1/3$. A public announcement of the true fact *A* updates this model to the new model $\mathbf{M}|A$:

$$p, q \overset{you}{\text{———————}} \neg p, \neg q$$

with φ true everywhere. In that new model, then, $P_{\mathbf{M}|A}(\varphi) = 1$. By contrast, an update that first announces φ takes $\mathbf{M}$ to the one-world model $\mathbf{M}|\varphi$:

$$\neg p, \neg q$$

Here *A* is false everywhere, and we get a probability $P_{\mathbf{M}|\varphi}(A) = 0$. Substituting all these values, we see that Bayes' Rule fails in its dynamic reading:

$$P(\varphi \mid A) = 1 \neq (0 \bullet (1/3))/(2/3)$$

More generally, in a dynamic perspective, order inversions are invalid, though they may work for special formulas. Still, Bayes' Rule has lived a useful life for centuries without logical blessing. Romeijn (2009) gives a modified Bayesian counter-analysis of the above reasoning.

Now we turn to our general analysis (van Benthem, Gerbrandy & Kooi (2009)).

Three sources of probability The preceding approaches have performed a two-fold 'probabilization' of *DEL* product update, distinguishing two factors:

(a) *prior probabilities of worlds* in the current epistemic probabilistic model **M**, representing agents' current informational attitudes,
(b) *occurrence probabilities for events* from the event model **E** encoding agents' views on what sort of process produces the new information.

But there is also a third type of uncertainty that plays in many realistic scenarios:

(c) *observation probability of events*, reflecting agents' uncertainty as to which event is actually being observed.

Recall the motivation for *DEL* in terms of observational access. I see you read a letter, and I know it is a rejection or an acceptance. You know the actual event (reading 'Yes', or reading 'No'), I do not. Here product update gives a new

[10] One can think of probabilities *P* for a third person *3* who holds all worlds equi-possible.

epistemic model without probabilities. To compute the latter, I may know about frequency of acceptance versus rejection letters, a type (b) occurrence probability. But there may also be more information *in the observation itself*! Perhaps I saw a glimpse of your letter – or you looked smug, and I think you were probably reading an acceptance. This would be an observation probability in sense (c). The latter notion is also known from scenario's motivating the Jeffrey Rule, where one is uncertain about evidence received under partial observation.[11]

A scenario where all three kinds of probability come together is this:

Example The hypochondriac.
You read about a disease, and start wondering. The chances of having the disease are slight, *1* in *100.000*. You read that one symptom is a certain gland being swollen. With the disease, the chance of this is *97%*, without the disease, it is *0*. You take a look. It is the first time you examine the gland and you do not know its proper size. You think chances are *50%* that the gland is swollen. What chance should you assign to having the disease? ■

We will now define a general mechanism for computing the answer.

8.4 Probabilistic product update

In what follows, our static epistemic probabilistic models **M** are as before, and so is our language. We will use the *DEL*-notation $[\mathbf{E}, e]\varphi$ from Chapter 4 to describe the effect of executing event model **E**, *e* in a current model **M**, *s*. Our event models are a bit special, to make them look like processes with uniformly specified occurrence probabilities:

DEFINITION Probabilistic event models.
Probabilistic event models are structures $\mathbf{E} = (E, \sim, \Phi, Pre, P)$ with (a) E a non-empty finite set of events, (b) $\sim$ a set of equivalence relations $\sim_i$ on E for each agent i, (c) Φ a set of pairwise inconsistent sentences ('preconditions'), (d) *Pre* assigns to each precondition $\varphi \in \Phi$ a probability distribution over E (we write $Pre(\varphi, e)$, the chance that e occurs given φ), and (e) for each i, P_i assigns each event e a probability distribution over E. ■

[11] Occurrence probability is often an objective frequency, and observation probability a subjective chance. Major views of probability co-exist in our setting.

The language for preconditions is given below: as in Chapter 4, there is a simultaneous recursion. Models work as follows. The *Pre* specifies occurrence probabilities of a process that makes events occur with probabilities depending on conditions Φ. Diseases and quizmasters are examples, with rules of the form 'if P holds, then do a with probability q', and so are Markov processes. Models also have observation probabilities, represented by the functions P_i. The probability $P_i(e)(e')$ is the probability assigned by an agent i to event e' taking place, given that e actually takes place. This adds probabilistic structure to the uncertainty relations $\sim_i$ in much the same way as happened in our static models.[12]

Our next goal is a dynamic update mechanism for these models. Merging input from all three sources of probability, it is a direct generalization of earlier rules:

Definition Probabilistic Product Update Rule.
Let **M** be an epistemic probabilistic model and let **E** be an event model. If s is a state in **M**, write $Pre(s, e)$ for the value of $Pre(\varphi, e)$ where φ is the unique element of Φ true at **M**, s. If no such φ exists, set $Pre(s, e) = 0$. The *product model* $\mathbf{M} \times \mathbf{E} = (S', \sim', P', V')$ is defined by:

(a) $S' = \{(s, e) \mid s \in S, e \in E \text{ and } Pre(s, e) > 0\}$
(b) $(s, e) \sim_i (s', e')$ iff $s \sim_i s'$ and $e \sim_i e'$
(c) $P'_i((s, e), (s', e')) := \dfrac{P_i(s)(s') \bullet Pre(s', e') \bullet P_i(e)(e')}{\Sigma_{s'' \in S, e'' \in E} P_i(s)(s'') \bullet Pre(s'', e'') \bullet P_i(e)(e'')}$
if the denominator >0 and 0 otherwise.
(d) $V'((s, e)) = V(s)$ ■

The new state space after the update consists of all pairs (s, e) where event e occurs with positive probability in s (as specified by *Pre*). The crucial part is the new probabilities $P'_i(s, e)$ for (s', e'). These are a product of the prior probability for s', the probability that e' actually occurs in s', and the probability that i assigns to observing e'. To get a proper probability measure, we normalize this value.[13] Here is how this rule works in practice:

[12] Each epistemic event model **E** in Chapter 4 can be expanded to a probabilistic event model.
[13] If the denominator in our rule sums to 0, we stipulated a total value 0. Thus $\mathbf{M} \times \mathbf{E}$ need not be a probabilistic epistemic model: $P_i(s, e)$ may assign probability 0 to all worlds. Bacchus (1990) has a probabilistic defense, and there are ways around this feature.

Example The Hypochondriac again.
The initial hypothesis about having the disease p is captured by a prior probability distribution

1/100.000   *99.999/100.000*
p .... $\neg p$

Then the hypochondriac examines the gland, with an occurrence probability (if he has the disease, the gland is swollen with probability *0.97*) and an observation probability (he thinks he is seeing a swollen gland with probability *0.5*) as stated in the given scenario. This is encoded in the following epistemic probabilistic event model:

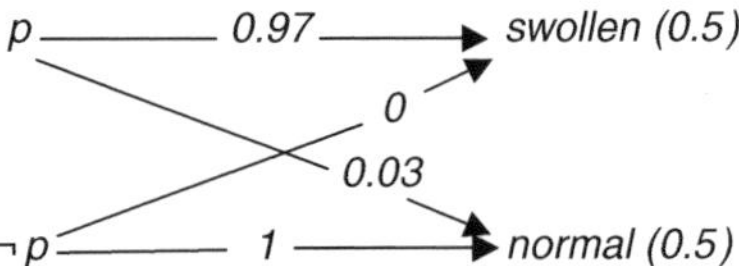

The product of our initial state with this model is as follows:

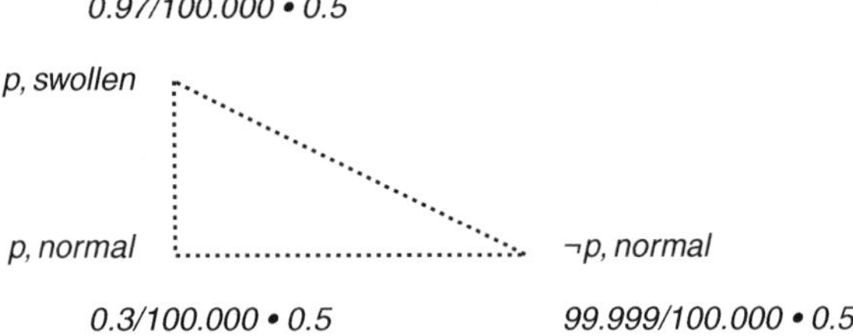

This diagram is our new information state after the episode. Renormalizing values, the new probability that the Hypochondriac should assign to having the disease is still *1* in *100.000*. His inconclusive observation has not produced any information about having the disease. Had he found it more probable that the gland was swollen, the probability for the disease would have come out higher than before by our product rule, and in the opposite case, that probability would have been lower. ■

We also see a typical *DEL* feature. Initially, we only had two options: having the disease or not. The update created three worlds, now with additional information if the gland is swollen or not.

8.5 Discussion and further developments

Systematic model construction Our epistemic probabilistic update rule starts from a simple probability space and, step by step, builds more complex product spaces with informational events encoded by event models. This control over probability spaces may be useful in practice, where management of relevant spaces, rather than applying the probability calculus, is the main difficulty in reasoning with uncertainty. Repeated over time, the new possibilities form all runs of a total informational process, linking up with more global epistemic probabilistic temporal logics (Chapter 11).

Model theory and probabilistic bisimulation The model theory of epistemic probabilistic logic can be developed like its earlier counterparts, using the earlier-mentioned epistemic-probabilistic bisimulation for our static language. It is easy to see that our product update rule respects such bisimulations between input models, and we are on our way.

Shifting loci of probabilistic information A natural question is if our three components: prior world probabilities, occurrence, and observation probabilities are really independent. In modeling real scenarios one can *choose* where to locate things. Van Benthem, Gerbrandy & Kooi (2009) give constructions on update models showing how under redefinition of events, occurrence probabilities can absorb observation probabilities, and vice versa. Such tricks do not endanger the intuitive appeal of our three-source scheme.

Processes and protocols Our mechanism is more powerful than may appear at first sight. Consider temporal processes over time (cf. Chapters 3, 4, 11). Many of their *protocols* are probabilistic: involving, say, an agent whose assertions have a certain probabilistic reliability.

Example Coins and Liars.
We know that some coin is either fair, or yields only heads. We represent observation of a throw of the coin with an initial model ***M*** with two options *Fair, Heads-only*, plus an event model **E** with two events *Heads* and *Tails*, related with the obvious probabilities:

Pre (Fair, Heads) = Pre (Fair, Tails) = 1/2,
Pre (Heads-only, Heads) = 1, Pre (Heads-only, Tails) = 0.

By our product update rule, one observation of *Tails* rules out the unfair coin, while each observation of *Heads* makes it more likely. In the same mood, we meet a stranger in a logic puzzle, who might be a *Truth Teller* or a *Liar*. We have to find out what is what. We encode the options as abstract *pair events* inside the event model (cf. Chapter 4):

(Truth Teller, !A), (Liar, !A)

encoding both the assertion made, and the type of agent making it. After that, update is exactly as in our earlier examples, and we read off the new values for the agent types. ■

A general construction for pairs *(Process type, Observed event)* is easily stated.

8.6 A complete dynamic probabilistic logic

Language and semantics To reason explicitly about probabilistic information change, we now extend static epistemic probabilistic logics with the dynamics of Chapters 3, 4.

DEFINITION Dynamic-epistemic-probabilistic language.
The *dynamic-epistemic-probabilistic language* extends our static language with a dynamic modality $[\mathbf{E}, e]\varphi$, with **E** a probabilistic event model, and *e* an event from its domain. ■

Note again the recursion: the formulas that define preconditions come from this language, but through the new dynamic modalities, such models themselves enter the language.

DEFINITION Semantics of probabilistic event models.
In an epistemic probability model $\mathbf{M} = (W, \sim, P, V)$ with $s \in W$, $\mathbf{M}, s \models [\mathbf{E}, e]\varphi$ iff for the unique $\psi \in \Phi$ with $\mathbf{M}, s \models \psi$, we have that $\mathbf{M} \times \mathbf{E}, (s, e) \models \varphi$ in the product model $\mathbf{M} \times \mathbf{E}$ as above. ■

A complete axiomatic system With all this in place, here is our main result:

THEOREM The dynamic-epistemic-probabilistic logic of update by probabilistic event models is completely axiomatizable over the chosen static logic.

Proof The core is the key recursion axiom for formulas $[\mathbf{E}, e]\varphi$. The following calculation is the heart of our reduction, with agent indices dropped for greater readability. Consider the value $P(\psi)$ of a formula ψ in a (pointed) product model *(M, s)* $\times$ *(E, e)*. We write *P(-)* in the initial model as P^M, $P^{M\times E}$ for values *P(-, -)* in the product model, and P^E for *P(-)* in the event model. For convenience, we use the existential dynamic modality *<E, e>*. For a start, if we have $\Sigma_{s''\in S,\ e''\in E} P^M(s'') \bullet Pre(s'', e'') \bullet P^E(e'') > 0$, then:

$$\begin{aligned} P^{M\times E}(\psi) &= \Sigma_{(s',e')\text{ in } M\times E:\ M\times E,(s',e') \models \psi} P^{M\times E}(s', e') \\ &= \Sigma_{s'\in S,\ e'\in E:\ M,\ s' \models <E,e'>\psi}\ P^{M\times E}(s', e') \\ &= \frac{\Sigma_{s'\in S,\ e'\in E:\ M,s' \models <E,e'>\psi}\ P^M(s') \bullet Pre(s', e') \bullet P^E(e')}{\Sigma_{s''\in S,\ e''\in E} P^M(s'') \bullet Pre(s'', e'') \bullet P^E(e'')} \end{aligned}$$

The numerator of this last equation can be written as

$$\Sigma_{\varphi\in\Phi,\ s'\in S,\ e'\in E:\ M,\ s' \models \varphi,\ M,\ s' \models <E,e'>\psi}\ P^M(s') \bullet Pre(\varphi, e') \bullet P^E(e')$$

which is equivalent to

$$\Sigma_{\varphi\in\Phi,\ e'\in E} P^M(\varphi \wedge <E, e'>\psi) \bullet Pre(\varphi, e') \bullet P^E(e')$$

We can then analyse the denominator of the equation in a similar way, and rewrite it as

$$\Sigma_{\varphi\in\Phi,\ e''\in E} P^M(\varphi) \bullet Pre(\varphi, e'') \bullet P^E(e')$$

So we can write the probability $P^{MxE}(\psi)$ in the new model as a term of the following form:

$$P^{M\times E}(\psi) = \frac{\Sigma_{\varphi\in\Phi,\ e'\in E}\ P^M(\varphi \wedge < E, e' >\psi) \bullet k_{\varphi,\ e'}}{\Sigma_{\varphi\in\Phi,\ e''\in E}\ P^M(\varphi) \bullet k_{\varphi,\ e''}}$$

where, for each φ and *f*, $k_{\varphi,\ f}$ is a constant, namely the value $Pre(\varphi, f) \bullet P^E(f)$.

Next, we enumerate the finite set of preconditions Φ and the domain of E as $\varphi_0, \ldots, \varphi_n, e_0, \ldots, e_m$. Then we rewrite any dynamic formula *<E, e>* $P(\psi) = r$ with *P* the probability after update to an equivalent equation in which *P* refers to probabilities in the prior model:

$$\frac{\Sigma_{1\le i\le n,\ 1\le j\le m} k_{\varphi i,\ ej} \bullet P(\varphi_i \wedge <E, e_j>\psi)}{\Sigma_{1\le i\le n,\ 1\le j\le m} k_{\varphi i,\ ej} \bullet P(\varphi_i)} = r$$

And the latter can be rewritten as a sum of terms:

$$\Sigma_{1\le i\le n,\ 1\le j\le m} k_{\varphi i,\ ej} \bullet P(\varphi_i \wedge <E, e_j>\psi) + \Sigma_{1\le i\le n, 1\le j\le m} - r \bullet k_{\varphi i,\ ej} \bullet P(\varphi_i) = 0$$

Now, to express these observations as one recursion axiom in our formal language, we need sums of terms. Our language with linear inequalities is up to just this job. But then, to restore the harmony of the total system, we must also find a reduction for inequalities:

$$[E, e]\alpha_1 \bullet P_i(\psi_1) + \ldots + \alpha_n \bullet P_i(\psi_n) \geq \beta$$

In this formula, we can replace separate terms $P(\psi_k)$ after the dynamic modal operator by their equivalents as just computed.[14] We then obtain an equivalent expression of the form

$$\Sigma_{1 \leq h \leq k,\ 1 \leq i \leq n,\ 1 \leq j \leq m}\ \alpha_h \bullet k_{\varphi i,\ ej} \bullet P(\varphi_i \wedge [E,\ e_j]\psi_h) + \Sigma_{1 \leq i \leq n,\ 1 \leq j \leq m} -\beta \bullet k_{\varphi i,\ ej} \bullet P(\varphi_i) \geq 0$$

This is still an inequality χ inside our language. The full axiom then becomes

$$([E, e]\ \alpha_1 \bullet P_i(\psi_1) + \ldots + \alpha_n \bullet P_i(\psi_n) \geq \beta) \leftrightarrow$$
$$((\Sigma_{1 \leq i \leq n,\ 1 \leq j \leq m}\ k_{\varphi i,\ ej} \bullet P(\varphi_i) > 0) \to \chi) \wedge ((\Sigma_{1 \leq i \leq n,\ 1 \leq j \leq m}\ k_{\varphi i,\ ej} \bullet P(\varphi_i) = 0) \to 0 \leq \beta)$$

This looks technical, but it can easily be computed in specific cases.

The other recursion axioms are as in Chapter 4, with preconditions $Pre_{E,\ e}$ of events e in our setting being the sentences $\vee_{\varphi \in \Phi,\ Pre(\varphi,\ e)\ \geq\ 0}\ \varphi$. Our proof concludes with the usual inside-out removal of dynamic modalities, effecting a reduction to the base logic. ■

Our relative style of axiomatization adding dynamics superstructure to a static base logic makes special sense in probabilistic settings, as it factors out the complexity of the underlying quantitative mathematical reasoning.

8.7 A challenge: weighted learning

Policies and weights Update may be more than our rule so far. Inductive logic (Carnap 1952), learning theory (Kelly 1996), and belief revision theory (Gärdenfors & Rott 1995) also stress *policies* on the part of agents. We have a probability distribution, we observe a new event. The new distribution depends on the *weights* agents assign to past experience versus the latest news. Different weights yield more radical or conservative policies.

[14] The denominator of the equation for the posterior probabilities must be > 0.

Example (adapted from Halpern 2003) The Dark Room.
An object can be light or dark. We start with the equiprobability distribution. Now we see that, with probability 3/4, the object is dark. What are the appropriate new probabilities? ■

Our earlier update rule weighs things here *equally*. We use signals '*Light*', '*Dark*', with occurrence probabilities 1 and 0 with the obvious Φ, and observation probabilities 1/4, 3/4. The new probability that the object is dark mixes these to a value between 1/2 and 3/4.

The $\alpha\beta\gamma$ formula Now, suppose agents give different weights to the three factors in our rule, say real values α, β, γ in $[0, 1]$. Here is a generalization:

DEFINITION Weighted Product Update Rule.[15]

$$P^{new}((s,e),(s',e')) :=$$

$$\frac{P(s)(s' \mid \varphi_{s'}) \bullet P(s)(s')^{\alpha} \bullet Pre(s',e')^{\beta} \bullet P(e)(e')^{\gamma}}{\Sigma_{\, s''\in S,\ e''\in E}\, P(s)(s'' \mid \varphi_{s''}) \bullet P(s)(s'')^{\alpha} \bullet Pre(s'',e'')^{\beta} \bullet P(e)(e'')^{\gamma}}$$

if the denominator is >0 – and 0, otherwise

Setting all three factors to 1 gives our original update. Setting $\alpha, \beta, \gamma = (0, 0, 1)$ is close to the earlier Jeffrey Update, mixing radicalism and conservatism.[16]

8.7 Conclusion

This chapter has linked dynamic-epistemic logic with the probabilistic tradition. We found a product update mechanism based on a principled distinction between prior world probability, occurrence probability, and observation probability. This provides a 'smooth' extension of the discrete update rules of earlier chapters, letting probabilities gently incorporate new information. Moreover, our mechanism supports a complete dynamic-epistemic probabilistic logic that can handle update with formulas of

[15] Cf. also Grunwald & Halpern (2003) on related forms of update, including Jeffrey Update.

[16] Technically, we have a pair (Φ, P) of a set of sentences partitioning the space and a probability distribution P over Φ. The Jeffrey Update of a prior P^{old} with the new information is then $P^{new}(s) = P^{old}(s \mid \varphi) \bullet P(\varphi)$. The new signal overrules prior information about the sentences in Φ, just like we had with belief revision policies like $\Uparrow P$, $\uparrow P$ in Chapter 7. For our Dark Room, Jeffrey Update makes the new probability of the object being dark 3/4, and of its being light 1/4. Thus we set new values for partition cells, but within these, relative probabilities of worlds remain the same.

arbitrary syntactic complexity. Thus dynamic logic and probability are compatible, and there may be an interesting flow of ideas across. Of course, the real task is now to extend the bridge head.

8.8 Further directions and open problems

Plausibility versus probability One obvious issue is how the plausibility models of Chapter 7 relate to a probabilistic approach. The style of thinking is different, in that most plausible worlds may ignore the cumulative probabilistic weight of the less plausible ones. A logical difference is that plausibility logics validate conjunction of beliefs: $(B\varphi \wedge B\psi) \rightarrow B(\varphi \wedge \psi)$, while probabilistic approaches typically do not, since the intersection $\varphi \wedge \psi$ may have lost probability mass below some threshold. This issue has been studied for belief revision with graded modalities (Spohn 1988; Aucher 2004). Can we find systematic transformations?

Surprises Our update rule treats zero denominators as a nuisance. But zero probability events represent a real phenomenon of true *surprises*. These have been studied in Aucher (2004) using infinitesimal numbers. An alternative is the use of conditional probability spaces in Baltag & Smets (2007) to represent surprise events and their epistemic effects.

Dynamifying probabilistic reasoning Our dynamic logic exemplifies our general programme of dynamifying existing systems. In probabilistic practice, however, our system may be too baroque, and well-chosen fragments are needed.[17] Probability theory also has challenges beyond our approach. One is the fundamental notion of *expected value*, a weighted utility over possible outcomes. This makes general sense for agency in the 'entanglement' of preference and belief studied in Chapter 9. To properly incorporate expected value, we need an extension of our dynamic logic with recursion axioms for new expected values after information has come in. Dealing with these steps may eventually also involve temporal extensions of dynamic-epistemic logics that can refer to the past (Chapter 11).[18]

[17] Cornelisse (2010) proposed an interesting subsystem closer to *PAL* with upgrade actions $\Uparrow P, r$ that reset a proposition P toward some new probability r.

[18] The linguistic analysis of questions in van Rooij (2003) and the game-theoretic one of Feinberg (2007) compare expected values in an old model and a new one after new information is received.

Philosophy of science In the philosophy of science, there is a flourishing literature on separate probabilistic update rules for popular scenarios, such as Sleeping Beauty and its ilk. It would be of interest to see if our logics can help systematize the area.

Postulates and Dutch Book arguments Laws of reasoning with probability are often justified by general postulates (cf. our discussion of postulational approaches in Chapters 3, 7). The most famous format are *Dutch Book Arguments* showing how the specific axioms of the probability calculus are the only ones that are fail-safe in multi-agent betting scenarios. Can we do similar analyses for the principles of our dynamic update logics?

8.9 Literature

There is a large literature linking probability, conditionals, and belief revision, for which the reader can consult standard sources. Kooi (2003) first merged probability with public announcement logic *PAL*. Van Benthem (2003a) extended this to public event models with occurrence probabilities. Van Benthem, Gerbrandy & Kooi (2009) has the system of this chapter. Aucher (2004) gives another analysis of epistemic-probabilistic update including surprise events, and Baltag & Smets (2007a) one more, using conditional probability spaces (Popper measures). Sack (2009) extends probabilistic *DEL* from finite models to infinite ones, using mathematical notions from probability theory. Grunwald & Halpern (2003) make proposals related to ours, though in another framework. Halpern (2003) is probably the major current source on logical approaches to reasoning with uncertainty.

9 Preference statics and dynamics

So far, we have shown how logical dynamics deals with agents' knowledge and beliefs, and informational events that change these. But as we noted in Chapter 1, agency also involves a second major system, not of information but of *evaluation*. It is values mixed with information that provide the driving force for rational action – and the colour of life. The barest record of evaluation are agents' *preferences* between worlds or actions. Thus, the next task in this book is dealing with preferences, and how they change under triggers like suggestions or commands. While this topic seems different in flavour from earlier ones, properly viewed, it yields to the same techniques as in Chapters 3, 7. Therefore, we will present our dynamic logics with a lighter touch, while emphasizing further interesting features of preference that make it special from a logical perspective.

9.1 Logical dynamics of practical reasoning

Consider how preferences function in scenarios of agency:

Example Decision problems.
A decision problem looks at actions available to agents, and asks what they will, or should, do based on their preference between the outcomes:

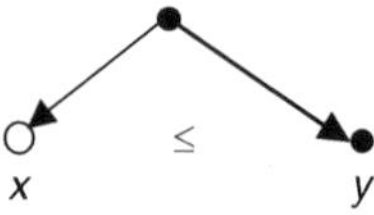

If an agent can choose between outcomes x and y, and she prefers y, then she should choose outcome y. ■

Information and evaluation also come together in *games*, looking at what players want, observe, and guess, and which moves are available to achieve their goals. In this setting, multi-agent interaction is essential, as my actions depend on what I think about yours.

Example Reasoning about interaction.
Here is an example that was already discussed in Chapter 1. In the following two game trees, preferences at end nodes are encoded in utility pairs *(value of A, value for E)*. The solution method of *Backward Induction* (cf. Chapter 10) is a typical piece of multi-agent reasoning about preference and belief. In the game to the right, essentially a single-agent decision problem, it tells player *A* to go right, where *E* takes both to the most desirable outcome *(99, 99)*. But interaction with more agents can be tricky. In the game on the left, Backward Induction tells *E* to turn left when she can, and then *A* (who realizes what *E* would do) will turn left at the start – where both suffer, since their pay-off is much lower than *(99, 99)*:

Why should players act this way? The relevant reasoning is a mixture of all notions so far. Player *A* turns left since she believes that *E* will turn left, and then her preference is for grabbing the value *1*. Thus, practical reasoning intertwines action, preference, and belief. ■

We will return to this scenario in Chapter 10, but for now, it shows the sort of preference ordering between options we want to study. In this chapter, we focus on preference alone, bringing in belief and action later. One area where this makes sense is *deontic logic*, the study of reasoning about agents' obligations. The latter are usually expressed in propositions that are true in the best of all worlds according to some moral authority. Moreover, the relevant 'better' order may change as commands come in, or new laws are enacted. Thus, deontic logic is a multi-agent preference logic, involving myself and one or more moral authorities.[1]

[1] The ideal situation might make my personal preference order coincide with the moral one, as in Kant's dictum that one should make duty coincide with inclination.

In these examples, we see an ordering of 'better', and we see maximizing along it in the notion 'best'. Thus, logics of preference can be designed in the style of plausibility models for belief (Chapter 7), while the methods of Chapters 3, 4 can deal with preference change. For a start, here are some ingredients from the literature:

Preference logics Von Wright (1963) proposed a logic with formulas $P\varphi\psi$ saying that every φ–situation is preferred over every ψ–situation 'ceteris paribus': a phrase to which we will return. Note that preference here is 'generic', running between propositions, i.e., sets of situations. Von Wright's calculus for reasoning with preference contains laws such as

$$P\varphi\psi \leftrightarrow P(\varphi \wedge \neg\psi)(\psi \wedge \neg\varphi)$$

This has led to current preference logics (cf. Hanson 2001), for which we will use a modal language below. Beyond this, there is a recent interest in *preference change* and its various triggers (Gruene-Yaroff & Hanson 2008). Our chapter will develop this evaluation dynamics in tandem with information dynamics, as we want to understand the entanglement.

9.2 Modal logic of betterness

Preference is multi-faceted: we can prefer one individual object, or one situation, over another – but preference can also be directed toward kinds of objects or generic situations, defined by propositions. A bona fide preference logic should do justice to both views. We start with a simple scenario on the object/world side, moving to generic notions later.

Basic models In this chapter, we start with a very simple semantic setting:

DEFINITION Modal betterness models.
Modal betterness models $\mathbf{M} = (W, \leq, V)$ have a set of worlds W,[2] a reflexive and transitive *betterness relation* $x \leq y$ ('world y is at least as good as world x'), and a valuation V for proposition letters at worlds (or equivalently, for unary properties of objects). ■

In practice, the relation may vary among agents, but we suppress subscripts $\leq_i$ for greater readability. We use the artificial term 'betterness' to stress

[2] These really stand for any sort of objects that are subject to evaluation and comparison.

that this is an abstract comparison, making no claim about the intuitive term 'preference', whose uses are diverse. These models occur in decision theory, where worlds are outcomes of actions, and game theory, where worlds are complete histories, with preferences for different players. The same kinds of orders were also used in Chapter 7 for relative plausibility as judged by an agent. While preference is not the same as plausibility, the analogy is helpful.

Modal languages Over our base models, we can interpret a standard modal language, and see which natural notions and patterns of reasoning can be defined in it.

DEFINITION Preference modality.
A modal formula $<\leq>\varphi$ makes the following local assertion at a world *w*:

$$M, w \models <\leq>\varphi \quad \text{iff} \quad \text{there exists a } v \geq w \text{ with } M, v \models \varphi$$

that is: there is a world *v* at least as good as *w* that satisfies φ. ■

In combination with other modal operators, this sparse betterness formalism can express many natural notions of preference-driven action.

Example Defining backward induction in preference action logic.
Finite game trees are models for a dynamic logic of atomic actions (players' moves) and unary predicates indicating players' turns at intermediate nodes and their utility values at end nodes (van Benthem 2002a). In Chapter 10, we will present a result from van Benthem, van Otterloo & Roy (2006) showing how the *backward induction solution* of a finite game[3] is the unique binary relation *bi* on the tree satisfying this modal preference-action law:

$$<bi>[bi^*](end \rightarrow \varphi) \rightarrow [move]<bi^*>(end \wedge <\leq>\varphi)$$

Here *move* is the union of all moves available to players, and * is reflexive-transitive closure. The formula says there is no alternative to the *BI*-move at the current node whose outcomes would be better than the *BI*-solution. ■

Thus, modal preference logic goes well with games. Boutilier (1994) showed how it can also define conditionals (Lewis 1973), analysing conditional logic in standard terms. We used this in Chapter 7. On finite reflexive and

[3] A famous benchmark example in the logical analysis of games; cf. Harrenstein (2004).

transitive orders, the following formula defines a conditional $A \Rightarrow B$ in the sense of 'B is true in all maximal A-worlds':

$U(A \rightarrow \langle\leq\rangle(A \wedge [\leq](A \rightarrow B)))$, with U the universal modality.[4]

The same expressive power will be relevant for preference.

Constraints on betterness Which properties should betterness have? *Total orders* satisfying reflexivity, transitivity, and connectedness are the norm in decision theory and game theory, as these properties relate to numerical utilities. But in the logical literature on preference or plausibility, even transitivity has been criticized (Hanson 2001). And in conditional logic, Lewis' totality is often abandoned in favour of *pre-orders* satisfying just reflexivity and transitivity, while acknowledging *four* irreducible basic relations:

$w \leq v, \neg\, v \leq w$	(often written as $w < v$)	*w strictly precedes v*
$v \leq w, \neg\, w \leq v$	(often written as $v < w$)	*v strictly precedes w*
$w \leq v, v \leq w$	(sometimes written as $w \sim v$)	*w, v are indifferent*
$\neg\, w \leq v, \neg\, v \leq w$	(sometimes written as $w \# v$)	*w, v are incomparable*

We prefer such a large class of models, with extra modal axioms if we want the relation to satisfy further constraints. The point of a logical analysis is to impose structure where needed, but also, to respect the right 'degrees of freedom' in an intuitive notion.

Further relations, further modalities? Given this, one can start with two relations: a weak order $w \leq v$ ('at least as good') and a strict order $w < v$ ('better'; $w \leq v \wedge \neg\, v \leq w$). Van Benthem, Girard & Roy (2009) axiomatize this language using separate modalities.

Example Frame correspondence for weak/strict betterness modalities.
By a standard modal semantic argument, the preference axiom $(\psi \wedge \langle\leq\rangle\varphi) \rightarrow (\langle<\rangle\varphi \vee \langle\leq\rangle(\varphi \wedge \langle\leq\rangle\psi))$ corresponds to the first-order frame property that $\forall x \forall y\, (x \leq y \rightarrow (x < y \vee y \leq x))$.

For much more on modal preference logic, see the dissertation Girard (2008).

[4] The modal language also easily defines variants, such as the existential 'each A-world sees *at least one* maximal A-world that is B'. Axiomatizing inference with these and other defined notions *per se* is the point of many completeness theorems in conditional logic. For preference, Halpern (1997) explicitly axiomatized a defined notion of preference of our later universal-existential type $\forall\exists$.

9.3 Defining global propositional preference

As we have said, a betterness relation need not yet determine what we mean by agents' preferences in a more colloquial sense. Many authors consider preference a generic relation between propositions, with von Wright (1963) as a famous example.[5]

Varieties of set lifting Technically, defining preferences between propositions calls for a comparison between sets of worlds. For a given relation $\leq$ among worlds, this may be achieved by *lifting*. One ubiquitous proposal in betterness lifting is the $\forall\exists$ stipulation that

a set Y is preferred to a set X if $\forall x \in X\ \exists y \in Y: x \leq y$

Van Benthem, Girard & Roy (2009) analyse von Wright's own view as the $\forall\forall$ stipulation that

a set Y is preferred to a set X if $\forall x \in X\ \forall y \in Y: x \leq y$,

and provide a complete logic. Liu (2008) provides a history of proposals for relation lifting in various fields (decision theory, philosophy, computer science), but no consensus has emerged. This may be a feature, rather than a bug. Preference between propositions may be genuinely different depending on the scenario, and then logic should not choose:

Example Different set preferences in games.
Comparing outcomes *A*, *B* reached by their moves, players have options:

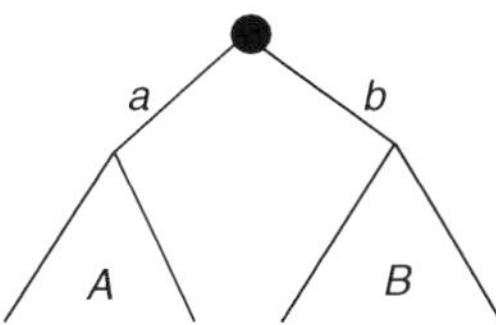

One might prefer a set of outcomes whose minimum utility value exceeds the maximum of another,

this is the $\forall\forall$ reading $max(A) < min(B)$

5 These differences are largely terminological – which is why debates are often bitter.

but one may also want the maximum of one set to exceed that of the other:

the $\forall\exists$ reading $max(A) < max(B)$

where each value in A has at least one higher value in B. Or, a pessimist might have the minimum of the preferred set higher than that of the other. There is simply no best choice. ■

Extended modal logics Many different liftings are definable in a modal base logic extended with a universal modality $U\varphi$: 'φ is true in all worlds' (cf. Chapter 2). This feature gives some additional expressive power without great cost in the modal model theory and the computational complexity of valid consequence. For instance, the $\forall\exists$ reading of preference is expressed as follows, with formulas for definable sets of worlds:

$$U(\varphi \rightarrow <\leq> \psi)$$

In what follows, we will use the notation $P\varphi\psi$ for lifted propositional preferences with no precise definition stipulated.[6]

9.4 Dynamics of evaluation change

But now for preference change. A modal model describes a current evaluation pattern for worlds, as seen by one or more agents. But the reality is that these patterns are not stable. Things happen that make us *change* our evaluations. This dynamics has long been in the air, witness our later references, leading up to mechanisms of relation change much like those discussed for plausibility in Chapter 7.[7] Realistic preference change has further features, seen with a deeper analysis of agents (Hanson 1995, Lang & van der Torre 2008). In this chapter, we show how the simpler version fits with the perspective of this book.

[6] One can also use stronger (first-order) logics to describe preferences. This is the balance in logic between illuminating definitions of key notions and argument patterns and computational complexity (cf. Chapter 2). Richer languages are fine, but as usual in this book, modal logic is a good place to start.

[7] We only discuss one logic strand here: cf. Hanson (1995) for a different point of entry.

9.5 A basic dynamic preference logic

We start with a very simple scenario from van Benthem & Liu (2007).

Dynamic logic of suggestions Betterness models are as before, and so is the modal base language with modalities $<\leq>$ and U. But the syntax now adds, for each formula of the language, a model-changing action $\#\varphi$ of 'suggestion',[8] defined as follows:

DEFINITION Ordering change by suggestion.
For each model **M**, w and formula φ, the *suggestion function* $\#\varphi$ returns the model $\mathbf{M}\#\varphi, w$ which is equal to **M**, w, but for the new betterness relation $\leq' = \leq - \{(x, y) \mid \mathbf{M}, x \models \varphi \;\&\; \mathbf{M}, y \models \neg\varphi\}$.[9] ■

Next, we enrich the language with action modalities interpreted as follows:[10]

$$\mathbf{M}, w \models [\#\varphi]\psi \quad \text{iff} \quad \mathbf{M}\#\varphi, w \models \psi$$

These talk about what agents will prefer after their comparison relation has changed. For instance, if you tell me to drink beer rather than wine, and I accept this, then I now come to prefer beer over wine, even if I did not do so before.

Now, as in dynamic-epistemic logic, the heart of the analysis is the recursion equation explaining when a preference obtains after an action. Here is the valid principle for suggestions, whose two cases follow the definition of the above model change:

$$<\#\varphi><\leq>\psi \leftrightarrow (\neg\varphi \wedge <\leq><\#\varphi>\psi) \vee (\varphi \wedge <\leq>(\varphi \wedge <\#\varphi>\psi))$$

THEOREM The dynamic logic of preference change under suggestions is axiomatized completely by the static modal logic of the underlying model class plus the following equivalences for the dynamic modality:

$$\begin{array}{lll}
[\#\varphi]p & \leftrightarrow & p \\
[\#\varphi]\neg\psi & \leftrightarrow & \neg[\#\varphi]\psi \\
[\#\varphi](\psi \wedge \chi) & \leftrightarrow & [\#\varphi]\psi \wedge [\#\varphi]\chi \\
[\#\varphi]U\psi & \leftrightarrow & U[\#\varphi]\psi \\
[\#\varphi]<\leq>\psi & \leftrightarrow & (\neg\varphi \wedge <\leq>[\#\varphi]\psi) \vee (\varphi \wedge <\leq>(\varphi \wedge [\#\varphi]\psi)).
\end{array}$$

8 This is of course just an informal reading, not a full-fledged social analysis of suggestion.

9 In this chapter '$\#\varphi$' stands for an act of suggesting that φ. Please do not confuse this with the notation '$\#\varphi$' for an act of *promotion* for φ in the syntactic awareness dynamics of Chapter 5.

10 Here the syntax is recursive: the formula φ may itself contain dynamic modalities.

Proof These axioms say that (i) upgrade for a suggestion does not change atomic facts, (ii) upgrade is a function, (iii) its modality is a normal one, (iv) upgrade does not change the set of worlds, and crucially, (v) the upgrade modality encodes the betterness effect of a suggestion. Applied inside out, these principles reduce any valid formula to an equivalent one without dynamic modalities, for which the base logic is complete by assumption. ■

This logic automatically gives us a dynamic logic of upgraded lifted propositional preferences.

Example Recursion laws for generic preferences.
Using the axioms, one computes how $\forall\exists$-type preferences $P\psi\chi$ change along:

$$\begin{aligned}
&[\#\varphi]P\psi\chi \leftrightarrow [\#\varphi]U(\psi \rightarrow <\leq>\chi) \leftrightarrow \\
&U[\#\varphi](\psi \rightarrow <\leq>\chi) \leftrightarrow U([\#\varphi]\psi \rightarrow [\#\varphi]<\leq>\chi) \leftrightarrow \\
&U([\#\varphi]\psi \rightarrow (\neg\varphi \wedge <\leq>[\#\varphi]\chi) \vee (\varphi \wedge <\leq>(\varphi \wedge [\#\varphi]\chi))) \leftrightarrow \\
&P([\#\varphi]\psi \wedge \neg\varphi)[\#\varphi]\chi \wedge P([\#\varphi]\psi \wedge \varphi)(\varphi \wedge [\#\varphi]\chi).
\end{aligned}$$

■

General relation transformers This result is just a trial run. Other relation transformers for betterness act on other triggers, and we aim for the same generality as in Chapter 7:

Example Drastic commands.
Let $\Uparrow\varphi$ be the radical trigger that makes all φ-worlds better than all $\neg\varphi$-ones, keeping the old order otherwise. This is stronger than a suggestion, making φ most desirable. Then we can use the axiom in Chapter 7 for safe belief, now using an existential modality E:

$$\begin{aligned}
[\Uparrow\varphi]<\leq>\psi \leftrightarrow (\neg\varphi \wedge E(\varphi \wedge [\Uparrow\varphi]\psi)) \vee (\neg\varphi \wedge <\leq>(\neg\varphi \wedge [\Uparrow\varphi]\psi)) \\
\vee (\varphi \wedge <\leq>(\varphi \wedge [\Uparrow\varphi]\psi))
\end{aligned}$$

The three clauses follow the three cases in the definition of radical upgrade. ■

These are just technical examples. Further betterness changes encode how people respond to what others claim or command, with a variety as with policies for belief revision – but we leave that to applications. A dynamic logic of preference should provide the right generality in triggers for upgrade. One general format is the *PDL* programs of Chapters 4, 7, involving *test*, *sequential composition*, and *union*, as in this earlier example:[11]

[11] Cf. also van Eijck's commentary on Sandu's chapter in Apt & van Rooij (2007).

Fact Suggestion is the map $\#\varphi(R) = (?\varphi\,;R\,;?\varphi) \cup (?\neg\varphi\,;R\,;?\neg\varphi) \cup (?\neg\varphi\,;R\,;?\varphi)$.

Constraints on betterness order once more Suppose that betterness satisfies constraints, will its transformed version still satisfy these? Indeed, the above suggestion actions take pre-orders to pre-orders, but they can destroy the *totality* of a betterness order:

Example Losing connectedness.
Suppose the agent prefers $\neg P$ over P as in the following connected model:

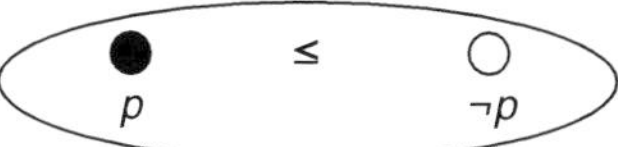

A suggestion $\#p$ will turn this into the following non-connected model:

■

Some people see this loss of basic properties as a drawback of relation transformers. But we feel that the situation is the other way around. The fact that some natural relation transformers break certain relational constraints on preference shows how fragile these constraints really are, and they provide natural scenarios for counter-examples.

Coda: what was versus what should become the case Especially in deontic logic, it is tempting to read upgrade commands $\#\varphi$ as 'come to prefer that φ', or 'your new duty will become to achieve φ'. Duties are often about what you should make the world look like. This is the *forward-oriented* view discussed in Chapters 3, 4, 11: one must produce a change making some postcondition on betterness true. Our approach, however, is *backward-oriented*, defining upgrades in terms of truth in the initial model – but the two perspectives co-exist peacefully for purely factual assertions. The contrast also comes up in related practical settings (cf. the notion of '*FIAT*' in Zarnic 1999), and hence *DEL* extended with factual change by stipulating postconditions (cf. Chapter 4) would also make sense here.

9.6 An alternative semantics: constraint-based preference

Now we come to a major feature of preference that was not part of our study of belief and plausibility order in Chapter 7. So far, we started from a betterness ordering of worlds, and then defined lifted notions of preference between propositions, i.e., properties of worlds. But another approach works in the opposite direction. Object comparisons are often made on the basis of *criteria*, and derived from how we apply these criteria, and prioritize them. Cars may be compared as to price, safety, and comfort, in some order of importance. On that view, criteria are primary, object order is derived. In our setting, criteria would be properties of worlds, expressed in propositions. This idea occurs in philosophy, economics (Rott 2001), linguistics, and cognitive science. We will now develop this alternative:

First-order priority logic A recent logic for this view of preference is found in de Jongh & Liu (2006). Take any finite linear *priority sequence* **P** of propositions, expressing the importance an agent attaches to the corresponding properties:

DEFINITION Object preference from priority.
Given a priority sequence **P**, the *derived object order* $x < y$ holds iff the objects x, y differ in at least one property in **P**, and the first $P \in \mathbf{P}$ where this happens is one with Py, $\neg Px$. ■

This stipulation is really a special case of the well-known notion of lexicographic ordering, if we view each property $P \in \mathbf{P}$ as inducing the following simple object order:[12]

$$x \leq^P y \text{ iff } (Px \rightarrow Py)$$

De Jongh and Liu give a complete first-order logic for induced preferences between objects. It hinges on the following representation result for object or world models:

THEOREM The orders produced via linear priority sequences are precisely the ones with *reflexivity*, *transitivity*, and *quasi-linearity*: $\forall xyz: x \leq y \rightarrow (x \leq z \vee z \leq y)$.

Liu (2008) extends this to betterness *pre-orders* induced, not by linear sequences but by the *priority graphs* of Andréka, Ryan & Schobbens (2002) (cf. Chapter 12). These can model more realistic situations where criteria may be incomparable. She also notes that there are many ways of defining object order from property order, that can be studied similarly. This diversity may be compared with that for lifting object order to world order.

[12] We will be free-wheeling in what follows between weak orders $\leq$ and strict ones $<$; but everything we say applies equally well to both versions and their modal axiomatizations.

This view makes sense much more generally than only for preference. One can also take a priority approach to belief, deriving plausibility order of worlds from an *entrenchment order* of propositions (cf. Gärdenfors & Rott 1995). Further interpretations will follow.

Dynamics This framework, too, facilitates preference change. This time, the priority order and set of relevant properties can change: a new criterion may come up, or a criterion may lose importance. Four main operations are *permuting* properties in a sequence, *prefixing* a new property, *postfixing* a new property, and *inserting* a property. De Jongh & Liu (2006) give a complete dynamic logic in our style, taking our methods to first-order logics. This can be generalized to non-linear priority graphs, leading to the algebra of graph operations in Andréka, Ryan & Schobbens (2002) (cf. Chapter 12), in particular, sequential and parallel composition.

Again, there are many interpretations for this dynamics. In a deontic setting, the priority graph may be viewed as a system of laws or norms, and changes in this structure model the adoption of new laws, or the appearance (or disappearance) of norms (van Benthem, Grossi & Liu 2010). Girard (2008) interprets priority graphs as agendas for investigation, and links agenda change to evolving research programmes in the philosophy of science.

Two-level connections The two views of preference are not rivals, but complementary. One either starts from a betterness relation between worlds and lifts this to propositional preference, or one starts from a importance order of propositions, and derives world order. One can combine these perspectives in interesting ways (Liu 2008):

DEFINITION Two-level structures.
Preferential *two-level structures* $(W, \leq, P, <)$ have worlds with a betterness order $\leq$ and 'important propositions' with a primitive priority order $<$:

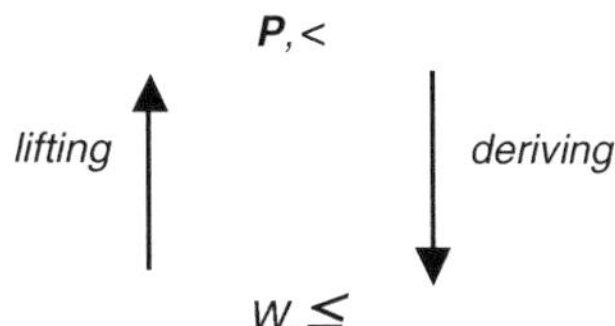

■

This picture suggests various interpretations, and many new questions.[13] We just state an elegant correspondence between the dynamics at the two levels. Prefixing propositions φ to a current priority sequence P has the same effect as the earlier relation transformer $\Uparrow\varphi$. More precisely, writing the lexicographic derivation of object order as a function *lex*:

FACT The valid identity $lex(\varphi\ ;\ P) = \Uparrow\varphi\ (lex(P))$ makes this diagram commute:

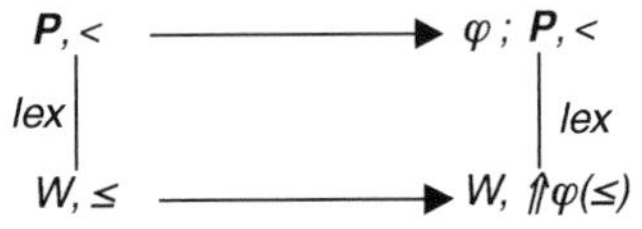

$$\begin{array}{ccc} P, < & \longrightarrow & \varphi\ ;\ P, < \\ lex \Big| & & \Big| lex \\ W, \leq & \longrightarrow & W, \Uparrow\varphi(\leq) \end{array}$$

Proof The definition of radical upgrade $\Uparrow\varphi$ really says that being φ is the first over-riding priority, while after that, the old order is followed.[14] ■

A general theory of inducing dynamics from one level to another is open.

9.7 Further aspects of preference: ceteris paribus logic

We have now discussed a basic modal approach to preference, as well as an alternative criteria-based view. Next comes one more major feature that presents a challenge to logic. Preferences usually hold only *ceteris paribus*, that is, under relevant circumstances. Van Benthem, Girard & Roy (2009) describe how to accommodate this in our logics:

Normality versus equality The term 'ceteris paribus' has two meanings. The *normality sense* says that we only make a comparison under normal circumstances. I prefer beer over wine, but not when dining at the Paris Ritz. This may be modelled by the 'normal' or most plausible worlds of a current model, as introduced in Chapter 7.

13 What happens when we derive betterness order from priority, and then lift it again – and vice versa? What happens when we treat the propositions in P as distinguished propositional constants in a modal language, and add a modal constraint logic?

14 This is an instance of the general algebraic calculus of priority graphs discussed in Chapter 12 – in particular, its law for sequential composition. For the special case of property graphs, there are also further laws. For instance, each such graph has a graph of *disjoint* properties generating the same object order (cf. Liu 2008). Rules for finding the latter merge Boolean algebra with preference logic.

Definition Preference under normal circumstances.
A global preference $P\varphi\psi$ holds in the *normality sense*, in any of the earlier lifted senses, if the latter holds when restricted to the doxastically most plausible worlds of the model. ■

If the set of normal worlds is definable by some proposition N, we can state this in our base logic as $P(N \wedge \varphi)(N \wedge \psi)$, using any relevant lifted preference P. But in general, we need both betterness and plausibility orders, as in Lang, van der Torre & Weydert (2003), with a matching combined logic of preference and belief. We will return to this in our next section.

In the *equality sense* of ceteris paribus, preference holds under the proviso that some propositions do not change truth values. You may prefer night over day, but only with 'work versus vacation' fixed (there may be vacation days that you prefer to work nights).

Definition Equality-based ceteris paribus preference.
A *ceteris paribus preference* for φ over ψ *in the equality sense* with respect to proposition A means that (i) among the A-worlds, the agent prefers φ over ψ, and (ii) among the $\neg A$-worlds, the agent prefers φ over ψ. Here, preference can be any of our lifted notions. ■

On this second account, cross-comparisons between the A and $\neg A$ worlds are irrelevant to a preference.[15] With a set of relevant propositions $\mathbf{A}$, one looks at the equivalence classes of worlds under the relation $\equiv_{\mathbf{A}}$ of sharing the same truth values on the A's.[16] This relation has also been studied as an account of *dependence* and independence of propositions (Doyle & Wellman (1994)). It also occurs in the semantics of questions in natural language (ten Cate & Shan (2002)), and with supervenience in philosophy.[17]

Equality-based ceteris paribus preference logic Van Benthem, Girard & Roy (2009) make equality-based ceteris paribus preferences an explicit part of the modal language.

[15] This is a conjunction of two normality readings: one with $N = A$, and one with $N = \neg A$.

[16] Von Wright (1963) kept one particular set $\mathbf{A}$ constant, viz. all proposition letters that do not occur in the φ, ψ in a statement $P\varphi\psi$. His preference logic has explicit rules expressing this special feature.

[17] For more general logics of dependence, cf. van Benthem (1996); Väänänen (2007).

DEFINITION Ceteris paribus modal preference logic.
The modal logic *CPL* extends basic model preference logic with *ceteris paribus* operators defined as

$$\begin{aligned}
&\mathbf{M}, s \models [\Gamma]\varphi \quad \text{iff} \quad \mathbf{M}, t \models \varphi \text{ for all } t \text{ with } s \equiv_\Gamma t \\
&\mathbf{M}, s \models [\Gamma]^{\leq}\varphi \quad \text{iff} \quad \mathbf{M}, t \models \varphi \text{ for all } t \text{ with } s \equiv_\Gamma t \text{ and } s \leq t \\
&\mathbf{M}, s \models [\Gamma]^{<}\varphi \quad \text{iff} \quad \mathbf{M}, t \models \varphi \text{ for all } t \text{ with } s \equiv_\Gamma t \text{ and } s < t
\end{aligned}$$ ■

Γ-equality-based ceteris paribus preference $P\varphi\psi$ can be defined as follows:

$$U(\varphi \to \langle\Gamma\rangle^{\leq}\psi)$$

In practice, the sets Γ are often finite, but the system also allows infinite sets, with even recursion in the definition of the ceteris paribus formulas. For the finite case, we have:

THEOREM The static logic of *CPL* is completely axiomatizable.

Proof The idea is this. All formulas in the new language have an equivalent formula in the base language, thanks to the basic laws for manipulating ceteris paribus clauses. The most important ones tell us how to change the sets Γ:

$$\begin{aligned}
&\langle\Gamma'\rangle^{\leq}\varphi \to \langle\Gamma\rangle^{\leq}\varphi \qquad \textit{if } \Gamma \subseteq \Gamma' \\
&(\alpha \wedge \langle\Gamma\rangle^{\leq}(\alpha \wedge \varphi)) \to \langle\Gamma \cup \{\alpha\}\rangle^{\leq}\varphi \\
&(\neg\alpha \wedge \langle\Gamma\rangle^{\leq}(\neg\alpha \wedge \varphi)) \to \langle\Gamma \cup \{\alpha\}\rangle^{\leq}\varphi
\end{aligned}$$

Applying these laws iteratively inside out will remove all ceteris paribus modalities until only cases $\langle\emptyset\rangle^{\leq}$ remain: that is, ordinary preference modalities from the base system. ■

The result is a practical calculus for reasoning with ceteris paribus propositions.[18]

9.8 Entanglement: preference, knowledge, and belief

Finally, we take up an issue that has come up at several places. We have analysed preference per se, but often it also has epistemic or doxastic aspects, being sensitive to changes in beliefs, and subject to introspection.

[18] This improves on logics like von Wright's where the set Γ is left implicit in context, that have tricky features of non-monotonicity and other surprising shifts in reasoning.

A standard approach would add epistemic structure to our models, and define richer preferences by combining earlier modalities for betterness, knowledge, and belief. Or should the marriage be more intimate?[19]

First degree: combining modalities Van Benthem & Liu (2007) give a logic of knowledge and preference with epistemic accessibility and betterness. Their language has modalities $<\leq>$, a universal modality, and modalities $K\varphi$. This can state things like

$KP\varphi\psi$	knowing that some generic preference holds,
$PK\varphi K\psi$	preferring to know φ over knowing ψ.[20]

The semantics allows for betterness comparisons beyond epistemically accessible worlds.[21] A language like this can change the earlier definition of lifted preferences $P\varphi\psi$ to epistemic variants like

$$K(\varphi \rightarrow <\leq>\psi)$$

Public announcements Preference dynamics now comes in two forms. There are direct betterness changing events as before, but preference may also change through *PAL*-style informative events $!\varphi$ as in Chapter 3. This is easily combined into one system:

THEOREM The combined logic of public announcement and suggestion consists of all separate principles for these operations plus two recursion axioms for betterness after update and knowledge after upgrade:

$$[!\varphi]<\leq>\psi \leftrightarrow (\varphi \rightarrow <\leq>(\varphi \wedge [!\varphi]\psi))$$
$$[\#\varphi]K_i\psi \leftrightarrow K_i[\#\varphi]\psi$$

Digression: upgrade versus update The *Art of Modelling* in Chapters 3, 4 returns here in the choice of initial models for the dynamics to start. Our system offers alternative models for the same scenario, trading preference change for information update (Liu 2008):

[19] Cf. Liu (2008). De Jongh & Liu (2006) make belief-based preference their central notion.

[20] This kind of statement raises tricky issues of future *learning*, that might work better in an epistemic or doxastic temporal logic of investigation (cf. Chapter 11).

[21] This can express a sense in which I prefer marching in the Roman Army to being an academic, even though I know that the former can never be.

Example Buying a house.
I am indifferent between a house near the park or downtown. Now I learn that a freeway will be built near the park, and I come to prefer the town house. This may be described as an initial *two-world* model

with a betterness indifference between the worlds. Taking a suggestion 'Town House' leaves both worlds, but removes a $\leq$-link, leaving a strictly better town house. But one can also describe the same scenario in terms of a *four-world* model with extended options

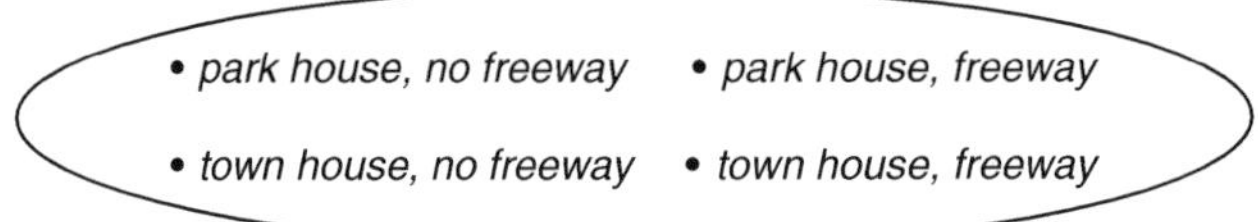

with obvious betterness relations between them. A public announcement of 'Freeway' now removes the two worlds to the left to get the model we got before by upgrading.[22] ■

Instead of knowledge, one can also merge the logics of belief of Chapter 7 with preference upgrade. Very similar points will apply as before, but the definable notions get more interesting. For instance, Lang & van der Torre (2008) discuss preference $P\varphi\psi$ as lifted betterness between only the *most plausible worlds* satisfying the propositions φ, ψ.

Second degree*: *intersections The logics so far may still miss a yet closer entanglement of preference and knowledge. An epistemic preference formula $K(\varphi \rightarrow <\leq>\psi)$, though subject to introspection, refers to ψ–worlds that are better than epistemically accessible φ–worlds. But there is no guarantee that these ψ–worlds *themselves* are accessible. But in our intuitive reading of the normality sense of ceteris paribus preference, we made the comparison *inside* the normal worlds, and likewise, we may want to make it inside the

[22] This example raises complex issues of language choice, and pre-encoding future events.

epistemically accessible worlds.[23] To do this, a modal language must talk about the *intersection* of the epistemic relation $\sim$ and betterness $\leq$:

DEFINITION Modal preference logic with epistemic intersection.
The *epistemic-preferential intersection modality* is interpreted as follows:

$$\mathbf{M}, s \models <\leq\cap\sim>\varphi \text{ iff there is a } t \text{ with } s \sim t \;\&\; s \leq t \text{ such that } \mathbf{M}, t \models \varphi \quad \blacksquare$$

Now we can define internally epistemized preference where each accessible φ-world sees an accessible ψ-world that is at least as good:

$$K(\varphi \rightarrow <\leq\cap\sim>\psi)$$

This new generic epistemic preference is no longer bisimulation-invariant (cf. Chapter 2), but it still allows for recursive analysis:

THEOREM The dynamic logic of epistemic-preferential intersection is completely axiomatizable, and its key axiom is the following equivalence:

$$<\#\varphi><\leq\cap\sim>\psi \leftrightarrow (\neg\varphi \wedge <\leq\cap\sim><\#\varphi>\psi) \vee (\varphi \wedge <\leq\cap\sim>(\varphi \wedge <\#\varphi>\psi))$$

Similar results hold with belief instead of knowledge. Dynamic events will produce both hard information and plausibility-changing soft information, both of which affect preference.

Third degree entanglement Still more intimately, preference and belief may even be taken to be interdefinable. Some literature on decision theory (cf. the survey in Pacuit & Roy 2006) suggests that we learn a person's beliefs from her preferences as revealed by her actions – or even that we learn preferences from beliefs (cf. Lewis 1988). In this book, entangled belief, preference, and action return in our study of games in Chapter 10. A rational player keeps all three aligned in a definite way, and different forms of entanglement defines different agent types that must reach equilibrium in a game.

9.9 Conclusion

Agents' preferences can be described in modal languages, just as knowledge and beliefs. These systems admit of dynamic logics that describe preference change triggered by events of accepting a command, suggestion, or more

[23] A similar entanglement is found in Baltag & Smets (2006).

drastic changes in normative structure. But preference is not just pure order of worlds, and we found interesting new phenomena that can be incorporated into logical dynamics, such as syntactic priority structure, management of ceteris paribus propositions, and entanglement with informational attitudes.

9.10 Further directions and open problems

Preference and actions Preferences hold between worlds or propositions, but also between *actions*. On an action-oriented view of ethics, if helping my neighbour is better than doing nothing, I must do it, whatever the consequences. How can we incorporate action-oriented preference? A formalism might merge preference logic with *propositional dynamic logic*. Since *PDL* has formulas as properties of states and programs as interstate relations, we can even put preference structure at two levels. One kind runs between states, the other between state transitions ('moves', 'events'), as in van der Meijden (1996), van Benthem (1999a).

Obligations and deontic logic This chapter obviously invites a junction with deontic logic and legal reasoning, including conditional obligations and global norms. There is a vast literature on deontic logic that we cannot reference here (cf. P. MacNamara's 2006 entry in the *Stanford On-Line Encyclopedia of Philosophy*, or Tan & van der Torre (1999) on how deontic logic enters computer science, a trend pioneered by J-J Meyer in the 1980s). Hansson (1969) is still relevant as a semantic precursor of the models used in this chapter, while Horty (2001) is a well-known philosophical-computational study of deontics and agency. For a state-of-the-art study, see Boella, Pigozzi & van der Torre (1999) on the logic of obligations, norms, and how these change. Liu (2008) is an attempt at creating links with dynamic-epistemic logic on topics like deontic paradoxes and norm change.

Art of modelling and thickness of worlds The example of Buying a House raised general issues of how much language to represent explicitly in the valuation of our models – and also, which possible future informational or evaluative events to encode in the description of worlds. On the whole, our dynamic logics are geared toward 'thin' worlds and light models, creating 'thickness' by further events that transform models. But there is always the option of making worlds in the initial model thicker from the start. Speaking generally, there is a *trade-off* here: the thicker the initial model, the simpler the subsequent events. I believe that every complex dynamic event, say, the *DEL*-style epistemic or doxastic updates of Chapters 4, 7, can be reduced to *PAL*-style public announcement by making worlds thicker. But I have never seen a precise technical formulation for this remodelling. Several technical facts in Chapter 10 on games and Chapter 11 on temporal trees seem relevant here, but I leave clarification of this issue to the reader.

Two-level preference logic The two-level view of object betterness plus priorities among properties raised many questions, such as harmony in dynamic operations or merging logics. The same issues arise if we would take the two-level view back to Chapter 7, and study relational belief revision in tandem with entrenchment dynamics for propositions. Van Benthem, Grossi & Liu (2010) shows how joint betterness/priority models for deontic logic with a matching modal language throw new light on the classical paradoxes of deontic reasoning.

Still further entanglement Entangled beliefs and preferences are the engine of decision theory (Hanson 1995; Bradley 2007). Can we add a qualitative logical counterpart to the fundamental notion of *expected value* to our logics, say based on degrees of plausibility? (A related probabilistic question occurred in Chapter 8.) Entanglement gets even richer when we add agents' *intentions*, as in Roy (2008) on the philosophy of action, or in *BDI* logics of agency in computer science (cf. Shoham & Leyton-Brown 2009).

Groups, social choice, and merge Preference logics with *groups* are a natural counterpart to epistemic and doxastic logics with groups (Chapters 2, 7, 12). Their dynamic versions describe group learning and preference formation through fact-finding and deliberation. Coalitional game theory and social choice theory (Endriss & Lang 2006) provide examples. A combined logic for actions and preferences for *coalitions* in games is given in Kurzen (2007). Group extensions of the logics in this chapter remain to be made, but see Grossi (2007) for a logical theory of institutions with normative aspects. Chapter 12 defines preference merge in structured groups, using Andréka, Ryan & Schobbens (2002).

Quantitative scoring There are also more quantitative versions of our preference logics, with *scoring rules* for worlds. Liu (2005) has models with numerical point values for worlds, and she gives a complete dynamic logic for a numerical *DEL*-style product update rule for worlds *(s, e)* using the separate values for *s*, *e*. Quantitative preference dynamics might be developed for most themes in this chapter, like we did in Chapter 8 for probability.

Dependence logic As we noted, ceteris paribus preference is related to general *logics of dependence* (van Benthem 1996; Väänänen 2007). Given the importance of the notion of dependence in games (Nash equilibrium involves the equality sense of ceteris paribus: cf. van Benthem, Girard & Roy 2009; players' behaviour depends on that of other players in extensive games) and many other disciplines, these links seem worth pursuing.

9.11 Literature

The literature on preference logic is surveyed in the handbook chapter Hanson (2001). Closer to this chapter, Van Benthem, van Eijck & Frolova (1993) proposed a dynamic logic for changing preferences triggered by actions including our 'suggestions'. Boutilier & Goldszmidt (1993) gave a semantics for conditionals in terms of actions that minimally change a given order so as to make all best antecedent worlds ones where the consequent is true. This upgrade idea was taken much further in Veltman (1996) on dynamic semantics of default reasoning, and in van der Torre & Tan (1999, 2001) on deontic reasoning and changing obligations. Zarnic (1999) studied practical reasoning with actions *FIAT* φ as changes in a preference ordering making the φ-worlds best. Yamada (2006, 2008) analysed acceptance of deontic commands as relation changers, and also gave the first complete dynamic-epistemic logics in our style. This chapter is based largely on Liu (2008), Girard (2008), van Benthem & Liu (2007), van Benthem, Girard & Roy (2009), whose themes have been explained in the text.

10 Decisions, actions, and games

We have now developed separate logics for knowledge update, inference, belief revision, and preference change. But concrete agency has all these entangled. A concrete intuitive setting where this happens is in *games*, and this chapter will explore such interactive scenarios. Our second reason for studying games here is as a concrete model of mid-term and long-term interaction over time (think of conversation or other forms of interactive agency), beyond the single steps that were the main focus in our logical systems so far.

This chapter is a 'mini-treatise' on logic and games introducing the reader to a lively new area that draws on two traditions: computational and philosophical logic. We discuss both *statics*, viewing games as encoding all possible runs of some process, and the *dynamics* when events change games. We start with examples. Then we introduce logics for static game structure, from moves and strategies to preferences and uncertainty. About halfway, we make a turn and start exploring what our dynamic logics add in terms of update and revision steps that change game models as new information arrives.[1] Our technical treatment is not exhaustive (we refer to further literature in many places, and van Benthem (to appearA) will go into more depth), but we do hope to convey the main picture.

10.1 Decisions, practical reasoning, and games

Action and preference Even the simplest scenarios of practical reasoning involve different logical notions at the same time. Recall this example from Chapter 8:

[1] This chapter is not a full survey of current interfaces between logic and game theory. For more, cf. Chapter 15, van der Hoek & Pauly (2006); van Benthem (1999a, to appearA).

Example One single decision.
An agent has two courses of action, but prefers one outcome to the other:

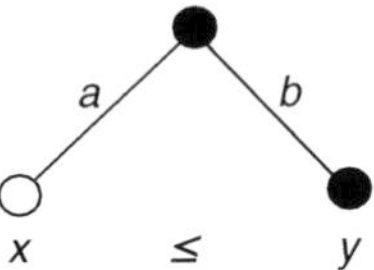

A typical form of reasoning here is the so-called Practical Syllogism: (i) the agent can choose either action *a* or *b*, (ii) the agent prefers the result of *a* over the result of *b*, and therefore, (iii) the agent will do [or maybe: should do?] *b*. ■

Choice of a best available option given one's beliefs is the basic notion of *rationality* in philosophy, economics, and many other fields. It can help predict agents' behaviour beforehand, or rationalize observed behaviour afterwards. And it intertwines all our notions so far: *preferences, actions*, and *beliefs* about what agents are going to do.

In decision theory, we see further roles of beliefs. Decisions may involve uncertainty about states of nature, and we choose an action with highest *expected value*, a weighted sum of utilities for the outcomes (cf. Chapters 8, 9). The probability distribution over states of nature represents our beliefs about the world. Here is a qualitative version:

Example Deciding with an external influence.
Nature has two moves *c, d*, and an agent has moves *a, b*. Now we get four combined moves:

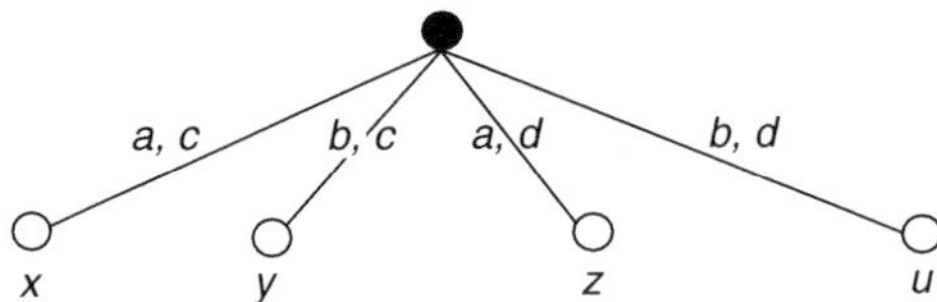

Suppose the agent thinks Nature's move *c* is more likely than *d*. This turns the outcomes into an epistemic-doxastic model (Chapter 7): the epistemic range has four worlds, but the most plausible ones are *x, y*, and the agent's beliefs only refer to the latter. ■

Thus we see entanglement of preference and belief as in Chapter 9, while also, as in Chapter 7, beliefs guide our actions. This mix becomes even more pronounced in games:

Solving games by backward induction In a multi-agent setting, behaviour is locked in place by mutual expectations. This requires an interactive decision dynamics, and standard game solution procedures like *Backward Induction* do exactly that. We will take this particular procedure as a running example in this chapter – not because we fully endorse it, but because it demonstrates many logical issues so beautifully:

Example Reasoning about interaction.
In the following game (cf. Chapters 1, 9), players' preferences are encoded in utility values, as pairs (*value for* A, *value for* E). Backward Induction tells player *E* to turn left at her turn, just as in our single decision case, which gives *A* a belief that this will happen, and so, based on this belief about his counter-player, *A* should turn left at the start:

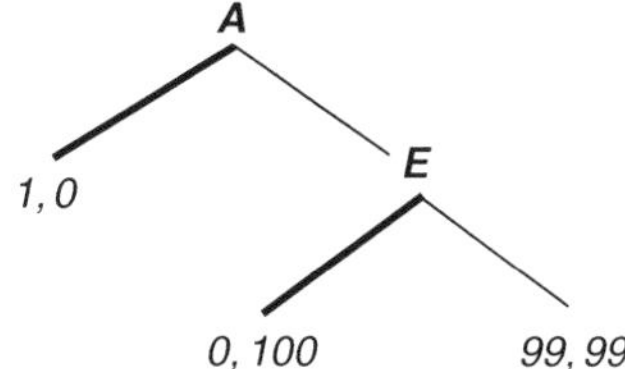

But *why* should players act this way? The reasoning is again a mixture of all notions so far. *A* turns left since she believes that *E* will turn left, and then her preference is for grabbing the value *1*. Once more, practical reasoning intertwines action, preference, and belief. ■

Here is the rule that captures all this, at least when preferences are encoded numerically:

Definition Backward Induction algorithm.
Starting from the leaves, one assigns values for each player to each node, using the rule

> Suppose *E* is to move at a node, and all values for daughters are known. The *E*-value is then the maximum of all the *E*-values on the daughters, and the *A*-value is the minimum of the *A*-values at all *E*-best daughters. The dual calculation for *A*'s turns is completely analogous. ■

This seems obvious and easy to apply, telling us players' best course of action (Osborne & Rubinstein 1994). And yet, it is packed with assumptions. We will perform a logical deconstruction later on, but for now, note that (a) the rule assumes the same reasoning by both players, (b) one makes worst-case assumptions about opponents, taking a minimum value when it is not our turn, (c) the rule changes its interpretation of values: at leaves they encode plain utilities; higher up in the game tree, they represent *expected utilities*. Thus, in terms of earlier chapters, Backward Induction is a mechanism for generating a *plausibility order* among histories, and hence, it relates our models in Chapter 7 to those of Chapter 9: betterness and plausibility order become intertwined in a systematic manner.

We will look at this reasoning in much more detail using our dynamic logics. But for now, we step back, and look at static logic of games *ab initio*. We first consider pure action structure, adding preference and epistemic structure for realistic games in due course.

10.2 Basic modal action logic of extensive games

In this chapter, we will view extensive games as multi-agent processes. Technically, such structures are models for a modal logic of computation in a straightforward sense:

DEFINITION Extensive game forms.
An *extensive game form* is a tree $\mathbf{M}$ = *(NODES, MOVES,* ***turn***, ***end***, *V)* with binary transition relations from the set *MOVES* pointing from parent to daughter nodes. Also, non-final nodes have unary proposition letters $\boldsymbol{turn}_i$ indicating the player whose turn it is, while ***end*** marks end nodes. The valuation *V* can also interpret other local predicates at nodes, such as utility values. ■

Henceforth, we will restrict attention to extensive games in the modal style.

Basic modal logic Extensive game trees support a standard modal language:

DEFINITION Modal game language and semantics.
Modal formulas are interpreted at nodes *s* in game trees **M**. Labelled modalities $<a>\varphi$ express that some move *a* is available leading to a next node in the game tree satisfying φ. Proposition letters true at nodes may be special-purpose constants for game structure, such as indications for turns and end-points, but also arbitrary local properties. ■

In particular, modal operator combinations now describe interaction:

Example Modal operators and strategic powers.
Consider a simple 2-step game like the following, between two players *A*, *E*:

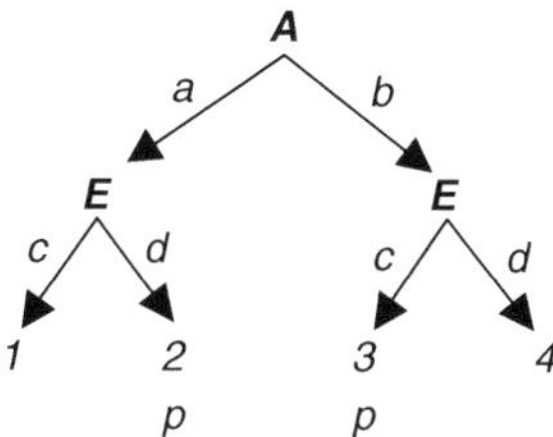

Player E clearly has a *strategy* for making sure that a state is reached where *p* holds, responding to whatever *A* does. This feature of the game is directly expressed by the modal formula $[a]\langle d\rangle p \wedge [b]\langle c\rangle p$. ■

More generally, letting *move* stand for the union of all relations available to players, in the preceding game, the modal operator combination

$$[\textit{move-A}]\langle \textit{move-E}\rangle \varphi$$

says that, at the current node, player E has a strategy for responding to A's initial move which ensures that the property expressed by φ results after two steps of play.[2]

Excluded middle and determinacy Extending this observation to extensive games up to some finite depth *k*, and using alternations $\Box\Diamond\Box\Diamond\ldots$ of modal operators up to length *k*, we can express the existence of winning strategies in fixed finite games. Indeed, given this connection, with finite depth, logical laws have immediate game-theoretic import. In particular, consider the valid *law of excluded middle* in the following modal form

$$\Box\Diamond\Box\Diamond\ldots\varphi \quad \vee \neg\Box\Diamond\Box\Diamond\ldots\varphi$$

or after some logical equivalences, pushing the negation inside:

$$\Box\Diamond\Box\Diamond\ldots\varphi \quad \vee \Diamond\Box\Diamond\Box\ldots\neg\varphi,$$

where the dots indicate the depth of the game. Here is its content:

FACT Modal excluded middle expresses the determinacy of finite games.

[2] Thus, one can express having winning strategies, losing strategies, etc. Links between such powers and logical formulas are crucial to *logic games* (van Benthem 2007d).

Here, *determinacy* is the fundamental property of many games that one of the two players has a winning strategy. This need not be true in infinite games (players cannot both have one, but maybe neither has), and Descriptive Set Theory has deep results in this realm.

Zermelo's theorem This brings us to perhaps the oldest game-theoretic result, predating Backward Induction, proved by Ernst Zermelo in 1913 for zero-sum games, where what one players wins is lost by the other ('win' versus 'lose' is the typical example):

THEOREM Every finite zero-sum 2-player game is determined.

Proof Here is a simple algorithm determining the player having the winning strategy at any given node of a game tree of this finite sort. It works bottom-up through the game tree. First, colour those end nodes *black* that are wins for player A, and colour the other end nodes *white*, being wins for E. Then extend this colouring stepwise as follows:

If all children of node s have been coloured already, do one of the following:

(a) if player A is to move, and at least one child is black:
colour s *black*; if all children are white, colour s *white*

(b) if player E is to move, and at least one child is white:
colour s *white*; if all children are black, colour s *black*.

This simplified Backward Induction eventually colours all nodes black where player A has a winning strategy, while colouring those where E has a winning strategy white. And the reason for its correctness is that a player has a winning strategy at one of her turns iff she can make a move to at least one daughter node where she has a winning strategy. ■

Zermelo's Theorem is widely applicable. Recall the Teaching Game from Chapter 1, our first example demonstrating the logical flavour of multi-move interaction:

Example Teaching, the grim realities.
A Student located at position S in the next diagram wants to reach the *escape* E below, while the Teacher wants to prevent him from getting there. Each line segment is a path that can be travelled. In each round of the game, the Teacher first cuts one connection, anywhere, and the Student must then travel one link still open at his current position:

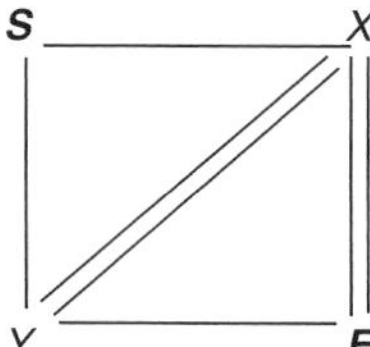

Education games like this arise on any graph with single or multiple lines.[3] ■

We now have a principled explanation why either Student or Teacher has a winning strategy, since this game is two-player zero sum and of finite depth.[4]

10.3 Fixed-point languages for equilibrium notions

A good test for logical languages is their power of representing basic proofs. Our modal language cannot express the above generic Zermelo argument. Starting from atomic predicates $\boldsymbol{win}_i$ at end nodes marking which player has won, we inductively defined new predicates WIN_i ('player i has a winning strategy at the current node' – here we use i, j to indicate the opposing players) through a recursion

$$WIN_i \leftrightarrow (\boldsymbol{end} \wedge \boldsymbol{win}_i) \vee (\boldsymbol{turn}_i \wedge \langle move\text{-}i \rangle WIN_i) \vee (turn_j \wedge [move\text{-}j]WIN_i)$$

Here *move-x* is the union of all moves for player x. This inductive definition for WIN_i is definable as a *fixed-point* expression in a richer system that we saw already in Chapter 4. The *modal μ-calculus* extends basic modal logic with operators for smallest and greatest *fixed-points* (cf. Bradfield & Stirling 2006; Blackburn, de Rijke & Venema 2000):

FACT The Zermelo solution is definable as follows in the modal μ-calculus:

$$WIN_i = \mu p \bullet (\boldsymbol{end} \wedge \boldsymbol{win}_i) \vee (\boldsymbol{turn}_i \wedge \langle move\text{-}i \rangle p) \vee (\boldsymbol{turn}_j \wedge [move\text{-}j]p)^5$$

[3] Gierasimczuk, Kurzen & Velázquez-Quesada (2009) give connections with real teaching.

[4] Zermelo's Theorem implies that in Chess, one player has a winning strategy, or the other a non-losing strategy, but almost a century later, we do not know which: the game tree is just too large. But the clock is ticking for Chess. Recently, for the game of Checkers, fifteen years of computer verification yielded the Zermelo answer: the starting player has a non-losing strategy.

[5] Crucially, the defining schema has only *positive occurrences* of the predicate p.

The μ–calculus has many uses in games: see below, and also Chapter 15.[6,7]

Other notions: forcing Winning is just one aspect. Games are all about control over outcomes that players have via their strategies. This suggests further logical notions:

DEFINITION Forcing modalities $\{i\}\varphi$.
$M, s \models \{i\}\varphi$ iff player i has a strategy for the subgame starting at node s which guarantees that only end nodes will be reached where φ holds, whatever the other player does. ■

FACT The modal μ–calculus can define forcing modalities for games.

Proof The modal fixed-point formula

$$\{i\}\varphi = \mu p \bullet (\mathbf{end} \wedge \varphi) \vee (\mathbf{turn}_i \wedge \langle move\text{-}i\rangle p) \vee (\mathbf{turn}_j \wedge [move\text{-}j]p)$$

defines the existence of a strategy for i making proposition φ hold at the end of the game, whatever the other player does.[8] ■

Analogously, just by shifting some modalities, the formula

$$COOP\varphi \leftrightarrow \mu p \bullet (\mathbf{end} \wedge \varphi) \vee (\mathbf{turn}_i \wedge \langle move\text{-}i\rangle p) \vee (\mathbf{turn}_j \wedge \langle move\text{-}j\rangle p)$$

defines the existence of a *cooperative outcome* φ.[9]

10.4 Explicit dynamic logic of strategies

So far, we left out a protagonist in our story. *Strategies* with successive interactive moves are what drives rational agency over time. Thus, it makes sense to move them explicitly into our logics, to state their properties and reason with long-term behaviour. Our discussion follows van Benthem (2007c),

[6] Our smallest fixed-point definition reflects the iterative equilibrium character of game solution (Osborne & Rubinstein 1994). In infinite games, we would switch to *greatest fixed-points* defining a largest predicate satisfying the recursion. This is also an intuitive view of strategies: they are not built from below, but can be used as needed, and remain at our service as fresh as ever next time we need them – the way we think of doctors. This is the perspective of *co-algebra* (Venema 2006). Greatest fixed-points seem the best match to the equilibrium theorems in game theory.

[7] As we shall see later, there are also other inductive formats for defining game solutions.

[8] We can also change the definition of $\{i\}\varphi$ to enforce truth of φ at all intermediate nodes.

[9] This is also definable in *PDL* by $\langle(?turn_i\,;\, move\text{-}i) \cup (?turn_j\,;\, move\text{-}j))^{*}\rangle(\mathbf{end} \wedge \varphi)$.

whose main tool is *propositional dynamic logic* (*PDL*), used in analysing conversation in Chapter 3 and common knowledge in Chapter 4. *PDL* is an extension of basic modal logic designed to study effects of imperative computer programs constructed using (a) sequential composition ;, (b) guarded choice *IF* ... *THEN* ... *ELSE* ..., and (c) guarded iterations *WHILE* ... *DO* ... For a start, we recall some earlier notions:

DEFINITION Propositional dynamic logic.
The *language of PDL* defines formulas and programs in a mutual recursion, with formulas denoting sets of worlds (local conditions on states of the process), while programs denote binary transition relations between worlds, recording pairs of input and output states for their successful terminating computations. Programs are created from

> atomic actions ('moves') $a, b, \ldots$ and tests $?\varphi$ for arbitrary formulas φ,[10] using the three operations of ; (interpreted as sequential composition), $\cup$ (non-deterministic choice), and * (non-deterministic finite iteration).

Formulas are as in our basic modal language, but with modalities $[\pi]\varphi$ saying that φ is true after every successful execution of the program π starting at the current world. ■

The logic *PDL* is decidable, and it has a simple complete set of axioms. This system can say much more about games. Admittedly, *PDL* focuses on terminating programs, and once more, we restrict attention to *finite games*.

Strategies as transition relations Strategies in game theory are partial functions on players' turns, given by instructions of the form 'if she plays this, then I play that'. More general strategies are *binary transition relations* with more than one best move. This view is like *plans* that agents have in interactive settings. A plan can be very useful when it only constrains my moves, without fixing a unique course of action. Thus, on top of the hard-wired move relations in a game, we now get defined further relations, corresponding to players' strategies, and these definitions can often be given explicitly in a *PDL* program format.

[10] Please note: these *PDL* tests inside one model are not the question actions of Chapter 6.

In particular, in finite games, we can define an explicit version of the earlier forcing modality, indicating the strategy involved – without recourse to the modal μ-calculus:

Fact For any game program expression σ, *PDL* can define an explicit forcing modality $\{\sigma, i\}\varphi$ stating that *σ is a strategy for player i forcing the game, against any play of the others, to pass only through states satisfying φ.*

Proof The formula $[((?turn_E ; \sigma) \cup (?turn_A ; move\text{-}A))^*]\ \varphi$ defines the forcing. ■

Here is a related observation (cf. van Benthem 2002a). Given relational strategies for two players **A**, **E**, we get to a substructure of the game described like this:

Fact Outcomes of running joint strategies σ, τ can be described in *PDL*.

Proof The formula $[((?turn_E ; \sigma) \cup (?turn_A ; \tau))^*]\ (\textbf{end} \rightarrow \varphi)$ does the job. ■

On a model-by-model basis, the expressive power of *PDL* is high (Rodenhäuser 2001). Consider any finite game **M** with strategy σ for player *i*. As a relation, σ is a finite set of ordered pairs *(s, t)*. Assume that we have an 'expressive model' **M**, where states *s* are definable in our modal language by formulas def_s.[11] Then we can define pairs *(s, t)* by formulas $def_s; a; def_t$, with *a* the relevant move, and take the relevant union:

Fact In expressive finite extensive games, all strategies are *PDL*-definable.[12]

The operations of *PDL* can also describe *combination* of strategies (van Benthem 2002a). In all, propositional dynamic logic does a good job in defining explicit strategies in simple extensive games. In what follows, we extend it to deal with more realistic game structures, such as preferences and imperfect information. But there are alternatives. Van Benthem (2007c) is a survey and defence of many kinds of logic with explicit strategies.[13]

[11] This can be achieved using temporal *past modalities* to describe the history up to *s*.

[12] As for *infinite games*, the modal μ-calculus is an extension of *PDL* that can express existence of infinite computations – but it has no explicit programs. One would like to add *infinite strategies* like 'keep playing *a*' that give infinite *a*-branches for greatest fixed-point formulas $\nu p\bullet <a>p$.

[13] For a temporal *STIT*-type logic of strategies in games with simultaneous actions, cf. Broersen (2009). (See also Herzig & Lorini (2010) on recent *STIT* logics of agents' powers.) Concurrent action without explicit strategies occurs in Alternating Temporal Logic

10.5 Preference logic and defining backward induction

Real games go beyond game forms by adding preferences or numerical utilities over outcomes. Defining the earlier Backward Induction procedure for solving extensive games has become a benchmark for game logics – and many solutions exist:

FACT The Backward Induction path is definable in modal preference logic.

Many solutions have been published by logicians and game-theorists, cf. de Bruin (2004); Harrenstein (2004); van der Hoek & Pauly (2006). We do not state an explicit *PDL* solution here, but we give one version involving the modal preference language of Chapter 9:

$<pref_i>\varphi$: player *i* prefers some node where φ holds to the current one.

The following result from van Benthem, van Otterloo & Roy (2006) defines the backward induction path as a unique relation σ by a frame correspondence for a modal axiom on finite structures. The original analysis assumed that all moves are unique:

FACT The *BI* strategy is the unique relation σ satisfying the following modal axiom for all propositions *p* – viewed as sets of nodes – for all players *i*:

$$(turn_i \wedge <\sigma^*>(\mathbf{end} \wedge p)) \rightarrow [move\text{-}i]<\sigma^*>(\mathbf{end} \wedge <pref_i> p)$$

Proof The argument is by induction on the depth of finite game trees. The crux here is that the given modal axiom expresses the following form of 'maximin' Rationality:

> No alternative move for the current player *i* guarantees a set of outcomes – via further play using σ – with a higher minimal value for *i* than the outcomes that result from playing σ all the way down the tree.

The typical picture to keep in mind here, and also in later analyses, is this:

(Alur, Henzinger & Kupferman (1997)) with an epistemic version *ATEL* in van der Hoek & Wooldridge (2003). Ågotnes, Goranko & Jamroga (2007) discuss *ATL* with strategies, van Otterloo (2005) highlights strategies in *ATEL*. An alternative might be versions of *PDL* with structured simultaneous actions, as suggested earlier.

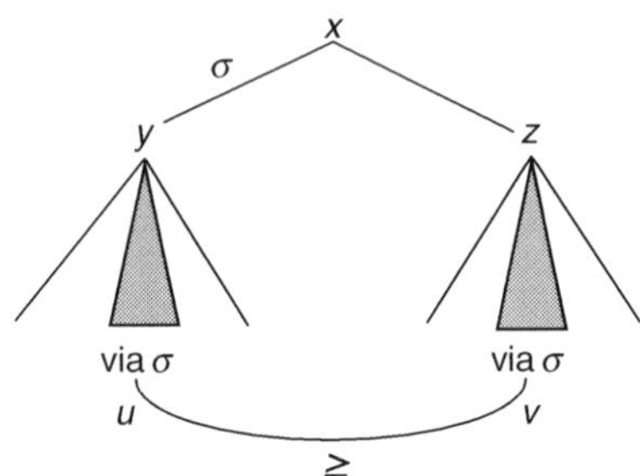

More precisely, the stated modal axiom is equivalent by a modal frame correspondence (cf. Chapter 2) to the following *confluence property* for action and preference:

$$\forall x\, \forall y\, ((turn_i(x) \wedge x\sigma y) \rightarrow \forall z\, (x \text{ } move \text{ } z \rightarrow \forall u\, ((end(u) \wedge y\sigma^* u) \rightarrow \exists v(end(v) \wedge z\sigma^* v \wedge v \leq_i u))))$$

This $\forall\forall\forall\exists$ format expresses the stated property about minimum values.[14] ■

The analysis also works for the natural relational version of Backward Induction, where the strategy connects a node to all daughters (one or more) with maximal values for the active player. Then σ is the *largest* subrelation of the *move* relation with the stated property.

Alternatives Our logical analysis can also deal with other relational versions of Backward Induction, that make alternative assumptions about players' behaviour. In particular, a more standard (though weaker) form of Rationality is 'avoiding dominance':

> No alternative move for the current player i guarantees outcomes via further play using σ that are all strictly better for i than the outcomes resulting from starting at the current move and then playing σ all the way down the tree.

This time, the modal axiom is

$$(turn_i \wedge <\sigma^*>[\sigma](\mathbf{end} \rightarrow p)) \rightarrow [move\text{-}i]<\sigma^*>(\mathbf{end} \wedge <pref_i>p)$$

[14] Confluence is reminiscent of the grid cells in Chapter 2, that could give combined modal logics high complexity. Little is known about the computational effects of rationality principles on our game logics.

and the corresponding formula is the ∀∀∃∃ form

$$\forall x\ \forall y\ ((turn_i(x) \wedge x\sigma y) \rightarrow \forall z(x\ move\ z \rightarrow \exists u\ \exists v\ (end(u) \wedge end(v) \wedge y\sigma^* v \wedge z\sigma^* u \wedge u \leq_i v)))$$

This version is technically a bit more convenient (van Benthem & Gheerbrant 2010). We will return to this scenario for game solution later on in this chapter, with a new dynamic analysis.

10.6 Epistemic logic of games with imperfect information

The next level of static structure gives up perfect information, a presupposition so far. In games with *imperfect information*, players need not know where they are in the tree. Think of card games, private communication, or real life with bounds on memory or observation. Such games have 'information sets': equivalence classes of epistemic relations $\sim_i$ between nodes that players i cannot distinguish, as in Chapters 2–5. Van Benthem (2001) points out how these games model an epistemic action language with knowledge operators $K_i\varphi$ interpreted in the usual manner as 'φ is true at all nodes $\sim_i$–related to the current one'.

Example Partial observation in games.
In the following imperfect information game, the dotted line marks player E's uncertainty about her position when her turn comes. She does not know the initial move played by A. Maybe A put his move in an envelope, or E was otherwise prevented from seeing:[15]

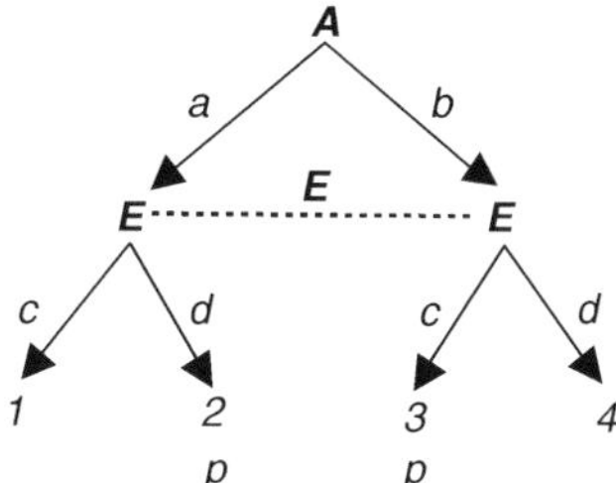

Structures like this interpret a combined *dynamic-epistemic* language. For instance, after A plays move b in the root, in both middle states, E knows

[15] An imperfect information game comes without an explicit scenario attached. We will return later to the natural issue how given epistemic game trees may *arise*.

that playing a or b will give her p – as the disjunction $<a>p \vee <b>p$ is true at both middle states:

$$K_E(<a>p \vee <b>p)$$

On the other hand, there is no specific move which **E** knows at this stage that will guarantee a p-outcome – and this shows in the truth of the formula

$$\neg K_E <a> p \wedge \neg K_E <b> p$$

Thus, **E** knows *de dicto* that she has a strategy which guarantees p, but she does not know, *de re*, any specific strategy that guarantees p. Such finer distinctions are typical for a language with both actions and knowledge for agents.[16]

Special games and modal axioms We can analyse special kinds of imperfect information game using modal frame correspondences. Recall the axiom $K_i[a]\varphi \rightarrow [a]K_i\varphi$ of Perfect Recall in Chapters 3, 4, that allowed an interchange of knowledge and action modalities. Again, it expresses a semantic confluence – this time, between *knowledge* and *action:*

FACT The axiom $K_i[a]p \rightarrow [a]K_ip$ holds for all propositions p iff

M satisfies $\forall xyz: ((x\ R_a\ y \wedge y \sim_i z) \rightarrow \exists u\ (x \sim_i u \wedge u\ R\ z))$.

Proof By standard correspondence techniques: cf. Blackburn, de Rijke & Venema (2000). The reader may want to check in a diagram how the given Confluence property guarantees the truth of $K_i[a]\varphi \rightarrow [a]K_i\varphi$, starting from an upper-left world x verifying $K_i[a]\varphi$:

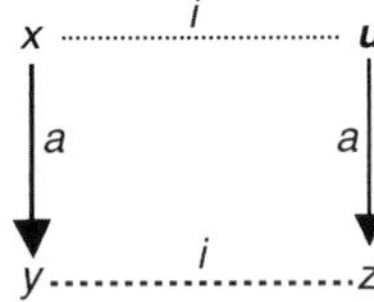

■

Perfect Recall says that, if the last observation has not provided differential information, present uncertainties must have come from past ones. Similar

[16] You may know that the ideal partner for you is around somewhere, but you might never convert this $K\exists$ combination into $\exists K$ knowledge which person is right for you.

analyses work for memory bounds, and further observational powers (van Benthem 2001). For instance, as a converse to Perfect Recall, agents satisfy *No Miracles* when their current epistemic uncertainty between worlds x, y can only disappear by observing subsequent events on x and on y that they can distinguish.[17] Incidentally, the game we just gave satisfies Perfect Recall, but No Miracles fails, since E suddenly knows where she is after she played her move. We will discuss these properties in greater generality in the epistemic-temporal logics of Chapter 11.

Uniform strategies Another striking feature of the above game with imperfect information is its *non-determinacy*. Player E's playing the opposite of player A was a strategy guaranteeing outcome p in the underlying perfect information game without an epistemic uncertainty in the middle – but it is useless now, since E cannot tell what A played. Game theorists allow only *uniform strategies* here, prescribing the same move at indistinguishable nodes. But then no player has a winning strategy in our game, when we interpret p as 'E wins' (and hence $\neg p$ as a win for A): A did not have one to begin with, while E loses hers.[18]

As for explicit strategies, we can add *PDL*-style programs to the epistemic setting. But there is a twist. We need the *knowledge programs* of Fagin *et al.* (1995), whose only test conditions for actions are knowledge statements. In such programs, moves for an agent are guarded by conditions that the agent knows to be true or false. Thus, knowledge programs define uniform strategies, where a player always chooses the same move at game nodes that she cannot distinguish epistemically. A converse also holds, given some conditions on expressive power of models (van Benthem 2001):

Fact On expressive finite games of imperfect information, the uniform strategies are precisely those definable by knowledge programs in epistemic *PDL*.

Several kinds of knowledge in games Imperfect information reflects limitations of players, resulting in ignorance of the past. But even perfect information

[17] This was noted by Halpern & Vardi in epistemic-temporal logic (Fagin *et al.* 1995).

[18] The game does have probabilistic solutions in *mixed strategies*: it is like Matching Pennies. Both players should play both moves with probability *1/2*, for an optimal outcome *0*.

games leave players ignorant of the *future*, perhaps because of ignorance about the kind of player one is up against. These richer views of knowledge will return later on in this chapter.

10.7 From statics to dynamics: *DEL*-representable games

Now we make a switch. So far, we have used static modal-preferential-epistemic logics to describe given game trees. Now we move to *dynamic* logics in the sense of this book, where games, or their associated models, can change because of triggering events. As a first step, we analyse how a given game might have come about through some dynamic process – the way we see a dormant volcano but imagine the tectonic forces that shaped it. We provide two illustrations, linking games of imperfect information first to *DEL* (cf. Chapter 4) and then to epistemic-temporal logics (cf. Chapter 11). After that, we will explore many further dynamic scenarios in the analysis of games, based on earlier chapters.

Imperfect information games and dynamic-epistemic logic The reader will long have seen a link with the logic *DEL* of Chapter 4. Which imperfect information games make sense in some underlying informational process, as opposed to mere sprinkling of uncertainty links over game trees? Consider any finite game as the domain of an event model, with preconditions on occurrence of moves encoded in the tree structure. Also assume that we are given observational limitations of players over these moves. We can then decorate the game tree with epistemic links through iterated product update, as in *DEL* update evolution:

Example Updates during play: propagating ignorance along a game tree.
When moving in the following game tree, players can distinguish their own moves, but not all moves of their opponents – as described in the accompanying event model:

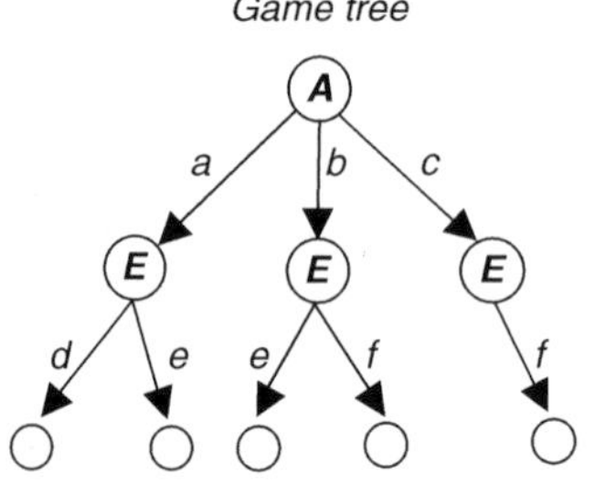

Here are the successive *DEL* updates that create the uncertainty links:

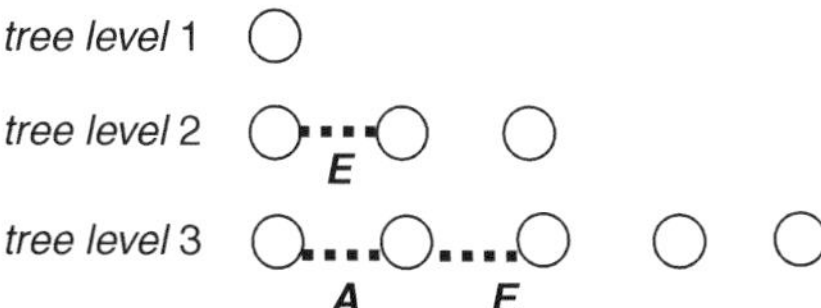

The resulting annotated tree is the following imperfect information game:

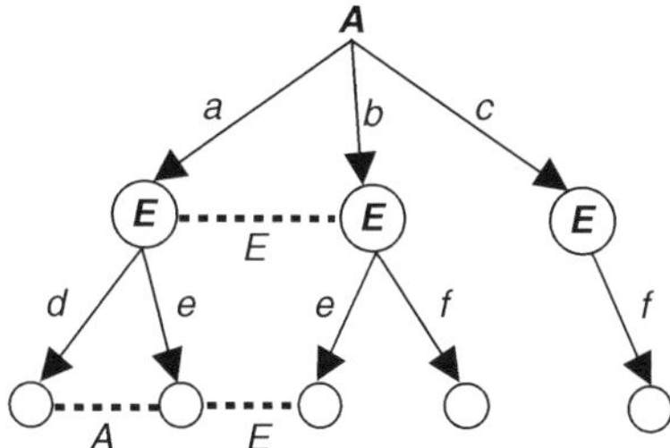

In Chapter 11, we will characterize the trees that arise from *DEL*-style update evolutions, in terms of properties whose definitions we postpone:

THEOREM An extensive game is isomorphic to an iterated update model *Tree*(*M*, E) over some epistemic event model E and initial model M iff it satisfies, for all players, (a) Perfect Recall, (b) No Miracles, (c) Bisimulation Invariance for the domains of moves.

10.8 Future uncertainty, procedural information, and temporal logic

As we have said, knowledge or ignorance in games has several senses. One is *observation uncertainty*: players may not be able to observe all moves completely, and hence need not know where they are in the game. This is the past-oriented view of *DEL*. But there is also future-oriented *expectation uncertainty*. In general, players have only limited procedural information about what will happen, and Chapters 5, 11 show that this, too, is a basic notion of information about the process agents are in. Future-oriented knowledge need not reduce to uncertainty between local nodes. It suggests uncertainty between whole histories, or between players' strategies: i.e., whole ways in which the game might evolve.[19]

Branching epistemic-temporal models The following structure is common to many fields (cf. Chapter 11 below for details). In tree models for branching time, legal histories h are possible evolutions of a game. At each stage, players are in a node s on some history whose past they know completely or partially, but whose future is yet to be revealed:

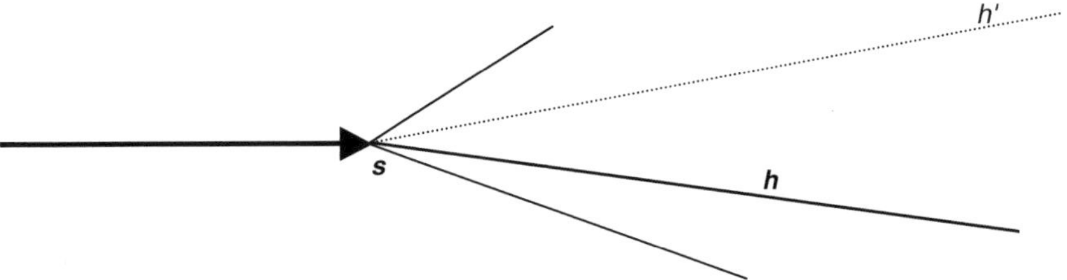

This can be described in an action language with knowledge, belief, and added temporal operators. We first describe games of perfect information (about the past):

(a) $\mathbf{M}, h, s \models F_a\varphi$ iff $s^\frown\langle a\rangle$ lies on h, and $\mathbf{M}, h, s^\frown\langle a\rangle \models \varphi$

(b) $\mathbf{M}, h, s \models P_a\varphi$ iff $s = s'^\frown\langle a\rangle$, and $\mathbf{M}, h, s' \models \varphi$

(c) $\mathbf{M}, h, s \models \Diamond_i\, \varphi$ iff $\mathbf{M}, h', s \models \varphi$ for some h' equal to h for i up to stage s.

Now, as moves are played publicly, players make public observations of them, leading to an epistemic-temporal version of our system *PAL* in Chapter 3:

FACT The following valid principle is the temporal equivalent of the key *PAL* recursion axiom for public announcement: $F_a\Diamond\varphi \leftrightarrow (F_aT \wedge \Diamond F_a\varphi)$.[20]

This principle will return as a recursion law for *PAL* with temporal protocols in Chapter 11.

Excursion: trading future for current uncertainty Again, there is a reconstruction closer to the local dynamics of *PAL* and *DEL*. Intuitively, each move by a player is a public announcement that changes the current game. Here is a folklore observation converting global uncertainty about the future into local uncertainty about the present:

[19] This section is from van Benthem (2004a) on update and revision in game trees.

[20] As in our earlier modal-epistemic analysis, this expresses a form of Perfect Recall.

Fact Trees with future uncertainty are isomorphic to epistemic tree models with current uncertainties.

Proof Given any game tree *G*, assign epistemic models M_s to each node *s* whose domain is the set of histories passing through *s* (all share the same past up to *s*), letting the agent be uncertain between all of them. Worlds in these models may be seen as pairs *(h, s)* with *h* any history passing through *s*. This will cut down the current set of histories in just the right manner. The above epistemic-temporal language matches this construction. ■

10.9 Intermezzo: three levels of game analysis, from 'thin' to 'thick'

At this point, it may be useful to distinguish three different levels at which games give rise to models for logics. All three come with their own intuitions, both static and dynamic.

Level One is the most immediate: extensive game trees are models for modal languages, with nodes as worlds, and accessibility relations for actions, preferences, and epistemic uncertainty. *Level Two* looks at extensive games as branching tree models with nodes and histories, supporting richer epistemic-temporal (-preferential) languages. The difference from Level One is slight in finite games, where histories correspond to end-points. But the intuitive step is clear, and Level Two cannot be reduced when game trees are infinite. But even this is not enough. Consider hypotheses about the future, involving procedural information about strategies. I may know that I am playing against either a simple automaton, or a sophisticated learner. Modelling this may go beyond epistemic-temporal models:

Example Strategic uncertainty.
In the following simple game, let *A* know beforehand that *E* will play the same move throughout:

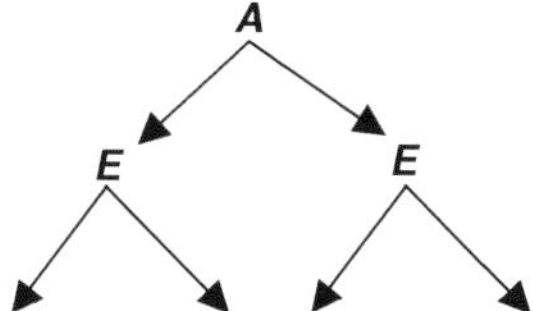

Then all four histories are still possible. But given his information, *A* only considers two *future trees*, viz.

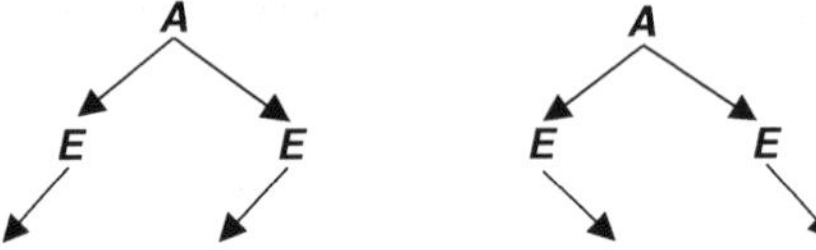

In longer games, this difference in modelling can be highly important, since observing only one move by E will tell A exactly what E's strategy will be in the whole game. ■

To model these richer settings, one needs *Level Three* game models:

DEFINITION Epistemic game models.
Epistemic game models for an extensive game G are epistemic models $M = (W, \sim_i, V)$ whose worlds carry local information about all nodes in G, plus strategy profiles: total specifications of each player's behaviour throughout the game. Players' information about structure and procedure is encoded by uncertainty relations $\sim_i$ between the worlds. ■

Partial knowledge about strategies is encoded in the set of profiles represented in such a model. And observing moves telling me which strategy you are following leads to dynamic update of the model, in the sense of our earlier chapters. Level-Three models are a natural limit for agency, and they are the usual models in the epistemic foundations of game theory (cf. Geanakoplos 1992; Battigalli & Siniscalchi 1999; Stalnaker 1999; Halpern 2003). Also, they are congenial to modal logicians because of their abstract possible worlds flavour.

Even so, our preference throughout this book has been for 'thin' epistemic models plus their explicit dynamics over 'thick' models encoding events and much else inside worlds (cf. Chapters 2, 4, 9). This lightness is indeed a major attraction of our dynamic framework for information flow. Hence, in the rest of this chapter we keep things simple, discussing issues at the thinnest level where they make sense: histories usually suffice.

10.10 Game change: public announcements, promises, and solving games

Now let us look at actual dynamic transformations of given games, and explicit triggers for them.

Promises and intentions One can break the impasse of a bad Backward Induction solution by changing the game through *promises* (van Benthem 2007g).

Example Promises and game change.
In the following game, discussed before, the bad Nash equilibrium *(1, 0)* can be avoided by *E*'s *promise* that she will not go left. This public announcement eliminates histories (we can make this binding by a fine on infractions)[21] – and the new equilibrium *(99, 99)* results, making both players better off:

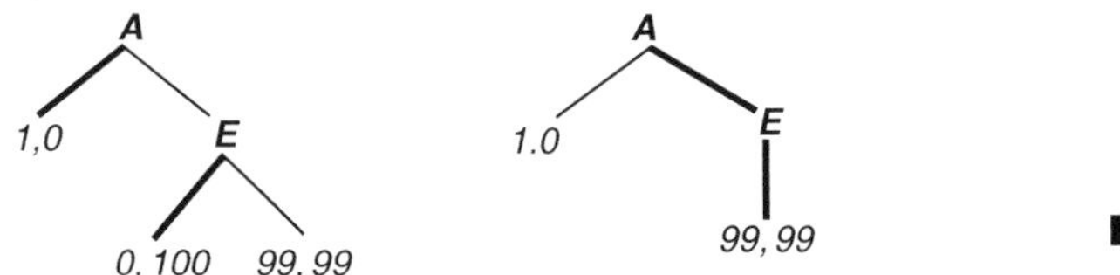

■

Van Otterloo (2005) has a dynamic logic of strategic enforceability, where games change by announcing intentions or preferences. Game theory has much more sophisticated analyses of such scenarios, including the study of 'cheap talk' (Osborne & Rubinstein 1994). One can also *add* new moves when trying to change a game in some desired direction, and indeed, there is a whole family of game transformations that make sense.[22]

Modal action logic and strategies Our methods from Chapters 3, 4 apply:

THEOREM The modal action logic of games plus public announcement is axiomatized by the modal game logic chosen, the recursion axioms of *PAL* for atoms and Booleans, plus the following law for the move modality:

$$<!P><a>\varphi \leftrightarrow (P \wedge <a>(P \wedge <!P>\varphi))$$

Using *PDL* for strategies in games as before, this leads to a logic $PDL + PAL$ adding public announcements *[!P]*. The following result uses the closure of *PDL* under relativization to definable submodels, both in its propositional

[21] Alternatively, this changes utilities and preferences to run Backward Induction anew.

[22] One might put all game changes beforehand in one grand *Super Game*, like the Supermodel of Chapters 3, 4 – but that would lose the flavour of what really happens.

and its program parts, with a recursive operation $\pi|P$ for programs π that wraps every atomic move a in tests to $?P; a; ?P$.

Theorem *PDL* + *PAL* is axiomatized by merging their separate laws while adding the following reduction axiom: $[!P]\{\sigma\}\varphi \leftrightarrow (P \rightarrow \{\sigma|P\}[!P]\varphi)$.[23]

There are also extended versions with epistemic preference languages.

Solving games by announcements of rationality Here is a more foundational use of logical dynamics, where public announcements act as 'reminders' driving a process of deliberation. Van Benthem (2007f) makes the solution process of extensive games itself the focus of a *PAL*-style analysis. Let us say that, at a turn for player i, a move a is *dominated* by a sibling b (a move available at the same decision point) if every history through a ends worse, in terms of i's preference, than every history through b. Now define:

Rationality No player chooses a strictly dominated move.

This makes an assertion about nodes in a game tree, viz. that they did not arise through a dominated move. Some nodes will satisfy this, others may not. Thus, announcing this formula as a fact about the players is informative, and it will in general make the current game tree smaller. But then we get a dynamics similar to that with the Muddy Children in Chapter 3. In the new smaller game tree, new nodes may become dominated, and hence announcing Rationality makes sense again, and so on. This process must reach a limit, a smallest subtree where no move is dominated any more. Here is how this works:

Example Solving games through iterated assertions of Rationality.
Consider a game with three turns, four branches, and pay-offs for A, E, resp.:

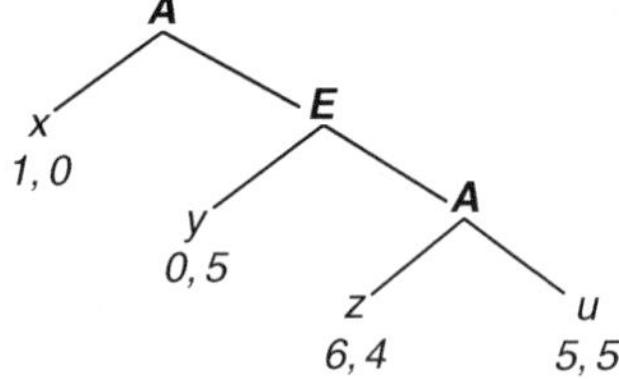

[23] The axiom derives what old plan I should have had in G to run a given plan in the new game G'. But usually, I have a plan σ *in* G to get effect φ. Now G changes to G' with more or fewer moves. How should I *revise* σ to get some related effect ψ in G'? This seems harder, and even the special case where G' is a subgame of G might be related to open technical problems like finding a complete syntactic characterization for the *PDL* formulas that are preserved under submodels.

Stage 0 of the procedure rules out point u (the only point where Rationality fails), Stage 1 then rules out z and the node above it (the new points where Rationality fails), and Stage 2 rules out y and the node above it. In the remaining game, Rationality holds throughout:

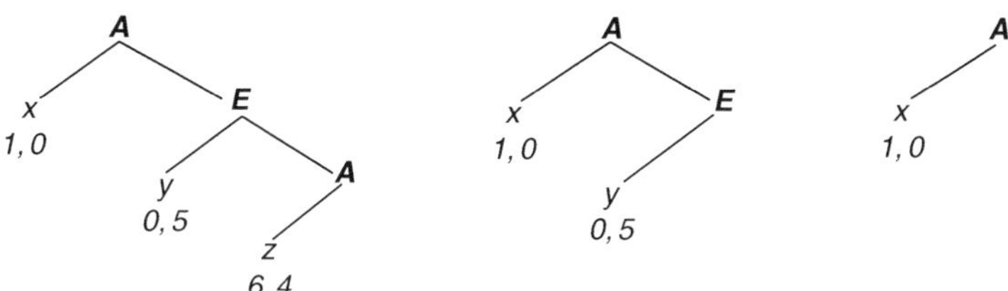

When Backward Induction assigns unique moves, we get this connection:

THEOREM The actual Backward Induction path for extensive games arises from repeated announcement of the assertion of Rationality to its limit.

Proof This can be proved by a simple induction on finite game trees.[24] ■

With general relational strategies, this particular iterated announcement scenario produces the earlier ∀∀∃∃ version of Backward Induction stated in Section 10.5.[25]

Dynamic instead of static foundations An important feature of our new scenario is this. In the terminology of Chapter 3, the above iterated announcement procedure for game trees (or game models) is *self-fulfilling*: it ends in non-empty largest submodels where players have *common knowledge of rationality*.[26] Thus, this dynamic style of game analysis is a big change away from the usual static characterizations of Backward Induction in the epistemic foundations of game theory (van Benthem 2007f):

> Common knowledge or common belief of rationality is not assumed, but *produced* by the logic.

[24] The repeated announcement limit yields only the actual *BI* path. To also get the *BI* moves at nodes x not reached by the latter, we must remember the stage for the subgame T_x with x as its root.

[25] We do not pursue such technical details here: the main point is the new scenario itself.

[26] We forego the issue of logical languages for explicitly *defining* the limit submodel.

In Chapter 15, we will look at similar iterated announcement procedures that can solve strategic games, working on game matrices or Level-Three models for games.[27]

The *PAL* style of game analysis also works with other assertions. Van Benthem (2004a, 2007f) consider alternatives to future-oriented rationality where players steer actions by considering the legitimate rights of other players because of their past merits.[28]

10.11 Belief, update, and revision in extensive games

So far, we have studied players' knowledge in games, observation-based or procedural. Next, consider the equally fundamental notion of *belief*, that has the same two aspects. Indeed, many foundational studies in game theory use belief rather than knowledge.

To deal with this, our earlier Level-One game models would add plausibility orders as in Chapter 7. A game tree annotated in this way records steps of knowledge update and belief revision as the game is played (cf. Board 1998), and we will use this view in Chapter 11. For the moment, we look at Level-Two branching trees, a vivid model of beliefs about the future:

Beliefs over time We now add binary relations $\leq_{i,\, s}$ of state-dependent *relative plausibility* between histories to branching models. As in Chapter 7, we then get a doxastic modality (an existential one here, for convenience), with absolute and conditional versions:

DEFINITION Absolute and conditional belief.
We set $\boldsymbol{M}, h, s \models \langle B, i\rangle\varphi$ iff $\boldsymbol{M}, h', s \models \varphi$ for some history h' coinciding with h up to stage s and most plausible for i according to the given relation $\leq_{i,\, s}$. As an extension, $\boldsymbol{M}, h, s \models \langle B, i\rangle^{\psi}\varphi$ iff $\boldsymbol{M}, h', s \models \varphi$ for some history h' most plausible for i according to the given $\leq_{i,\, s}$ among all histories coinciding with h up to stage s and satisfying $\boldsymbol{M}, h', s \models \psi$. ■

[27] There are also other ways to go. Baltag, Smets & Zvesper (2008) show how strong belief in 'dynamic rationality' defines epistemic-doxastic models whose plausibility relations encode the Backward Induction strategy, in a way that avoids the usual 'Paradox of Backward Induction' (see below) by assuming 'incurable optimism' about players' reverting to Rationality later on in the game.

[28] Cf. the contrast in social philosophy between fairness as desired future income distribution and entitlement views calling a society fair if it arose from a history of fair transactions.

Suppose we are at node s in the game, and move a is played in public. Epistemically, this just eliminates some histories from the current set. But there is now also belief revision, as we move to a new plausibility relation $\leq_{i,\ s^\wedge a}$ describing the updated beliefs:

Hard belief update With *hard information*, as in Chapter 7, the new plausibility relation is the old one, restricted to a smaller set of histories. Here are matching recursion laws, where the temporal operator $F_a\varphi$ says that a is the next event on the current branch, and that φ is true on the current branch immediately after a has taken place:

FACT The following principles are valid for hard revision along a tree:

$$F_a <B,i> \varphi \leftrightarrow (F_aT \wedge <B,i>^{FaT} F_a\varphi)$$
$$F_a <B,i>^{\psi} \varphi \leftrightarrow (F_aT \wedge <B,i>^{Fa\psi} F_a\varphi)$$

This is like the dynamic doxastic recursion axioms for hard information in Chapter 7.[29] There are also analogies with temporal protocol axioms for belief revision (Chapter 11, and Dégrémont 2010). As for concrete temporal scenarios driven by hard update, Dégrémont & Roy (2009) present a beautiful analysis of disagreement between agents (cf. Aumann 1976) via iterated announcements of conflicts in belief: Chapter 15 has more details.

Soft game update So far, our dynamics is driven by hard information from observed moves. An analogue for soft information and *plausibility changes*, the more exciting theme in Chapter 7, calls for new events as soft triggers – say, signals that players receive in the course of a game. Recursion axioms will then reflect revision policies.[30]

Our final illustration is such a soft reconstruction of our running example. Again we give just one illustration, leaving deeper logical theory for another occasion.

Backward induction in a soft light So far we gave two analyses of Backward Induction. The first in Section 10.5 was a modal frame correspondence with a rationality assertion. The second in Section 10.10 was dynamic, capturing

[29] This is a principle of *Coherence*: the most plausible histories for i at h', s' are the *intersection* of B_t with all continuations of s'. Cf. Bonanno (2007) on temporal *AGM* theory.

[30] In Chapter 11, we study belief revision under soft information in branching trees.

the *BI*-path as a limit of successive hard announcements *!RAT* of Rationality. But perhaps the most appealing take on the *BI* strategy uses soft update (van Benthem 2009b). All information produced by the algorithm is in the binary *plausibility relations* created inductively for players among end nodes in the game.[31] For concreteness, consider our running example once more:

Example The bad Nash equilibrium, hard and soft.
The hard scenario, in terms of events *!RAT* removes nodes from the tree that are strictly dominated by siblings as long as this can be done, resulting in the following stages:

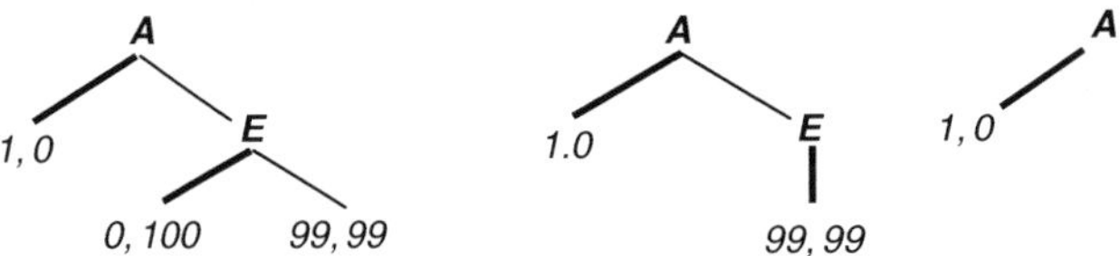

By contrast, a soft scenario does not eliminate nodes but *modifies the plausibility relation*. We start with all end-points of the game tree incomparable (other versions would have them equiplausible). Next, at each stage, we compare sibling nodes, using the following notion:

> A turn *x* for player *i dominates* its sibling *y in beliefs* if the most plausible end nodes reachable after *x* are all better for the active player than all the most plausible end nodes reachable after *y*.

*Rationality** (*RAT**) is the assertion that no player plays a move that is dominated in beliefs. Now we perform a relation change that is like a radical upgrade $\Uparrow$*RAT** as in Chapter 7:

> If *x* dominates *y* in beliefs, we make all end nodes from *x* more plausible than those reachable from *y*, keeping the old order inside these zones.

This changes the plausibility order, and hence the dominance pattern, so iteration can start. Here are the stages for this procedure in our example:

[31] There is an easy correspondence between strategies as subrelations of the total *move*-relation and 'move-respecting' total orders of endpoints. For details, cf. van Benthem & Gheerbrant (2010).

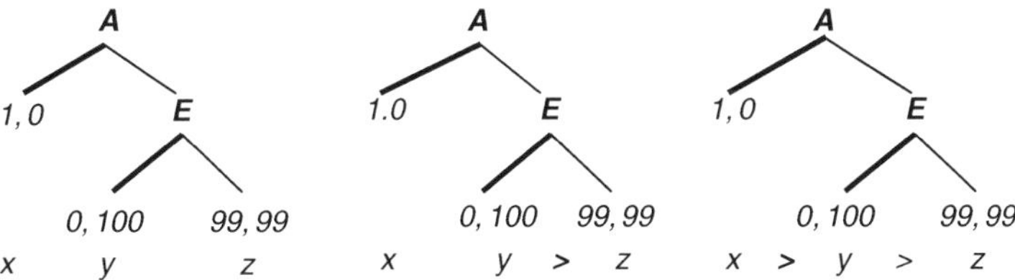

In the first game tree, going right is not yet dominated in beliefs for *A* by going left. *RAT** only has bite at *E*'s turn, and an upgrade takes place that makes *(0, 100)* more plausible than *(99, 99)*. After this upgrade, however, going right has now become dominated in beliefs, and a new upgrade takes place, making *A*'s going left most plausible. ∎

THEOREM On finite trees, the Backward Induction strategy is encoded in the plausibility order for end nodes created by iterated radical upgrade with rationality-in-belief.

At the end, players have acquired *common belief in rationality*.

Other revision policies Backward Induction is just one scenario for creating plausibility in a game. To see alternatives, consider what has been called a paradox in its reasoning. Assuming the above analysis, we expect a player to follow the *BI* path. So, if she does not, we must revise our beliefs about her reasoning, so why would we assume that she will play *BI* later on? Backward Induction seems to bite itself in the tail. This is not a total inconsistency: the dynamic scenarios that we have given make sense. Even so, consider a concrete example:

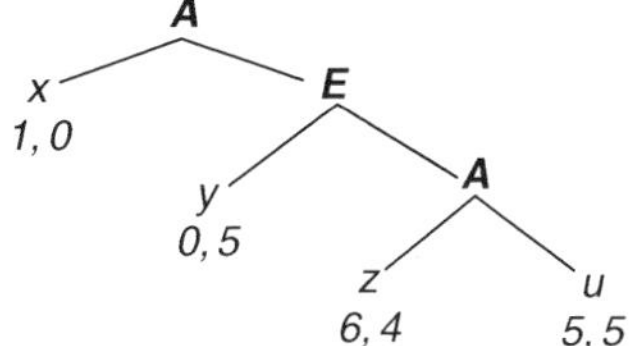

Backward Induction tells us that *A* will go left at the start. So, if *A* plays *right*, what should *E* conclude? There are many different options, such as ‘it was just an error, and *A* will go back to being rational’, ‘*A* is trying to tell me that he wants me to go right, and I will surely be rewarded for that’, ‘*A* is an automaton with a general rightward tendency’, and so on. Our dynamic logics do not choose for the agent, and support many policies, in line with Stalnaker's view on rationality in games as tied to players' policies of belief revision (Stalnaker 1999).[32]

10.12 Conclusion

We have seen how games invite logics of action, knowledge, belief, and preference. In particular, we showed how, once the static base has been chosen, our dynamic logics can deal with entanglement of all these notions, as they play in interactive agency, not just locally, but also through time. To make these points, we gave pilot studies rather than grand theory – but some of our themes will return in Chapters 11, 15.

Even at this stage, we hope to have shown that logical dynamics and games are a natural match. Indeed, we found many interfaces between logic and game theory that may lead eventually to something new, a *Theory of Play*.

10.13 Coda: game logics and logic games

What is the point of logical definitions for game-theoretic notions? The way I see it, logic has the same virtues as always. By formalizing a practice, we see more clearly what makes it tick. In particular, logical syntax is a perspicuous high-level way of capturing the gist of complex game-theoretic constructions. For instance, our modal Zermelo formula gave the essence of the recursive solution process in a nutshell. Zvesper (2010) has other examples where simple modal fixed-point rules summarize sophisticated complex game-theoretic proofs. Of course, there is no unique best game logic, and first-order logic makes sense, too. Moreover, a logical stance also suggests new notions about games, as it goes far beyond specific examples (cf. Chapter 15). Finally, logical results about expressive power, completeness, and complexity can be used for model checking, proof search, and other ways of analysing players. Still, in this book, games are only one example of rational agency, and our logic style of analysis has broader aims than serving game theory.

But there is also another connection: less central to our study of information dynamics, but quite important in general. As we saw in Chapter 2, basic notions of logic themselves have a game character, with logic games of

[32] I admit that one natural reaction to these surprise events is a switch to an entirely *new reasoning style* about agents. That might require more finely grained syntax-based views.

argumentation, evaluation, or model comparison. Thus, logic does not just *describe games*, it also *embodies games*. Only in this dual perspective, the true grit of the logic and games connection becomes clear.[33]

10.14 Further directions and open problems

We list a few topics broadening the interface of logical dynamics and games.

Designing concrete knowledge games The games studied in this chapter are all abstract. But many epistemic and doxastic scenarios in earlier chapters also suggest concrete games, say, of being the first to know certain important facts. It would be of interest to have such 'gamifications' of our dynamic logics: Ågotnes & van Ditmarsch (2009) is a first study.

Dynamics of rationalization We are often better at rationalizing behaviour afterwards than in predicting it. Given minimally rational strategies, there is an algorithm for creating preferences that make these the Backward Induction outcome – and even with given preferences, one can postulate beliefs that make the behaviour rational (van Benthem 2007g). One would like to understand this dynamics, too, in terms of our logics.

Dynamics in games with imperfect information This chapter distinguished two kinds of knowledge: one linked to past observation in imperfect information games, the other to uncertainty about how a game will proceed. We did not bring the two views together in our dynamic perspective. Now, imperfect information games are a difficult area of study. We just illustrate the task ahead with two simple scenarios (with the order '*A-value, E-value*'):[34]

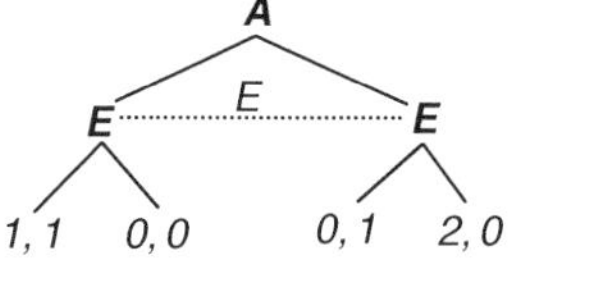

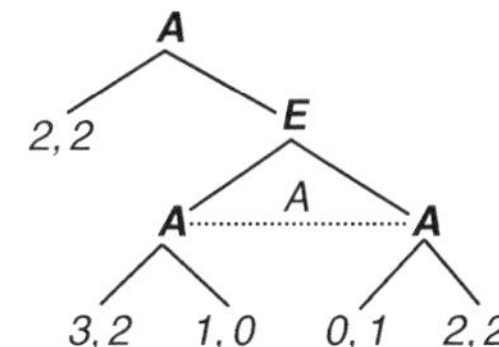

The game to the left seems a straightforward extension of our techniques with dominance, but the one to the right raises tricky issues of what *A* would

[33] This duality is the main thrust of the companion volume van Benthem (to appearA).

[34] The tree to the right is slightly adapted from an invited lecture by Robert Stalnaker at the *Gloriclass Farewell Event*, ILLC Amsterdam, January 2010.

be telling *E* by moving right. We leave the question of what should or will happen in both games to the reader. Dégrémont (2010), Zvesper (2010) have more extensive discussion of what *DEL* might say here.

Groups and parallel action We have not discussed *coalitions* and group agency in games. A very simple desideratum here would be a treatment of games with *simultaneous moves*. Nothing in our dynamic logics prevents either development. In fact, it is easy to extend *DEL* and *PDL* with compound actions that are tuples of acts for each player, adding some appropriate vocabulary accessing this structure. It just has not been done yet.

Bounded rationality We characterized the *DEL*-like games of imperfect information in terms of players with Perfect Recall observing public moves. But there are many other types of player, and the treatment of this chapter needs to be extended to cope with *agent diversity* in powers of memory (cf. Chapters 3, 4) and inferential resources (cf. the syntax-based dynamics of Chapter 5) for choosing their strategies toward others.

Infinite games All games in this chapter are finite. Can we extend our dynamic logics to deal with *infinite games*, the paradigm in evolutionary game theory (Hofbauer & Sigmund 1998) and much of computer science? A transition to the infinite was implicit in the iterated announcement scenarios of Chapter 3, and it also arises in modal process theories (Bradfield & Stirling 2006). See also Chapter 11, where we will fit *DEL* into Grand Stage views of temporally evolving actions, knowledge, and belief.

Games equivalence and other logics Beyond extensive games, there are *strategic games* with only strategy profiles and their outcomes. A deeper study would raise the question *when two games are the same* (van Benthem 2002). Consider the following two games:

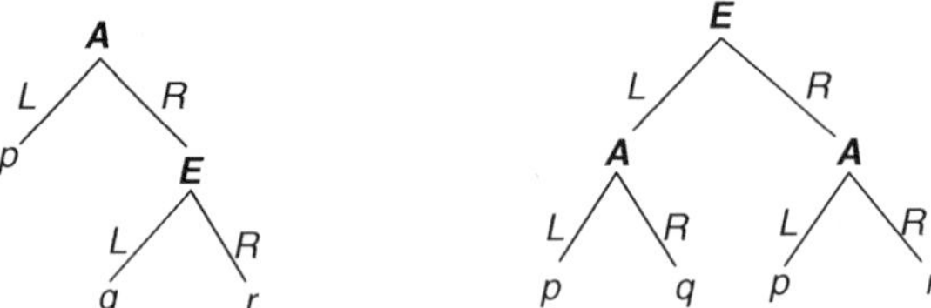

Are these the same? As with computational processes, the answer depends on our level of interest. If we focus on turns and *local moves*, then the games are not equivalent – and some sort of bisimulation would be the right invariance, showing a difference in some matching modal language. But if we focus on players' *powers*, the two games are the same. In both, using their strategies players can force the same sets of outcomes. Player *A* can force the outcome to fall in the sets $\{p\}$ or $\{q, r\}$, *E* can force the outcome to fall in the sets $\{p, q\}$ or $\{p, r\}$. This time, the right invariance is *power bisimulation*, and the matching language uses the global forcing modalities defined earlier in this chapter (van Benthem 2001).[35] All these issues become more delicate in the presence of *preferences*. Van Benthem (2002a) looks at intuitions of preference-equivalence, or supporting the same equilibria. Thus, this chapter has only told part of a much larger story of logic and games. Chapter 15 is a case study of logical dynamics on games in strategic form.

Model theory, fixed-point logics, and complexity There are many questions about the games in this chapter that invite standard mathematical logic. For instance, little is known about preservation behaviour of assertions about players and strategies across various *game transformations*, including just going to subgames. Say, if we know what equilibria exist in one game, what can we say systematically about those in related games?

Another source of logical questions are solution procedures. For instance, our hard and soft dynamic scenarios for Backward Induction raise the issue of how their limits are definable. Working on finite games, van Benthem & Gheerbrant (2010) define the update steps by means of suitable formulas, connect our various construals, and show that sometimes *monotone fixed-point logic LFP(FO)* can define limits, while for most limits of iterated hard or soft update, *inflationary fixed-point logic IFP(FO)* is more natural. (Gurevich & Shelah (1986), Kreutzer (2004) show that these logics have the same expressive power – but their styles differ. Cf. Chapter 15 for related issues with strategic games.) They also give an alternative analysis, pointing out how the notions of rational game solution in this chapter define unique relations by recursion on a well-founded order on finite trees, namely, the composition of the *sibling* and *tree dominance* relations. The dissertation Gheerbrant (2010) has further technical details.

One important issue is *which fragments* of such strong logics for inductive definition are needed in the theory of game solution. One intriguing

[35] Parikh (1985) has a complete modal game logic based on power structures.

question here concerns the role of *Rationality*, usually seen as a way of simplifying theory. Its logical syntax seems significant, since it gives a Confluence diagram with moves and preference. This suggests a similarity with the Tiling problems of Chapter 2 that generated high *complexity* of logics. Are logics for games with rational players essentially complex?

The gist of it all: modal logics of best action Our final open problem is about a modal logic, just a fragment of sophisticated fixed-point logics. In the spirit of mathematical abstraction, it would be good to extract a simple surface logic for reasoning with the ideas in this chapter, while hiding most of the machinery. Now, when all is said and done, game solution procedures take a structure with actions and preferences, and then compute a new relation that may be called *best action*. And the latter, of course, is the only thing that we agents are really after (cf. the 'logic of solved games' in van Otterloo 2005):

> *Open problem* Can we axiomatize the modal logic of finite game trees with a *move* relation and its transitive closure, turns and *preference* relations for players, and a new relation *best* as computed by Backward Induction?

My guess would be that we get a simple modal logic for the moves plus a preference logic as in Chapter 9, while the modality $<best>$ would satisfy one major bridge axiom that we have seen earlier:

$$(turn_i \wedge <best> [best^*](\mathbf{end} \rightarrow p)) \rightarrow [move\text{-}i] <best^*> (\mathbf{end} \wedge <pref_i> p).$$[36]

10.15 Literature

This chapter stands in a vast literature. Just a tiny sample of congenial papers in game theory, philosophy, and computer science are Aumann (1976), Geanakoplos & Polemarchakis (1982), Battigalli & Siniscalchi (1999), Bonanno (2001), Stalnaker (1999), Halpern (2003), and Brandenburger & Keisler (2006). Van der Hoek & Pauly (2006) is a handbook survey of interfaces of games with modal logic. Other good sources are the dissertations Pauly (2001), de Bruin (2004), Harrenstein (2004), Dégrémont (2010), Zvesper (2010), and the

[36] It may be technically necessary here to add a 'conditionalized' version of the *best* action modality, like we did for common knowledge in Chapter 3.

textbook Pacuit & Roy (2010). The material in this chapter is based largely on van Benthem (2001, 2002a, 2004a, 2009b). Further *DEL* perspectives on extensive games are found in Baltag (2001), Baltag (2002), Baltag, Smets & Zvesper (2009). Finally, van Benthem (1999a) is a survey of both game logics and logic games, and van Benthem (to appearA) is a monograph on recent interfaces between logic and game theory.

11 Processes over time

The preceding chapters took our study of rational agency from single update steps to mid-term activities like finite games that mix agents' actions, beliefs, and preferences. In the limit, this leads to long-term behaviour over possibly infinite time, that has many features of its own. In particular, in addition to information about facts, agents can now have procedural information about the process they are in. This chapter makes a junction between dynamic epistemic logic and temporal logics of discrete events, occurring in philosophy, computer science, and other disciplines. We prove semantic representation theorems, and show how dynamic-epistemic languages are fragments of temporal ones for the evolution of knowledge and belief. Amongst other things, this gives a better understanding of the balance between expressive power and computational complexity for agent logics. We also show how these links, once found, lead to merges of ideas between frameworks, proposing new systems of *PAL* or *DEL* with informational protocols.

11.1 Dynamic-epistemic logic meets temporal logics

The Grand Stage The following global view has surfaced at various places in Chapters 4, 10, in branching tree-like pictures for agents over time:

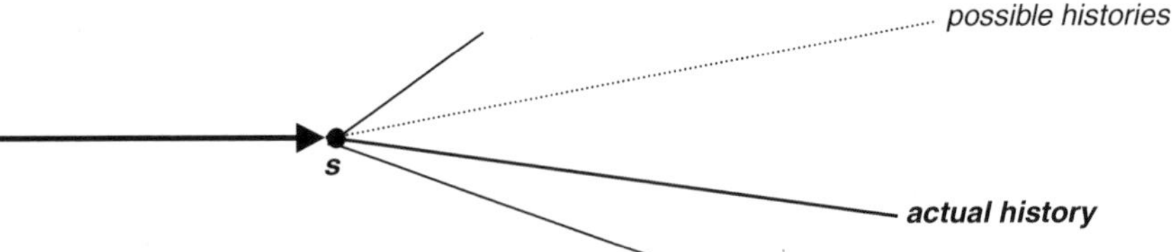

Branching temporal models are a Grand Stage view of agency, with histories as complete runs of some information-driven process, described by languages with epistemic and temporal operators.[1] The Grand Stage is a natural habitat for the local dynamics of *DEL*, and this chapter brings the two views together. Temporal trees can be created through constructive unfolding of an initial epistemic model M by successive product updates $M \times E$ with event models E (cf. the 'update evolution' of Chapter 4), and we will determine which trees arise in this way. Thus, *DEL* adds fine-structure to temporal models. We will use this to connect facts about epistemic-temporal logics and our findings about *DEL*.

Protocols Linking frameworks leads to flow of ideas. Our key example will be *protocols*, constraints on possible histories of an agent process. Message-passing systems may demand that only true information is passed, or each request is answered. In conversation, some things cannot be said, and there are rules like 'do not repeat yourself', 'let others speak in turn'. Restricting the legitimate sequences of announcements affects our logics:

Example Contracting consecutive assertions.
A nice *PAL*-validity in Chapter 3 stated that the effect of two consecutive announcements $!P$, $!Q$ is the same as that of the single announcement $!(P \wedge [!P]Q)$. This equivalence may fail in protocol-based models, as the latter trick assertion may not be an admissible one. ■

Protocols occur in puzzles (the Muddy Children made only epistemic statements), games, and learning. Physical experiments, too, obey protocols, in line with our broader view of *PAL* and *DEL* as logics of observation. Finally, knowing a protocol is a new form of *procedural information* (cf. Chapter 5) beyond information about facts and other agents.

11.2 Basics of epistemic-temporal logic

Temporal logics come in flavours (cf. Fagin *et al.* 1995; Halpern, van der Meyden & Vardi 2004; Hodkinson & Reynolds 2006; van Benthem & Pacuit 2006). Chapter 9 used complete branches (perhaps infinite) for actual histories plus finite stages on them. In this chapter, we use only finite histories as indices of evaluation, living in a modalized future of possible histories extending the current one.

[1] This view underlies frameworks like Interpreted Systems (Fagin *et al.* 1995), Epistemic-Temporal Logic (Parikh & Ramanujam 2003), *STIT* (Belnap, Perloff & Xu 2001), or Game Semantics (Abramsky 2008).

Models and language Take sets $\mathbb{A}$ of agents and $\mathbb{E}$ of events (usually finite). A *history* is a finite sequence of events, and E^* is the set of all histories. Here *he* is history *h* followed by event *e*, representing the unique history after *e* has happened in *h*. We write $h \leq h'$ if *h* is a prefix of *h'*, and $h \leq_e h'$ if $h' = he$. Our first semantic notion represents protocols.[2]

DEFINITION *ETL* Frames.

A *protocol* is a set of histories $\mathbb{H} \subseteq E^*$ closed under prefixes. An *ETL frame* is a tuple $(E, H, \{\sim_i\}_{i \in A})$ with a protocol $\mathbb{H}$, and accessibility relations $\sim_i$. An *ETL-model* is an *ETL*-frame plus a valuation map *V* sending proposition letters to sets of histories in $\mathbb{H}$. ■

An *ETL* frame describes how knowledge evolves over time in some informational process. The relations $\sim_i$ represent uncertainty of agents about how the current history has gone, due to their limited powers of observation or memory. Thus, $h \sim_i h'$ means that from agent *i*'s point of view, the history *h'* looks the same as the history *h*.

An *epistemic temporal language* L_{ETL} for these structures extends *EL* from Chapter 2 with event modalities. It is generated from a set of atomic propositions *At* by the following syntax:

$$p \mid \neg\varphi \mid \varphi \wedge \psi \mid [i]\varphi \mid \langle e \rangle \varphi \qquad \text{where } i \in A, e \in E \text{ and } p \in At.$$

Here $[i]\varphi$ stands for $K_i\varphi$. Booleans, and dual modalities $\langle i \rangle$, $[e]$ are as usual.[3]

DEFINITION Truth of L_{ETL} formulas.

Let $\mathbf{M} = (E, H, \{\sim_i\}_{i \in A}, V)$ be an *ETL* model. The truth of a formula φ at a history $h \in \mathbb{H}$, denoted $\mathbf{M}, h \models \varphi$, is defined inductively as usual, with the following key clauses:

(a) $\mathbf{M}, h \models [i]\varphi$	iff	for each $h' \in \mathbb{H}$, if $h \sim_i h'$, then $\mathbf{M}, h' \models \varphi$
(b) $\mathbf{M}, h \models \langle e \rangle \varphi$	iff	there exists $h' = he \in \mathbb{H}$ with $\mathbf{M}, h' \models \varphi$ ■

[2] In what follows, a protocol is a family of finite histories. A more general setting would allow for *infinite histories*, where protocols need not reduce to such finitely presented ones.

[3] We can add group operators of distributed or common knowledge, as in earlier chapters. In temporal logic, such extensions can have dramatic effects on the complexity of validity: as we shall see below.

Agent properties Further constraints on models reflect special features of agents, or of the informational process of the model.[4] These come as conditions on epistemic and action accessibility, or as epistemic-temporal axioms matched by modal frame correspondences. Here are some examples from earlier chapters (we suppress indices for convenience):

FACT The axiom $K[e]\varphi \rightarrow [e]K\varphi$ corresponds to *Perfect Recall*:
if $he \sim k$, then there is a history h' with $k = h'e$ and $h \sim h'$.[5]

This says that agents' current uncertainties can only come from previous uncertainties: a strong form of perfect memory. An induction on distance from the root then derives

Synchronicity: uncertainties $h \sim k$ only occur between h, k at the same tree level.

Weaker forms of Perfect Recall in game theory lack Synchronicity, allowing uncertainty links that cross between tree levels. Note that the axiom presupposes perfect observation of the current event e: in *DEL*, it would not hold, as uncertainty can also be created by the current observation, when some event f is indistinguishable from e for the agent. This point will return in our analysis below. In a similar fashion, we have a dual fact:

FACT The axiom $[e]K\varphi \rightarrow K[e]\varphi$ corresponds to *No Miracles*:
for all ke with $h \sim k$, we also have $he \sim ke$.[6]

This principle is sometimes called 'No Learning', but its content is rather that learning can take place, but only by observing new events to resolve current uncertainties.

Epistemic-temporal languages also describe other agents. Take a *memory-free* automaton that only remembers the last-observed event, making any two histories he, ke ending in the same event epistemically accessible. Then, with finitely many events, knowledge of the automaton can be defined in the temporal part of the language. Using backward modalities P_e plus a universal modality U over all histories, we have the equivalence

[4] The border line can be vague: am I clever as an agent, or thanks to the process I am in?

[5] The elementary proof uses a simple modal substitution argument. Details simplify by assuming, as in our tree models, that transition relations for events e are *partial functions*.

[6] When we write h, he, etc., we always assume that these histories occur in the protocol.

$$K\varphi \leftrightarrow \vee_e(<e^{\cup}>T \wedge U(<e^{\cup}>T \rightarrow \varphi))$$

Similar ideas work for bounded memory in general (Halpern & Vardi 1989; Liu 2008). Thus, properties of processes and types of agents meet with epistemic-temporal languages.[7]

11.3 A basic representation theorem

Now we can state how *DEL* and *ETL* are related. Recall the scenario of *Update Evolution* in Chapter 4: some initial epistemic model **M** is given, and it then gets transformed by the gradual application of event models $E_1, E_2, \ldots$ to form a sequence

$$\mathbf{M}_0 = \mathbf{M},\ \mathbf{M}_1 = \mathbf{M}_0 \times E_1,\ \mathbf{M}_2 = \mathbf{M}_1 \times E_2, \ldots$$ [8]

It helps to visualize this in trees, or rather forest pictures like the following:

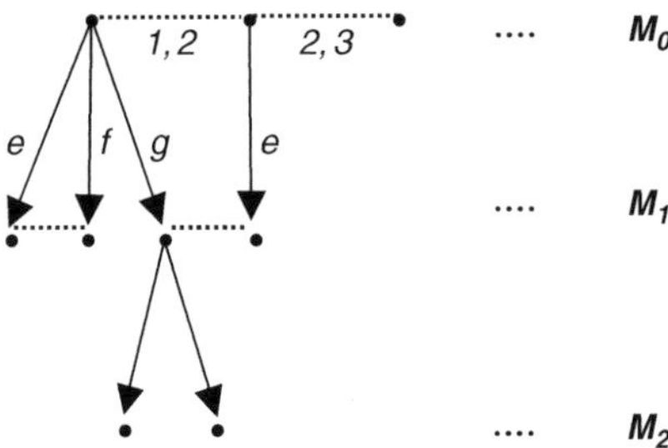

where stages are horizontal, while worlds may extend downward via 0, 1, or more event successors. Through product update, worlds in these models arise from successive pair formation, forming finite sequences starting with one world in the initial epistemic model **M** followed by a finite sequence of events that were executable when their turn came. But that means that these worlds are just histories in the sense of the above *ETL* models.

DEFINITION Induced *ETL* forests.
Given an epistemic model **M** and a finite or countable sequence of event models *E*, the *induced ETL-model Forest(***M***, E)* has as its histories all finite

[7] Further questions arise with common knowledge. For instance, if we assume Perfect Recall for all individual agents separately, it also holds for the group: $C_G[e]\varphi \rightarrow [e]C_G\varphi$ becomes valid.

[8] An important special case had one event model *E* throughout. Van Benthem & Liu (2004) suggest that the latter simple format suffices for mimicking the more general approach here.

sequences $(w, e_1, \ldots, e_k)$ produced by successive product update, with accessibility relations and valuation as in *DEL*. ■

Drawing pictures shows how this works. In particular, induced *ETL*-models have a simple protocol $\mathbb{H}$ given by all the finite sequences that pass the local requirements of the update rule. Accordingly, they have three striking properties making them stand out:

FACT ETL-models **H** of the special form *Forest*(**M**, **E**) satisfy the following three principles, where quantified variables h, h', k, ... range only over histories that are present in **M**:

(a) If $he \sim k$, then there is some f with $k = h'f$ and $h \sim h'$ — *Perfect Recall*
(b) If $h \sim k$, and $h'e \sim k'f$, then $he \sim kf$ — *Uniform No Miracles*
(c) The domain of any event e is definable in the epistemic base language. — *Definable Executability*

Now, the crucial point is that this can be converted to a representation result for *DEL* inside *ETL* (van Benthem 2001; van Benthem & Liu 2004):

THEOREM For *ETL* models **H**, the following two conditions are equivalent:

(a) **H** is isomorphic to some model *Forest*(**M**, **E**),
(b) **H** satisfies Perfect Recall, Uniform No Miracles, and Definable Executability.

Proof The direction from (a) to (b) is the preceding Fact. Conversely, consider any *ETL*-model **H** satisfying the three conditions. We define an update sequence as follows:

(i) **M** is the set of histories in **H** of length 1, copying their given epistemic accessibilities and valuation,
(ii) $\mathbf{E}_k$ is the set of events occurring at tree level $k+1$ in **H**, setting $e \sim f$ if there exist histories h, k of length k with $he \sim kf$ in **H**. Definability of preconditions is the Definable Executability.

We prove by induction that the tree levels $\mathbf{H}_k$ at depth k of the *ETL* model **H** are isomorphic to the epistemic models $\mathbf{M}_k = \mathbf{M} \times \mathbf{E}_1 \times \ldots \times \mathbf{E}_{k-1}$. The crucial fact is this, using our definition and the first two given properties (here (s, e) is the same history as 'se'):

$$(s, e) \sim_{H_k} (t, f) \quad \text{iff} \quad (s, e) \sim_{\mathbf{M}_k} (t, f)$$

From left to right. By Perfect Recall, $s \sim t$ in H_{k-1}, and so by the inductive hypothesis, $s \sim t$ in $\mathbf{M}_{k-1}$. Also, by our definition, $e \sim f$ holds in E_k. Then by the forward half of the Product Update rule, we have that $(s, e) \sim_{\mathbf{M}_k} (t, f)$. *From right to left.* By the other half of Product Update, $s \sim t$ in $\mathbf{M}_{k-1}$, and by the inductive hypothesis, $s \sim t$ in H_{k-1}. Next, since $e \sim f$, by our definition, there exist histories i, j with $ie \sim jf$ in H_k. By Uniform No Miracles then, $se \sim tf$ holds in H. ■

This result stipulates definability for preconditions of events e, i.e., the domains of the matching partial functions in the tree H. Here is a more purely structural version:

THEOREM The preceding theorem still holds when we replace Definable Executability by *Bisimulation Invariance*: that is, closure of event domains under all purely epistemic bisimulations of the *ETL*-model H.

The proof follows from two facts in Chapter 2: (a) epistemically definable sets of worlds are invariant for epistemic bisimulations, and (b) each invariant set has an explicit definition in the *infinitary* epistemic language.[9] Our two results tell us how special *DEL* update is as a mechanism generating epistemic-temporal models. It is about idealized agents with perfect memory and driven by observation only, while their informational protocols involve only local epistemic conditions on executability.

Variations This is just a starting point. In particular, a mild relaxation of the definability requirement for events allows more general preconditions referring to the epistemic past beyond local truth. Think of conversation with no repeated assertions: this needs a memory of what was said that need not be encoded in a local state. Also, other styles of representations make sense, representing *ETL*-models only up to epistemic-temporal *bisimulation*. Finally, our proof method can also characterize effects of other update rules, such as the earlier $(s, e) \sim (t, f)$ *iff* $e \sim f$ for memory-free agents.

[9] While this only guarantees finite epistemic definitions for preconditions on finite models, we feel that further tightening of conditions adds no further insight.

11.4 Temporal languages: expressive power and complexity

DEL as a temporal language The conditions in our representation theorem for *DEL* evolution as *ETL* models suggest definability in matching epistemic-temporal languages. We saw how Perfect Recall corresponds to a syntactic operator switch between knowledge and action modalities.[10] Indeed, what is the language of *DEL* in Chapter 4 and following when viewed as an epistemic-temporal formalism? This is quite easy to answer:

> *DEL* is a static knowledge language for individual and collective epistemic agents, plus a *one-step future operator* saying what is going to hold after some specified next event.

This is less than what can be said in general epistemic-temporal logics. The latter also have *past* operators, as we saw with event preconditions for extended protocols.[11] And typically also, a temporal language can talk about the whole *future*, and effects of arbitrary finite sequences of evens from now on. This is crucial in specifying how an information process is to behave, in terms of 'safety' and 'liveness' properties. A complete repertoire of temporal modalities will depend on how one views the model, and agents living inside it. For instance, with the earlier Synchronicity, it also makes sense to have a modality $<=>\varphi$ of *simultaneity* saying that φ is true at some history of the same length.

The balance with complexity But then we meet the Balance discussed in Chapter 2. Increases in expressive power may lead to upward jumps in computational *complexity* of combined logics of knowledge and time. The first investigation of these phenomena was made in Halpern & Vardi (1989). Here is a table with a few observations from their work showing where the dangerous thresholds lie for the complexity of validity:

	K, P, F	K, C_G, F_e	K, C_G, F_e, P_e	K, C_G, F
All *ETL* models	*decidable*	*decidable*	*decidable*	*RE*
Perfect Recall	*RE*	*RE*	*RE*	Π^1_1-complete
No Miracles	*RE*	*RE*	*RE*	Π^1_1-complete

[10] Facts like these suggest a general modal correspondence theory on *ETL*-frames.

[11] A one-step past modality Y also occurred in Chapter 3, where a public announcement $!\varphi$ achieved common knowledge that φ was true just before the event: $[!\varphi]C_G Y\varphi$.

Here complexities run from decidable through axiomatizable (*RE*) to Π^1_1-complete, which is the complexity of truth for universal second-order statements in arithmetic.[12] The latter complexity is often a worst case for modal logics, witness Chapter 2. Van Benthem & Pacuit (2006) is a survey of expressive power and complexity in connection with *DEL*, citing much relevant background in work on tree logics and products of modal logics.

Dangerous agent properties As we just saw, epistemic-temporal logic over all *ETL*-models is simple even with rich vocabularies, but things change with special assumptions on agents such as Perfect Recall. The technical explanation is *grid encoding* (Chapter 2). Essentially, Perfect Recall[13] makes epistemic accessibility and future moves in time behave like a grid model of type $IN \times IN$, with cells enforced by a confluence property that we have seen already (Chapters 3, 10), as pictured in the uncertainty-action diagram

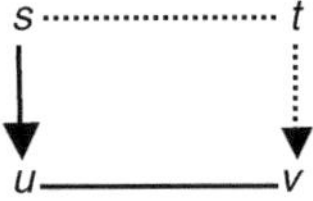

Satisfiability in the language can then express the Recurrent Tiling Problem that is known to have $\sum^1_1$-complete complexity. But to really encode the tiling argument, the language needs sufficient expressive power, in particular, a *universal quantifier* ranging over all points in the grid. This can be supplied by combining an unbounded future modality in the tree, plus a common knowledge modality accessing all reachable points at the same horizontal tree level. If one of these resources is not available, however – say we have common knowledge but no unbounded future, complexity may drop, as shown in the table.

Technical points The balance in these results is subtle, partly because of a tension between two pictures in Chapter 2. As trees, our models should have simple logics, by Rabin's Theorem on the decidability of the monadic second-order logic of trees with the relation of initial segment and partial successor

12 There seems to be a gap between the complexities *RE* and Π^1_1-*complete*, few epistemic-temporal logics fall in-between. This also occurs with extensions of first-order logic, where being able to *define a copy of the natural numbers IN* is a watershed. If you cannot, like first-order logic, complexity stays low: if you can, like first-order fixed-point logics or second-order logic, complexity jumps.

13 Similar observations to all that follows hold for the converse principle of No Miracles.

functions. Indeed, pure process theories can be simple in temporal logic. But the process theory of epistemic agents that handle information adds a second relation, of epistemic accessibility, and then a tree may carry grid structure after all. Van Benthem & Pacuit (2006) explain how Perfect Recall supports tiling proofs even though it just requires basic cell structure downward in a tree.[14] Also, there is a difference between our *forest* models, where the first level may have many starting points (the worlds in the initial epistemic model ***M***), and *trees* with just one root. To get grids in trees with single roots and perhaps finite horizontal levels, we must create cells by a trick:

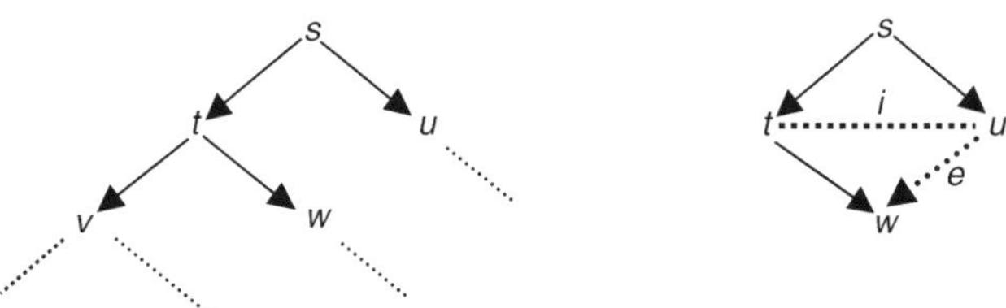

Here we reach the key bottom corner of a cell by an epistemic move plus an event move, as shown to the right. To make use of the latter, the language needs to mix epistemic and temporal steps in patterns $(\sim_i ; e)^*$. This requires a *propositional dynamic logic* PDL_{et} with both epistemic accessibility and temporal event moves as basic transition relations:

THEOREM The validity problem for PDL_{et} is Π^1_1-complete.

The proof is in van Benthem & Pacuit (2006). This is one of the many points in this book where *PDL* program structure makes sense.

DEL as an ETL-logic Against this background, we can now place *DEL* and understand its behaviour in this book. Its language is the *K*, C_G, F_e slot in the earlier table, over models satisfying Perfect Recall and No Miracles. Thus, there is grid structure, but the expressive resources of the language do not exploit it to the full, using only one-step future operators <*!P*> or <***E***, *e*>. If we add unbounded future, the same complexity arises as for *ETL*. Indeed,

[14] Models with Perfect Recall suffice for tiling *arbitrary finite submodels* of N × N – and, by Koenig's Lemma, the latter suffices for the existence of a complete tiling. Some tricks are needed placing literals p, $\neg p$ to ensure existence of enough branches.

Miller & Moss (2005) show that the logic of just public announcement with common knowledge and Kleene iteration of assertions *!P* is Π^1_1-complete.[15]

But more interesting is a language extension that came up earlier. It seems obvious that adding one-step *past* does not endanger decidability. But even beyond, we have this

Open Problem Does *DEL* stay decidable over *ETL* tree models when we add an *unbounded past* operator that can only go back finitely many steps to the root?

Discussion: complexity of agents How do complexity results for logics relate to agency? They tell us delicate things about richness of behaviour. Take the following paradox(ette). How can the logic of ideal agents with perfect memory be so highly complex, while the logic of arbitrary agents is simple, witness the first line in the above Table? A moment's reflection dissolves the paradox. The general logic describes what is true for all agents: a simple story.[16] For some special kinds of agent, that logic stays simple, as we saw with bounded memory, whose epistemic-temporal logic was embedded in the pure temporal logic of our models. But agents with perfect memory are so regular that an *ETL*-record of their activities shows grid patterns that encode arithmetical computation – and the Π^1_1-completeness says that understanding this behaviour requires substantial mathematics.

To us, this computational perspective on agency is more than a formal tool. In Chapter 14, we discuss real analogies between computation and conversation. For some remaining worries on the computational complexity of agency itself: see the end of this chapter.

11.5 Adding protocols to dynamic-epistemic logic

Now that we have things in one setting, we can go further and create merges, transferring ideas from one framework to another. The preceding sections showed how *DEL* adds fine-structure to *ETL*. Product update is a mechanism for creating temporal models, and *DEL* imports some of that model structure into the object language, where it becomes subject to explicit manipulation.

15 The Miller & Moss result leaves a loophole. It is not known what happens exactly to the logic over families of finite models and their submodels reachable by public announcements.

16 This no longer holds if we can explicitly define *agent types* inside the language.

In doing so, *DEL* type languages suggest new fragments of *ETL* and other process languages, providing concrete new systems for investigation.

Protocols In the opposite direction, a notion missing in *DEL* is that of a temporal *protocol* defining or constraining the informational process agents are in. Clearly, this procedural information (cf. Chapters 3, 13) is crucial to agency. Now *DEL* does have *preconditions* constraining which events are executable where, cutting down on possible histories. Van Benthem & Liu (2004) suggest that this can represent most natural protocols, especially if we go a bit beyond preconditions in the pure epistemic language. But this approach only works for protocols that are locally defined restrictions on events.[17]

DEL protocol models Much more can be learnt by also having *DEL* protocols in a straightforward *ETL* style, an idea first proposed in Gerbrandy (1999a). What follows is based on the results in van Benthem *et al.* (2009):

Definition *DEL* protocols.
Let **E** be the class of all pointed event models. A *DEL protocol* is a set $\mathbf{P} \subseteq \mathbf{E}^*$ closed under taking initial segments. Let **M** be any epistemic model. A *state-dependent DEL protocol* is a map **P** sending worlds in **M** to *DEL* protocols. If the protocol assigned is the same for all worlds, the state-dependent *DEL* protocol is called *uniform*. ■

In Chapters 3, 4, the dynamic modalities of *PAL* and *DEL* were interpreted in the total universe of all possible epistemic models, representing total freedom of information flow. But now, protocols in the epistemic-temporal sense restrict the range of reachable models. Though details of this require some technical care, the idea is very simple. We extend the earlier notion of update unfolding as in the following illustration:

Example *ETL* model generated by a uniform *PAL* protocol.
We use a public announcement protocol for graphical simplicity. Consider the following epistemic model **M** with four worlds and agents *1*, *2*:

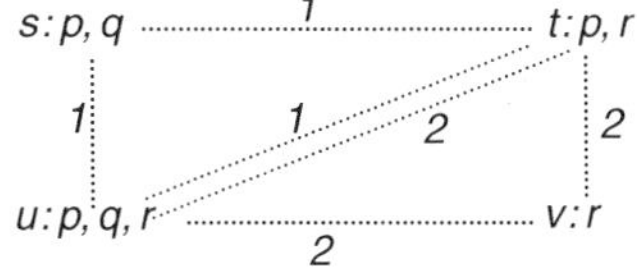

[17] Chapter 8 did a bit more, shifting protocol information into the *definition of the events*, using 'thick events' of forms like *Liar says P* instead of bare public announcements *!P*.

with a protocol **P** = {<!p>, <!p, !q>, <!p, !r>} of the available sequences of announcements or observations. The *ETL* forest model in the following diagram is the obvious update evolution of **M** in the earlier sense, with histories restricted by the event sequences in **P**. Note how some worlds drop out, while others 'multiply':

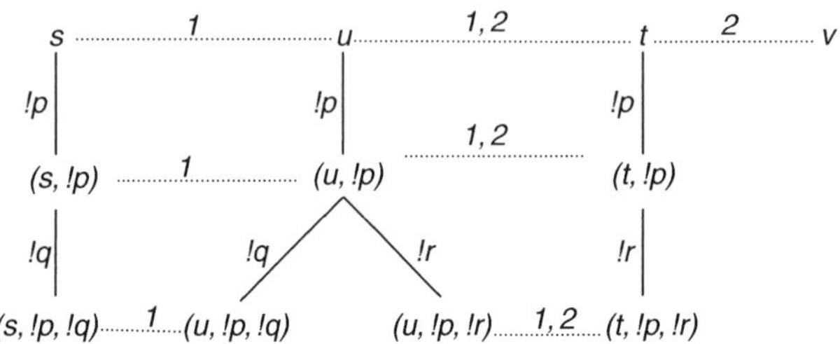

Remark This example also shows the syntactic nature of our protocols. Different formulas *P, Q* lead to different events, even when semantically equivalent. This syntactic flavour, used in Hoshi (2009) for analysing inferential information (cf. Chapter 5), is a disadvantage in some settings – but we leave more semantic definitions as a desideratum.

Now comes the general notion behind this, suppressing some technicalities:[18]

DEFINITION Generated *ETL* models from *DEL* protocols.
Let **M** be an epistemic model and **P** a state-dependent *DEL* protocol on **M**. The *induced ETL* (forest) *model Forest(**M**, **P**)* is defined as follows. Starting from **M** as the first layer, one computes the finite update evolutions of **M** containing only sequences $(w, e_1, \ldots, e_k)$ where preconditions are satisfied as earlier, while now also the event sequence $(e_1, \ldots, e_k)$ must be in the local protocol assigned by **P** to *w*. The complete epistemic tree or forest model is then generated as a straightforward union of these finite stages.[19] ■

State-dependent protocols are very flexible. Uniform protocols are the same in every world, making them *common knowledge*: the more usual assumption in epistemic-temporal logic.[20] But when procedural information is agent-dependent, there are finer distinctions:

[18] The precise definitions in van Benthem *et al.* (2009) run through two pages.

[19] One might also allow protocol constraints at later stages in the update evolution.

[20] This formulation makes more sense when our language can *define* protocols.

Example A protocol that is not common knowledge.
Let the model **M** have two worlds *s*, *t* as depicted here:

$s : p$ .... $t : p, q$

The state-dependent protocol **P** assigns $\{<!p>\}$ to *s* and $\{<!q>\}$ to *t*. In the induced model the formula $<!p>T$ is true at *s*, meaning that the information that *p* holds can be revealed by the protocol. But the agent does not know this procedural fact, since $<!p>T$ is false at *t*, and hence the epistemic-dynamic formula $K<!p>T$ is false at *s*. By contrast, with uniform protocols, a true formula $<!p>T$ is common knowledge: $<!p>T \rightarrow C_G<!p>T$ is valid. ■

Representation revisited Which *ETL*-models are induced by epistemic models with a state-dependent protocol? For the universal protocol of *all possible* finite sequences of event models, we get trees of all models reachable by update from some fixed model **M**. In Section 11.3, we characterized the induced class for uniform protocols, with one sequence of event models without branchings, using Perfect Recall, Uniform No Miracles, and Bisimulation Invariance for epistemic bisimulations. Van Benthem *et al.* 2008 extend this analysis to possibly branching state-dependent protocols.[21]

Translation Finally, our current perspective may also be viewed as a syntactic translation from *DEL* to *ETL* in the following sense. Suppose we are working with the full protocol $\textbf{Prot}_{DEL}$ of all finite sequences of event models. Then we have this equivalence:

FACT For any epistemic model **M**, world *w*, and any formula φ in *DEL*,

$$\mathbf{M}, w \models \varphi \text{ iff } Forest(M, \mathbf{Prot}_{DEL}), <w> \models \varphi.$$

11.6 Determining the logic of *PAL* protocols

Adding *DEL* protocols raises new questions for dynamic-epistemic logic itself.

PAL protocols A telling example is information flow by public announcement. The earlier definitions specialize to protocols for conversation, or experiments where only a few things can be measured, in certain orders.

[21] Local versions of the earlier conditions work, linking histories only to histories reachable by epistemic accessibility. Extracting event models from *ETL* forests takes extra care.

What is the logic of epistemic models plus models reachable by some announcement protocol? Note that *PAL* itself no longer qualifies. It was the logic of arbitrary models subjected to the universal protocol of all announcement sequences. But when the latter are constrained, two axioms from Chapter 3 will fail.

Note In the rest of this chapter, for convenience, protocols only involve pure epistemic formulas without dynamic announcement modalities. This restriction was lifted in Hoshi (2009), and most assertions below can be extended to the full language of *PAL*. We will use existential action modalities for their greater vividness in a procedural setting:

Example Failures of *PAL* valididities.
PAL had a valid axiom $<!P>q \leftrightarrow P \wedge q$. As a special case, this implies

$$<!P>T \leftrightarrow P$$

From left to right, this is valid with any protocol: an announcement *!P* can only be executed when *P* holds. But the direction from right to left is no longer valid: *P* may be true at the current world, but the protocol need not allow public announcement of this fact at this stage. Next, consider the crucial knowledge recursion law, in its existential version

$$<!P><i>\varphi \leftrightarrow (P\wedge<i><!P>\varphi)$$

This, too, fails in general from right to left. Even when *P* is true right now, and the agent thinks it possible that *P* can be announced to make φ true, she need not know the protocol – and a state-dependent protocol need not allow action *!P* in the actual world. ■

The point is this: assertions $<!P>T$ now come to express genuine *procedural information* about the informative process agents are in, and hence, they no longer reduce to basic epistemic statements. Stated more critically, *PAL* only expressed factual and epistemic information, but left no room for genuine procedural information. We now remedy this:

Definition The logic *TPAL*.
The *logic TPAL of arbitrary announcement protocols* has the same language as *PAL*, and its axioms consist of (a) the chosen static epistemic base logic, (b) the minimal modal logic for each announcement modality, and (c) the following modified recursion axioms:

$$<!P>q \leftrightarrow <!P>T \wedge q \qquad \textit{for atomic facts } q$$
$$<!P>(\varphi \vee \psi) \leftrightarrow (<!P>\varphi \vee <!P>\psi)$$
$$<!P>\neg\varphi \leftrightarrow <!P>T \wedge \neg<!P>\varphi$$ [22]
$$<!P>K_i\varphi \leftrightarrow <!P>T \wedge K_i(<!P>T \rightarrow <!P>\varphi)$$

It is easy to verify that these modified recursion laws hold generally:

FACT The axioms of *TPAL* are sound on all *PAL* protocol models.

Proof We explain the validity of the crucial axiom $<!P>K_i\varphi \leftrightarrow <!P>T \wedge K_i(<!P>T \rightarrow <!P>\varphi)$. From left to right, if $<!P>K_i\varphi$ is true at world s in a model $\mathbf{M}$, then $!P$ is executable, i.e. $<!P>T$ holds at s. Moreover, $K_i\varphi$ holds at $(s, !P)$ in the updated model $\mathbf{M}|P$. Next, for each $\sim_i$-successor t of s where $<!P>T$ holds, the world $(t, !P)$ makes it into $\mathbf{M}|P$ as a $\sim_i$-successor of $(s, !P)$. But then φ holds at $(t, !P)$, by the truth of $K_i\varphi$ at $(s, !P)$, and $<!P>\varphi$ holds at t. From right to left, let world s in the model $\mathbf{M}$ satisfy $<!P>T \wedge K_i(<!P>T \rightarrow <!P>\varphi)$. By the executability of $!P$, the world $(s, !P)$ is in $\mathbf{M}|P$. Now consider any of its $\sim_i$-successors there: it must be of the form $(t, !P)$ with $s \sim_i t$. But then, $\mathbf{M}, t$ satisfies $<!P>\varphi$, and therefore $(t, !P)$ satisfies φ in the updated model $\mathbf{M}|P$. ■

Note that our original method of finding *recursion axioms* for effects of informational events still works. But now, it has been decoupled from the more drastic reduction to the pure epistemic form that held for *PAL* in its original version. Also, earlier schematic validities are typically going to fail now, such as the statement composition law

$$<!P><!Q>\varphi \leftrightarrow <!(P\wedge <!P>Q)>\varphi$$

The protocol need not allow any compound statements $!(P \wedge <!P>Q)$ at all.[23]

In-between It is also of interest to compare the above *TPAL* axiom for knowledge $<!P>K_i\varphi \leftrightarrow (<!P>T \wedge K_i(<!P>T \rightarrow <!P>\varphi))$ with the stronger variant

$$<!P>K_i\varphi \leftrightarrow <!P>T \wedge K_i(P \rightarrow <!P>\varphi)$$

Setting $\varphi = T$, this implies $<!P>T \rightarrow K_i(P \rightarrow <!P>T)$, saying that agents know which statements are currently available for announcement. This is an intermediate requirement on protocols, in-between the most general setting and the full protocol.

[22] A useful effect of the negation axiom is this: $[!P]\varphi \leftrightarrow <!P>\varphi \leftrightarrow (<!P>T \rightarrow <!P>\varphi)$.

[23] *TPAL* still allows for a sort of normal form, since every formula is equivalent to a purely epistemic combination of proposition letters p plus procedural atoms $<!P>T$.

Theorem The logic *TPAL* is complete for the class of all *PAL* protocol models.

Proof Here is just a brief outline, to show the difference in labour required with our earlier fast completeness proofs: details are in van Benthem *et al.* (2008). No reduction argument of the earlier kind works. Instead, we need to do a variant of a standard modal Henkin construction (cf. Blackburn, de Rijke & Venema 2000):

(a) Start from the canonical model of all *TPAL* maximally consistent sets, with epistemic accessibility defined as usual (that is, $w \sim_i v$ if for all $K_i\varphi \in w$: $\varphi \in v$) as the initial level of worlds w.[24]

Then create further levels through finite sequences of announcements with maximally consistent sets at each stage. Suppose that we have reached stage Σ in such a sequence:

(b) The successors to the sequence are created as follows. Take any formula $<!P>T$ in Σ, and note that in the canonical model, by the Boolean axioms of *TPAL*, the set $\Sigma^P = \{\alpha \mid <!P>\alpha \in \Sigma\}$ is itself maximally consistent. Now, add a $!P$ move to the current sequence, and place the latter set right after it:

The successive levels contain epistemic models derived in this way from a previous stage by the same announcement action. Their ordering for worlds is copied from that of the preceding level. The following picture may help visualize this:

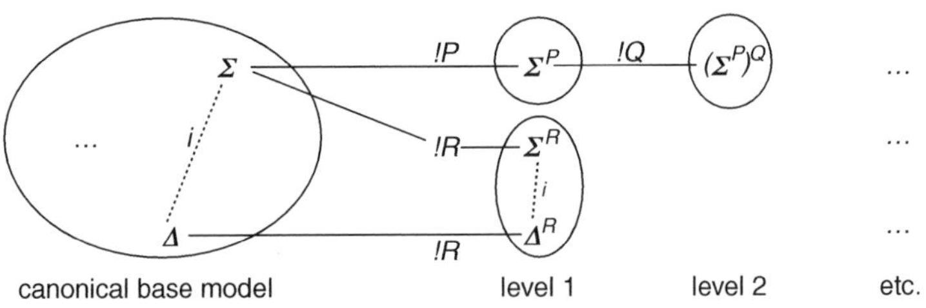

This construction yields a matching semantic forest model whose worlds are initial worlds followed by sequences of announcement events, where levels reflect construction stages. As usual, the heart of the matter is harmony between syntax and semantics:

[24] As we shall see, this level already pre-encodes the further protocol through its true assertions involving (stacked) procedural formulas of the form $<!P>T$.

Truth Lemma A formula φ belongs to the last set on a finite sequence h iff φ is true at that sequence h viewed as a history in the epistemic-temporal model matching the construction.

The proof reduces reasoning about formulas φ in worlds w at finite levels k to reasoning about stacked dynamic formulas $<!P_1>\ldots<!P_k>\varphi$ in the initial canonical model, where $P_1, \ldots, P_k$ reflects the unique construction of w. Repeatedly applying the recursion axiom $<!P>K_i\varphi \leftrightarrow (<!P>T \wedge K_i(<!P>T \rightarrow <!P>\varphi))$ then shows that the epistemic relations at the final level are inherited from the model at the initial stage, as they should.

The final stage of the construction is a routine verification that the ad-hoc model matching the construction is indeed a protocol model of the sort we have been considering. ■

Uniform protocols The construction gets slightly simpler with uniform protocols assigning the same set of sequences to each initial world. Their axiomatization works best with a universal modality U over all worlds, stipulating axioms for recursion over announcement actions, plus a strong form of common knowledge of the protocol:

(a) $<!P>U\varphi \leftrightarrow (<!P>T \wedge U(<!P>T \rightarrow <!P>\varphi))$
(b) $<!P>T \rightarrow U(P \rightarrow <!P>T)$

Decidability and complexity Analysing the completeness argument in more detail shows the following result:

THEOREM Validity in *TPAL* is decidable.

But the additional expressive power has a price:

Open Problem What is the precise computational complexity of *TPAL*?

There is no obvious reduction of *TPAL* to *PAL*, whose complexity was *Pspace*-complete (Chapter 3). So, how do *TPAL* and *PAL* compare as logics? Here is the only result so far:

THEOREM There exists a faithful polynomial-time embedding for validity from *PAL* into the logic *TPAL* extended with the universal modality.

Again, the proof is in van Benthem *et al.* (2008). The results of this section have been extended from *PAL* to the full system *DEL* in Hoshi (2009) with syntax-induced event models and a construction that carefully tracks preconditions Pre_e.

11.7 Language extensions

The preceding results were for the epistemic base language, occasionally with a universal modality. Realistic scenarios for agency with protocols will bring in other operators:

Group modalities We have no complete logic for *TPAL* yet that includes *common knowledge* $C_G\varphi$. Also, several scenarios in Chapters 3, 12, 15 turn implicit factual knowledge of groups into common knowledge through iterated public announcement of what agents know to be true. This calls for languages with the *distributed knowledge* modality $D_G\varphi$. It seems quite plausible that, like *PAL* itself, *TPAL* can be extended to deal with such extensions, by adding principles like

$$<!P>C_G{}^{\psi}\varphi \leftrightarrow (<!P>T \wedge C_G{}^{<!P>\psi}<!P>\varphi)$$
$$<!P>D_G\varphi \leftrightarrow (<!P>T \wedge D_G[P]\varphi)$$

Logics of protocols Concrete protocols for communication or experiment are a natural complement to our earlier local analysis of update steps in agency. The best examples so far are from epistemic-temporal logics, cf. Fagin *et al.* (1995). One key result is the problem of the Byzantine Generals who must coordinate their attacks: if a communication channel is not known to be reliable, no new common knowledge arises from communication.[25] Such special protocols validate axioms beyond *TPAL*. For instance, consider the epistemic-temporal assertion that implicit knowledge eventually turns into common knowledge:

$$D_G\varphi \rightarrow FC_G\varphi$$

This is not a general law of epistemic-temporal logic: its truth depends on the type of statement φ and the communication available in the channel. Chapters 3, 12 show that $D_G\varphi \rightarrow F\, C_G\varphi$ holds with simultaneous announcement, and even in some sequential settings. Gerbrandy (1999b), Hoshi (2009) study laws for special protocols such as the Muddy Children. Chapter 15 in

[25] Gerbrandy (1999a) analyses such arguments in dynamic-epistemic logics. Cf. also Roelofsen (2006) and van Eijck, Sietsma & Wang (2009) on channels and protocols in *DEL*.

this book adds logics for protocols with iterated statements about knowledge and belief that arise in solution procedures for games.

11.8 Beliefs over time

One test for the epistemic-temporal analysis in this chapter is how it generalizes to other attitudes that drive rational agency, such as belief. Epistemic-doxastic-temporal *DETL models* are branching event trees as before, with nodes in epistemic equivalence classes now also ordered by *plausibility relations* for agents (connected orders, for convenience). These tree or forest models interpret belief modalities at histories, in the style of Chapter 7. But they are very general, and as with knowledge, we ask which of them arise as traces of some systematic update scenario.

For this purpose, we take epistemic-doxastic models **M** and plausibility event models **E** to create products **M** × **E** whose plausibility relation obeys (cf. Baltag & Smets 2006):

Priority Rule $(s,e) \leq (t,f)$ iff $(s \leq t \wedge e \leq f) \vee e < f$

Van Benthem & Dégrémont (2008) extend the representation theorems of Sections 11.3, 11.5 to link belief updates to temporal forests. Let update evolution take place from some initial model along a sequence of plausibility event models in some uniform protocol.[26] Here are the relevant properties:

Fact The histories *h, h', j, j'* arising from iterated Priority Update satisfy the following two principles for any events *e, f*:

(a) whenever $je \leq j'f$, then $he \geq h'f$ implies $h \geq h'$ *Plausibility Revelation*

(b) whenever $je \leq j'f$, then $h \leq h'$ implies $he \leq h'f$ *Plausibility Propagation*

Representation Together, these two principles express the revision policy in the Priority Rule: its bias toward the last-observed event, but also its conservativity with respect to previous worlds whenever possible given the former priority. Here is the key result:

[26] The cited reference also analyses the case of pre-orders, and of state-dependent protocols.

THEOREM A *DETL* model can be represented as the update evolution of an epistemic-doxastic model under a sequence of epistemic-plausibility updates iff it satisfies the structural conditions of Section 11.3, with Bisimulation Invariance now for epistemic-doxastic bisimulations, plus Plausibility Revelation and Propagation.

Proof The idea of the proof is as before. Given a *DETL*-model H, we say that

> $e \leq f$ in the model E_k if the events e, f occur at the same tree level k, and there are histories h, h' (of length $k-1$) with $he \leq_H h'f$.

One can then check inductively, making crucial use of Priority Update plus Plausibility Revelation and Propagation in H, that the given plausibility order in H matches the one computed by sequences of events in the update evolution stages

$$M_H \times E_1 \times \ldots \times E_k$$

starting from the epistemic plausibility model M_H that was put at the bottom level of the tree. ■

Languages, logics, and long-term information scenarios Next, one can introduce doxastic temporal languages over our tree models, extending dynamic-doxastic logic to a temporal setting. Van Benthem & Dégrémont (2008) uses the safe belief modality of Chapter 7 to state correspondences with agent properties. Dégrémont (2010) proves completeness for the logics, comparing them with the postulational analysis of Bonanno (2007). He also adds doxastic protocols, linking up with game theory and learning theory. Further doxastic protocols occur in Baltag & Smets (2009a, b) on long-term belief update in groups via iterated soft announcements of the form $\Uparrow\varphi$ of Chapter 7. See Chapter 15 for some concrete examples.

11.9 Conclusion

We have linked the dynamic logics of agency in earlier chapters to long-term temporal logics of knowledge and belief, in a precise technical sense. We found that this is a natural combination, where *DEL* describes fine-structure of widely used *ETL*-style branching tree models, and so we made a contribution to framework convergence in a crowded area. In the process, we also saw how ideas flow across frameworks, resulting in an interesting new version of

PAL and *DEL* with protocols, incorporating genuine procedural information, and shedding the 'fast reduction' ideology of our earlier chapters where needed.

11.10 Further directions and open problems

This chapter has made a first connection, but it leaves many desiderata:

Groups and preferences Given the social nature of agency (and interactive protocols), we need extensions to *group notions* like common and distributed knowledge and belief. Also, since evaluation works over time as well, we need temporal versions of the *preference logics* of Chapter 9. But our major concerns at the end of this chapter are the following:

Logics with explicit protocols Like *ETL*, our logics leave protocols implicit in models, with only muted impact in the language via assertions $<!P>T$ or $<E, e>T$. The latter are just one-step local events for *PAL* or *DEL*, whereas we also want to bring out agents' long-term behaviour explicitly, like we did with strategies in Chapter 10.[27] We want to define this behaviour explicitly, and talk and reason about it. Van Benthem & Pacuit (2007) propose a version of epistemic *PDL* for this purpose, akin to the knowledge programs of Fagin *et al.* (1995). Van Eijck, Sietsma & Wang (2009) use *PDL*-definable protocols to explore concrete communication scenarios (Wang 2010 develops further theory). Explicitness becomes particularly pressing when we realize that a protocol itself can be subject to dynamic change, just as games could change in Chapter 9. How to best model *protocol change*?

Joining forces with learning theory Temporal information mechanisms and long-term convergence to knowledge drive Formal Learning Theory (Kelly 1996, 1998; Hendricks 2003). What is the connection with our logics? *DEL* describes local learning of facts, but not long-term identification of the total history one is on. The latter suggests the branching temporal models of this chapter that are also the habitat of learning theory. One can learn long-term properties then, say about strategies of other players, depending on what one observes, and what one knows about the protocol. A logical perspective

[27] Strategies and protocols are not the same intuitively, but they are formally similar.

can distinguish a whole range of possible learning goals, and our languages for epistemic-temporal models do just that:

> $FK\varphi$ or modalized variants express for suitable φ that there comes a stage where the agent will know that φ. Stronger variants are $FGK\varphi$ or $F(GK\varphi \vee GK\varphi)$.

Even more distinctions arise by adding *beliefs* that fit the ambient hypotheses in learning scenarios. What such assertions say depends on one's epistemic-doxastic-temporal models: versions with and without explicit histories both make sense. In fact, our epistemic-doxastic-temporal languages can express success principles for learning such as

> $F(B\psi \rightarrow \psi)$, or $F(B\psi \rightarrow K\psi)$ saying that my beliefs will eventually be true, or even, that they will turn into knowledge.[28]

Chapters 3, 10, 12, 15 have examples of learning by the special mechanism of iterated announcement for the same assertion. The dynamics for that can be updates with either hard or soft information (cf. Chapter 7, and Baltag & Smets 2009). Gierasimczuk (2009), Dégrémont (2010), Gierasimczuk & de Jongh (2009). As for further links between Formal Learning Theory, *PAL*, and *DEL*, Kelly (2002) suggests that learning theory is a natural extension of belief revision, separating bad policies from optimal ones. This seems attractive, also from a logical point of view. Gierasimczuk (2010) develops all these links (based on the colloboration reported in Baltag, Gierasimczuk & Smets 2010) finding many surprising connections with the analysis of limit behaviour of iterated epistemic and doxastic assertions in Chapters 10, 15.

Complexity of logics: further sources Perfect Recall created grid patterns in trees, and hence the logic of agents with perfect memory turned out to be complex. What about complexity effects of the doxastic agent properties in this chapter? Structurally, Plausibility Revelation for belief seems to act like Perfect Recall. And what about still more complex entangled properties with preference, such as Rationality of agents in games?

Complexity of agents versus complexity of logics Here is a major worry, resuming an earlier discussion in this chapter. Does the high complexity of epistemic-temporal logics really mean that the tasks performed by, say, agents

[28] Learning theory studies finite identification of a history, but also, and more typically, identification in the limit.

with Perfect Recall are complex? Logic complexity lives at a meta-level, while task complexity lives at an object-level. The two differ: the theory of a simple activity can be complex. Object-level complexity of tasks for agents performing them might call for a major reworking of our logical analysis.[29]

Agents and automata It has been noted many times that *DEL* or *ETL* do not provide an explicit account of *agents* by themselves, their abstract models only show the 'epistemic traces' of agents in action. Though this seems a well-chosen abstraction level, ignoring details of implementation, Ramanujam (2008) makes a strong plea for linking *DEL* with *Automata Theory*, where agents are automata with concrete state spaces and memory structure. This seems a congenial and correct idea, but: it has to be carried out. Ramanujam (2010) is a masterful illustration for the phenomenon of memory.

Outreach: agency, process algebra, and dynamical systems We conclude by repeating a few lines of outreach that started in Chapter 4. Our systems interface with temporal logics in AI such as the Situation Calculus (cf. van Benthem 2009b; van Ditmarsch, Herzig & de Lima 2007; Lakemeyer 2009), or *STIT*-based formalisms for agency and games as studied in Toulouse and Utrecht (cf. Broersen, Herzig & Troquard 2006; Balbiani, Herzig & Troquard 2007; Herzig & Lorini 2010; Xu 2010). Another relevant line is epistemic versions of Alternating Temporal Logic (Van der Hoek & Wooldridge 2003; Van Otterloo 2005; Ågotnes, Goranko & Jamroga 2007; Goranko & van Drimmelen 2006). Computer science, too, has elegant calculi of process construction, such as Process Algebra (Milner 1999; Bergstra, Ponse & Smolka 2001) and Game Semantics (Abramsky & Jagadeesan 1992). With an explicit calculus of event models, *DEL* and *ETL* link such systems to modal languages describing properties of internal system states as a process unfolds. Is there a useful merge? An upcoming issue of the *Journal of Logic, Language and Information* (van Benthem & Pacuit 2010) on temporal logics of agency brings some of these systems together.[30]

[29] High complexity for agent tasks may be a good thing in rational agency, say, when a chairman finds it too hard to manipulate the agenda for her own purposes.

[30] And to repeat an issue from earlier chapters, now that we have a temporal logic in place, what is the connection between our discrete framework and continuous ones like evolutionary game theory or the mathematical theory of dynamical systems?

11.11 Literature

Two key references on epistemic-temporal logics in different guises are Fagin *et al.* (1995), and Parikh & Ramanujam (2003), both reporting work going back to the 1980s. Other temporal frameworks for agency are *STIT* (Belnap, Perloff & Xu 2001), and the Situation Calculus (McCarthy 1963; Reiter 2001). Van Benthem & Pacuit (2006) give many further references, also to the computational tradition in temporal logic going back to Rabin's theorem (cf. Thomas 1992). For natural extensions of *DEL* to temporal languages, see Sack (2008), Hoshi & Yap (2009). The representation theorems in this chapter are from van Benthem (2001), van Benthem & Liu (2004), van Benthem & Dégrémont (2008). They are extended to partial observation, belief, and questions in Hoshi (2009), Dégrémont (2010), and van Benthem & Minica (2009). Gerbrandy (1999a) introduces *DEL* protocols, van Benthem *et al.* (2009) has the main technical results reported here.

12 Epistemic group structure and collective agency

While we have looked extensively at individual agents and their interaction, a further basic feature in rational agency is the formation of collective entities: groups of agents that have information, beliefs, and preferences, and that are capable of collective action. Groups can be found in social epistemology, social choice theory, and the theory of coalitional games. Some relevant notions occurred in the preceding chapters, especially common knowledge – but they remained a side theme. Indeed, the logical structure of collectives is quite intricate, witness the semantics of plurals and collective expressions in natural language, which is by no means a simple extension of the logic of individuals and their properties. This book develops no systematic theory of collective agents, but this chapter collects a few themes and observations, connecting logical dynamics to new areas such as social choice.

12.1 Collective agents in static logics

Groups occur in the epistemic logic of Chapter 2 with knowledge modalities such as $C_G\varphi$ or $D_G\varphi$. But the logic had no explicit epistemic laws for natural group forming operations such as $G_1 \cup G_2$, $G_1 \cup G_2$.[1] Actually, two logics in this book did provide group structure. One is the epistemic version *E-PDL* of propositional dynamic logic in Chapter 4, where epistemic program expressions defined complex 'collective agents' such as $i\ ;\ (?p\ ;\ j \cup k)^*$. Another was mentioned in Chapter 2: the combined topologies of van Benthem & Sarenac (2005). Even so, epistemic logic still needs a serious extension to collective agents: adding common or distributed knowledge is too timid. Here, group

[1] The indices in the standard notation are always concrete sets of agents. Adding an explicit abstract *group algebra* to epistemic logic would be an interesting generalization.

structure and information may be intertwined: for instance, membership of a group seems to imply that one knows this.[2] Moreover, groups also have beliefs and preferences, and they engage in collective action. In all these cases, generalization may not be straightforward. Collective attitudes or actions may reduce to behaviour of individual group members, but they need not. One sees this variety with collective predicates in natural language, such as 'Scientists agree that the Earth is warming', or 'The sailors quarrelled.' There is no canonical dictionary semantics for what these things mean in terms of individual predication. Finally, there is a temporal aspect. Groups may change their composition, so their structure may be in flux, with members entering and leaving – and then, dynamic and temporal logics come into play.

12.2 Group knowledge and communication

One dynamic perspective on groups is that they must *form* and stay together by being involved in shared activities. This fits well with logical dynamics, and we will pursue some illustrations in this chapter, first with knowledge, then with belief.

For a start, the recurrent static notion of common knowledge in this book is not heaven-sent: it is the result of doing work, such as making public announcements. Chapter 3 raised the issue of what information groups can achieve through internal communication. We discuss this a bit further here, though a general theory is still beyond our reach:

'Tell All': maximal communication Consider two epistemic agents in an information model **M**, at an actual world *s*. They can tell each other things they know, cutting down the model. Suppose they are cooperative. What is the best correct information they can give?

Example The best agents can do by internal communication.
What is the best that can be achieved in the following model?[3]

[2] Also, groups are often held together by *trust* between agents, a delicate epistemic feature. Cf. Holliday (2009) for a *DEL*-style analysis of the dynamics of reported beliefs, testimony, and the building of trust, based on soft upgrades beyond those studied in Chapter 7.

[3] To make things more exciting, one can mark worlds with unique proposition letters.

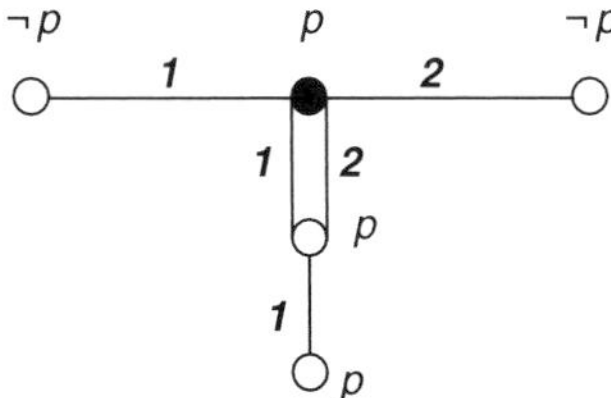

Geometrical intuition suggests that this must be:

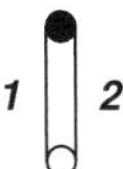

This is correct. For instance, *1* might say 'I don't know if *p*', ruling out the rightmost world, and then *2* 'I don't know either', ruling out the leftmost and bottom worlds. They could also say these things simultaneously and get the same effect. ■

For simplicity, in what follows, we stick to *finite models* **M** where each epistemic relation is an equivalence relation. Clearly, any sequence of updates where agents say all they know must terminate in some submodel that can no longer be reduced. This is reached when everything each agent knows is true in every world, and hence common knowledge. It is not clear that there is a unique 'communication core' (henceforth, the *Core*) to which this must converge, but van Benthem (2000) proposes the set of worlds reachable from the actual world in **M** via each uncertainty link. These interpret distributed knowledge in the sense of Chapter 2 – so we want to know when this is reached. A related issue was raised in Chapter 11 as a property of informational protocols, namely, when it holds that

communication turns distributed knowledge into common knowledge.

When we try to make all this more precise, there are complications:

Example Problems with bisimulation.
In the following model, the communication core is just the actual world **x**, but all worlds satisfy the same epistemic formulas:

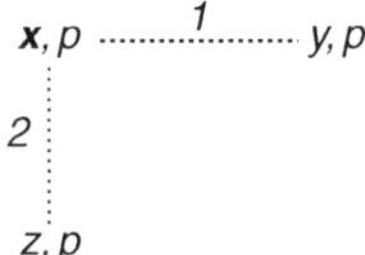

The reason is that there is a bisimulation contraction (cf. Chapter 2) to a one-world model. This may not look bad yet, but worse scenarios will follow. ■

Communication does not get us to the communicative core here. An immediate response might be to reject such inflated models, working with bisimulation contractions only. This improves things to a certain extent:

Proposition On finite bisimulation-contracted models, the Core is reached by one simultaneous announcement of what each agent knows to be true.

Proof By the contraction under bisimulation, all worlds t in the model satisfy a unique defining epistemic formula δ_t, as shown in Chapter 2. Each agent can now communicate all she knows by stating the *disjunction* $\vee\, \delta_t$ for all worlds t she considers indistinguishable from the actual one. This simultaneous move cuts the model down to the actual world plus all worlds reachable from it in the intersection of all $\sim_i$-alternatives. ■

As with the Muddy Children puzzle, things get more complicated when we let agents speak sequentially. Still, we do have one positive result:[4]

Proposition On bisimulation-contracted finite models with two agents, the Core is reached by two single-agent announcements plus a contraction.

Proof Let agent *1* state all she knows as before. This reduces the initial model $\boldsymbol{M}, s$ to a model $\boldsymbol{M}_1, s$ with just the actual world plus its $\sim_1$-successors. Now we want to let agent *2* define the set of remaining $\sim_2$-successors in a similar fashion. But there is a difficulty: a new model after a *PAL*-update need not be contracted under bisimulation, as we saw in Chapter 3. And if we contract first, to make things right, we may not get to the Core, since different worlds in the Core may now contract to one (after the first update, they may have come to verify the same epistemic formulas in $\boldsymbol{M}_1$).

To get around this, let agent *2* first state the best she can. This is the set of formulas true in all her $\sim_2$-successors from s (in the Core) plus all worlds having the same epistemic theory as one of these in $\boldsymbol{M}_1$. One formula suffices

[4] The reader can skip the following passage without loss of continuity.

for defining such a finite set (we omit the simple argument). The result is a submodel M_{12}, s whose domain consists of the Core plus perhaps some 'mirror worlds' that were modally equivalent to some world in the Core in the model M_1. Now, our claim is this:

Lemma Taking the identity on the Core and connecting each mirror world in M_{12} to all the Core worlds satisfying the same epistemic formulas in M_1 is an epistemic bisimulation.

Here is a picture (the relation is total between the two models):

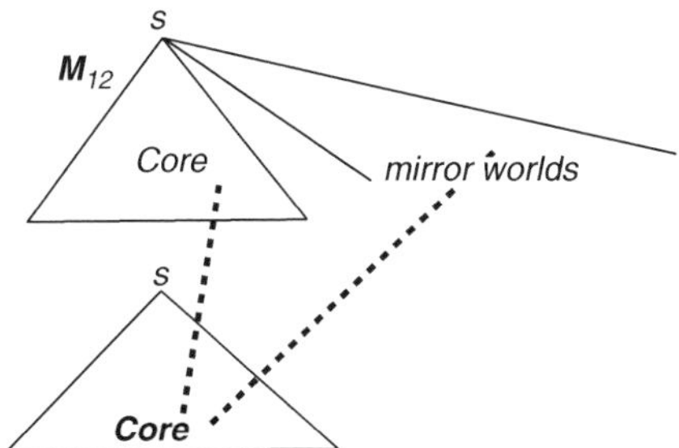

We have to check that the given map is a bisimulation from M_{12} to the free-standing model ***Core*** with just the core worlds. The atomic clause is clear by definition. As for the zigzag clauses, the relation $\sim_1$ is total in both models, and nothing has to be proved for it. Next consider the relation $\sim_2$, with a match of world x in M_{12} with a world y in ***Core***. Since $\sim_2$ is total in ***Core***, linking any given $\sim_2$-successor for x in M_{12} to a suitable $\sim_2$-successor of y is automatic. In the opposite direction, there are two cases. If x was in the Core, and y has an $\sim_2$-successor z in ***Core***, that same z also serves for x in M_{12}. But if x was a mirror world, we argue differently. By the definition of our relation, x satisfied the same epistemic formulas in M_1 as the Core world y. Now, since $y \sim_2 z$ in the Core, $M_1, y \models <2>\varphi_z$ where φ_z defined all worlds sharing z's epistemic theory in M_1. But then x satisfied $<2>\varphi_z$ as well, giving it a $\sim_2$-successor w in M_1 satisfying φ_z. But then w was a mirror world for z in the Core, and so it stayed in M_{12}. Hence it is the $\sim_2$-successor for x as needed, that got mapped to z. ■

The very complexity of this argument for such a simple conclusion is ominous. Indeed, with three agents, the core need not be reached at all, unless we are careful:

Example (Jelle Gerbrandy, p.c.) Different conversation, different information. Consider the following model, with three agents, and actual world *1*:

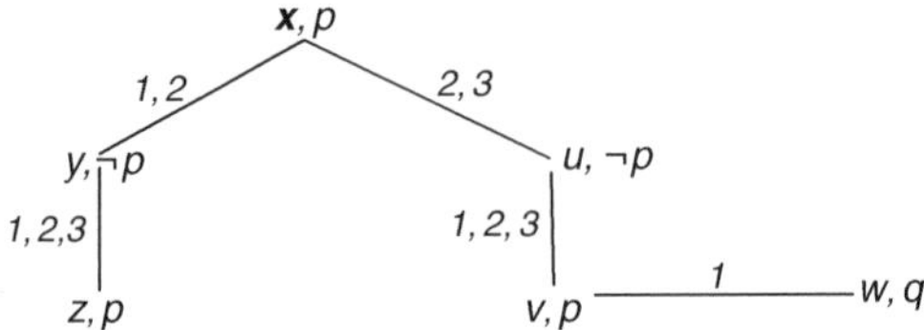

Each world is defined by a unique epistemic formula. The Core is just $\{x\}$, and it is reached if first *1* and then *3* state all they know. But if *3* starts by announcing $\neg q$, this rules out world w that made the difference for *1, 2* on the left and right. The new model arising then is easily seen to be bisimilar to the much simpler

1, 2, 3

p ——————— $\neg p$

and no further announcements will help.

Discussion: what is implicit knowledge of a group? There are subtleties here. In the final model ***Core***, worlds may satisfy formulas different from their epistemic theory in the initial model: the D_G/C_G conversion that we were after applies at best to factual formulas.[5] Also the use of bisimulation contraction works for our standard epistemic language, whereas the modality $D_G\varphi$ itself was not invariant for bisimulation (cf. Chapter 2): one more case of imbalance.

But one can also argue that all this rather speaks *against distributed knowledge*, as failing to capture the dynamic intuition of knowledge that one makes explicit by communication. Our dynamic scenarios *themselves* might be taken as yielding a more appropriate notion of *implicit knowledge* that agents can obtain by communication and contraction.

Constrained assertions The preceding scenarios were open, as long as agents said things they knew to be true. But in the puzzle of the Muddy Children of Chapters 1, 3, children could only state their knowledge or ignorance of their mud status, and in the end common knowledge resulted. This is a frequent

[5] It may be distributed knowledge that no one knows where the treasure is, but inside the Core the location may be common knowledge. Compare quantifier restriction versus relativization: $D_G\varphi$ looks only at worlds in the Core, but it evaluates the formula φ there *in the whole model*. By contrast, internal evaluation in the Core behaves like totally relativized statements $(\varphi)^{Core}$.

scenario: maximal communication *with one specific assertion*. Chapters 10, 15 use this to solve games by repeated announcements of players' rationality. As we will see in Chapter 15, the limit of repeatedly announcing a formula φ (even in infinite models) is a submodel where one of two things has happened: φ is now common knowledge ('self-fulfilling'), or $\neg\varphi$ is common knowledge ('self-refuting').

Beliefs and soft information Communication scenarios can play with beliefs just as well as knowledge. Even further surprising phenomena come to light then. Chapter 15 has some recent examples from Dégrémont & Roy (2009) showing how announcing differences in beliefs can switch who believes what – though in the limit, agreement will result. It also has results from Baltag & Smets (2009a, 2009b) on repeated announcement of soft information (cf. Chapter 7), where no worlds are eliminated, but a plausibility order gets modified.[6] A more modest aim of communication might be creating a shared plausibility pattern.

Dynamics of communication Groups form and persist through actions of communication, and hence group knowledge and belief are fundamentally linked to this dynamics. This view uncovers a wide range of complex phenomena, of which we have only seen a few.[7] Our tentative tour suggests many new logical problems, but right now we move on.

12.3 Belief and preference merge for groups

Our topic so far was modification of information and plausibility for single agents in an interactive process of communication. But a more radical step can be made. Consider the formation of *collective beliefs* or preferences, say, when a group forms in an encounter. Technically, group merge requires integration of separate epistemic-doxastic models that may contradict each other or involve different languages. Belief merge and social choice are two particular headings where this issue arises. Here we will just present one way to go.

Andréka, Ryan & Schobbens (2002) (*ARS*) propose a model that fits well with our logics. For a start, they note that the usual accounts of creating collective relations from individual ones work with input that is too poor,

6 But there are again ugly surprises, such as *cycles* when updating plausibility orders.

7 Apt, Witzel & Zvesper (2009) study group communication with explicit channel structure.

viz. just a set of relations. In general, we need richer input: a graph of dominance order in the group. For this to work well, as in Chapters 7, 9, relations are reflexive transitive pre-orders, not necessarily connected:

DEFINITION Prioritized relation merge.
Given an ordered *priority graph* $\mathbf{G} = (G, <)$ of indices for individual binary relations that may have multiple occurrences in the graph, the *merged group priority relation* is:[8]

$x \leq_G y$ iff for all indices $i \in \mathbf{G}$, either $x \leq_i y$, or
there is some $j > i$ in $\mathbf{G}$ with $x <_i y$ ■

Example Merging via simple graphs.
Consider the following two simple graph pictures with two relations each:

Putting the relation R above S, the merged group priority orders objects lexicographically: $x \leq y$ iff either $x\ R\ y \wedge x\ S\ y$ or there is a difference in the S relation and $x\ R^+\ y$ with R^+ the strict version of R. Putting R alongside S leads to the intersection $x\ R\ y \wedge x\ S\ y$. ■

ARS merge already occurred in Chapter 9. There, perhaps confusingly, it was used to find a preference relation for a single agent that has to integrate an ordered family of criteria. The latter were propositions P, and so relations in the graph were of the form

$x \leq_P y$ iff $(Px \rightarrow Py)$

Girard (2008), Liu (2008) show how this subsumes belief merge, priority-based preference, ceteris paribus logic, and agendas in inquiry (cf. Chapter 6).[9]

As for dynamics, there are various natural operations that change and combine priority graphs:

[8] Thus, either x comes below y, or if not, y 'compensates' for this by doing better on some comparison relation in the set with a higher priority in the graph.

[9] Andréka, Ryan & Schobbens (2002) prove that priority graphs are universal as a preference aggregation procedure, and give a complete graph algebra. Girard (2008) has an alternative modal analysis.

sequential composition $G_1 ; G_2$ (putting G_1 on top of G_2, retaining the same order inside) and *parallel composition* $G_1 \mid\mid G_2$ (disjoint union of graphs).

As we just saw, $G_1 \mid\mid G_2$ defines intersection of the separate relations for the graphs G_1, G_2, and $G_1 ; G_2$ defines lexicographic order (cf. the radical upgrade of Chapter 7). *ARS* is a step towards an abstract logic of group agency, and we will see one particular use in the next section.

12.4 Belief change as social choice

Groups are composed of agents, but agents themselves may also be groups of entities when we analyse their structure in detail. To show how this can be illuminating, we return to the belief revision rules of Chapter 7, and ask for a more principled analysis than what was offered there. More concretely, we will analyse belief revision as a process of group merge for 'signals'. We will show how the basic rule of Priority Update is social choice between relations in an initial model **M** and an event model **E**, where the relation in **M** × **E** results from either treating the two as equally important, or taking them in a hierarchy.

Abstract setting: ordering pair objects given component orders Two pre-orders (A, R) and (B, S) are given, with possibly different domains A, B: for instance, think of a doxastic model **M** and an event model **E** with their separate domains and plausibility orders. Now we seek to order the product $A \times B$ by a relation $O(R, S)$ over pairs (a, b).[10]

The main analogy The Priority Update Rule took the event model **E** to rank above the doxastic model **M** in terms of authority, defining the following order in **M** × **E**:[11]

$$(s,e) \leq (t,f) \text{ iff } (s \leq t \wedge e\, I\, f\,) \vee e < f$$

With pre-orders, we must state intuitions for four cases $x < y$, $y < x$, $x \sim y$ (indifferent), and $x \# y$ (incomparable). For vividness, we mark these cases graphically as $\rightarrow$, $\leftarrow$, $\sim$, and #.

Intuitive conditions on plausibility update What sort of process are we trying to capture? I first choose a very restrictive set to zoom in exclusively on Priority

[10] In general, we only need to order a *subset* of this full product, as with *DEL* in Chapter 4.

[11] Here $e\, I f$ stands for *indifference*: $e \leq f \wedge f \leq e$. By a simple computation on the rule, we get the version for strict order: $(s, e) < (t, f)$ iff $(s < t \wedge e \leq f) \vee e < f$.

Update. Later on I relax this, to get greater variety in update rules. The first condition says that the choice should not depend on individual features of objects, only their ordering pattern:

Condition (a) Permutation invariance

Consider any two permutations of A and B. Thinking of A, B as disjoint sets, without loss of generality, we can see this as one permutation π. We require:

$$O(\pi[R], \pi[S]) = \pi[O(R, S)]$$

This standard condition imposes a strong uniformity on possible formats of definition (cf. the accounts of logicality in van Benthem 2002b).

Here is one more constraint:

Condition (b) Locality

$$O(R, S)((a, b), (a', b')) \text{ iff } O(R|\{a, a'\}, S|\{b, b'\})((a, b), (a', b'))$$

Thus, we only order using the objects occurring in a pair, a form of context-independence akin to the well-known Independence from Irrelevant Alternatives in social choice.[12]

Table format Taken together, Permutation Invariance and Locality force any operation O to be definable by its behaviour in the following 4×4-table:

	S on b, b'	$\rightarrow$	$\leftarrow$	$\sim$	#
R on a, a'	$\rightarrow$	-	-	-	-
	$\leftarrow$	-	-	-	-
	$\sim$	-	-	-	-
	#	-	-	-	-

Here entries stand for the four isomorphism classes on two objects: all that matters given the above invariance constraint. Under certain conditions, some parts of this table are even forced. We will fill in the same four types of entry in the table, subject to further conditions.[13]

[12] Locality holds for the *radical update* $\Uparrow A$ of Chapter 7, that can be modelled by Priority Update using a two-point event model with an A-signal more plausible than a $\neg A$-signal. But Locality fails for *conservative update* $\uparrow A$ where we place only *the best A-worlds* on top in the new ordering. Checking if worlds are *maximal in A* requires running through other worlds.

[13] *Caveat.* Strictly speaking, one might just want to put *YES/NO* in the slots marking whether the relation $\leq$ holds between the pairs. In using the four types, strictly speaking, one should check that all intuitions to be stated hold for Priority Product Update as defined above.

Choice conditions on the aggregation procedure Now we state some conditions on how the component relations are going to be used in the final result. Even though we will only be using them for a choice with two actors, they make sense more generally. The names have been chosen for vividness, but nothing is claimed for them in naturalistic terms:

Condition (c) Abstentions

If a subgroup votes indifferent ($\sim$), then the remaining agents determine the social outcome.

Condition (d) Closed agenda

The social outcome always occurs among the opinions of the voters.

This implies *Unanimity*: 'if all members of a group agree, then take their shared outcome', but it is much stronger. Finally, consider agents who care, and are not indifferent about outcomes. An 'over rule' is a case where one opinion wins over the other.

Condition (e) Overruling

If an agent's opinion ever overrules that of another, then her opinion *always* overrules that other agent.

This goes against the spirit of democracy and letting everyone win once in a while, but we should not hide that this is what the Priority Rule does with its bias toward the last event.

Our main result now captures Priority Update, though with a twist. We derive that input order must be hierarchical. But we do not force the authority to be the second argument – say, the event model E.[14] Thus our result speaks of 'a', not 'the', Priority Update:

THEOREM A preference aggregation function is a Priority Update iff it satisfies the conditions of Permutation Invariance, Locality, Abstentions, Closed Agenda, and Overruling.

Proof First, Priority Update satisfies all stated conditions. Here one needs to check that the original formulation boils down to the case format in our Table. For instance, if event arguments are incomparable, this will block any comparison between the pairs, whence the last column. Also, if $e < f$, and $s < t$, it is easy to check that then $(s, e) < (t, f)$. Etcetera.

[14] The other option of giving priority to the first argument (the initial model M) is a conservative anti-Jeffreyan variant (cf. Chapter 8) where little learning takes place.

Conversely, we analyse possible Table entries subject to our conditions. Here the diagonal is clear by Unanimity, and the row and column for the indifference case by Abstentions:

	S on b, b'	$\rightarrow$	$\leftarrow$	$\sim$	#
R on a, a'	$\rightarrow$	$\rightarrow$	1	$\rightarrow$	2
	$\leftarrow$	3	$\leftarrow$	$\leftarrow$	4
	$\sim$	$\rightarrow$	$\leftarrow$	$\sim$	#
	#	5	6	#	#

This leaves six slots to be filled. But there are really only three choices, by simple symmetry considerations. E.g., an entry for $\rightarrow \leftarrow$ automatically induces one for $\leftarrow \rightarrow$.

Now consider slot 1. By Closed Agenda, this must be either $\rightarrow$ or $\leftarrow$. Without loss of generality, consider the latter: *S* overrules *R*. Using Overruling to fill the other cases with *S*'s opinion, and applying Permutation Invariance, our table is this:

	S on b, b'	$\rightarrow$	$\leftarrow$	$\sim$	#
R on a, a'	$\rightarrow$	$\rightarrow$	$\leftarrow$	$\rightarrow$	#
	$\leftarrow$	$\rightarrow$	$\leftarrow$	$\leftarrow$	#
	$\sim$	$\rightarrow$	$\leftarrow$	$\sim$	#
	#	$\rightarrow$	$\leftarrow$	#	#

It is easy to see that this final diagram is precisely the one for Priority Update in its original sense. The other possible case would give preference to the ordering on **M**.[15] ■

Weaker conditions: additional update rules Now we relax conditions to allow democratic variants where arguments count equally – in a special case, a flat epistemic product update of **M** and **E** where $(s, e) \leq (t, f)$ iff $s \leq t$ *and* $e \leq f$.[16] Now Closed Agenda fails: with this rule, the above clash case $\rightarrow \leftarrow$ ends up in #. Instead, we state two new principles:

Condition (f) Unanimity

If voters all agree, then their vote is the social outcome.

15 Both are instances of the basic 'But' operator of *ARS*, i.e., sequential graph composition.

16 Intersection of relations was *ARS*-style parallel composition: 'And', instead of 'But'.

Condition (g) Alignment

If anyone changes their vote to get closer to the group outcome, the group outcome does not change.

THEOREM A preference merge function satisfies Permutation Invariance, Locality, Abstentions, Overruling, Unanimity, and Alignment iff it is either (a) a priority update, or (b) flat product update.[17]

Proof The crucial step is that, without Closed Agenda, Slot 6 in our diagram

S on b, b'		→	←	∼	#
R on a, a'	→	→	6	→	2
	←	3	←	←	4
	∼	→	←	∼	#
	#	5	6	#	#

may also have entries ∼ or #. But Alignment rules out ∼. If *S* changes its vote to ∼, the outcome should still be ∼, but it is in fact →. So, the entry must be #. But then, using Alignment once more for both voters (plus Permutation Invariance), all remaining slots are #:

S on b, b'		→	←	∼	#
R on a, a'	→	→	#	→	#
	←	#	←	←	#
	∼	→	←	∼	#
	#	#	#	#	#

This is clearly the table for the flat update. ∎

Variations Our results are just a start. For instance, dropping Overruling allows for mixtures of influence for ***M*** and *E*.[18] Also, other themes from social choice theory make sense. For instance, what corresponds to commonly made restrictions on the individual preference profiles fed into the

[17] There are analogies with May's Theorem on majority voting: cf. List & Goodin (2006).

[18] More might also be said about relative power of update rules in creating new relational patterns. Compare Priority Update versus Flat Product Update. Which rule is more general if we allow re-encoding of the relational arguments that provide their inputs?

rule?[19] But perhaps the main benefit is the view itself. I am intrigued by the idea that 'I' am 'we': the social aggregate of all signals in my life.

12.5 Further directions: dynamics of deliberation

This chapter consisted of some observations showing how logical dynamics interfaces with groups as entities in their own right. Many themes need to be elaborated, but also, the list we considered is far from complete.[20] [21] Here is one further direction to be explored which I find particularly appealing from the general viewpoint of logical dynamics:

Dynamic epistemic logic fits well with social choice theory, and it can provide two things: *informational structure*, and more finely grained *procedure*. For the first, think of Arrow's Theorem, and the horror of a dictator whose opinions are the social outcome. But even if it is common knowledge 'de dicto' that there is such a dictator, this does no harm if there is no person whom we *know 'de re'* to be the dictator. Not even the dictator herself may know that she is one. To see the real issues of democracy, we need social choice plus epistemic logic.

As for the second aspect, social choice rules seem far removed from the practice of rational communication and debate, and intuitive notions of fairness having to do with these. One would want to study in detail how groups arrive at choices by *deliberating*, and ways in which agents then experience preference changes extending the concerns of Chapter 9.[22] Perceived fairness resides as much in the process as in a final act of voting. But this is not to say that our dynamics has all the answers. To the contrary, dynamic-epistemic logic should start looking at discussion and debate, since

[19] My answer: assumptions on the *continuity* of the information streams we encounter in the world. We only learn well if the universe is kind enough to us.

[20] A deeper study might use the *linguistic semantics* of individual and collective predicates. See the chapters on Temporality and Plurals in van Benthem & ter Meulen (1997).

[21] Cf. also Dégrémont (2010), Kurzen (2010) on *computational complexity* for group activities.

[22] This includes two processes: adjustment of individual preferences through social encounters, and joining in the formation of new groups with preferences of their own.

it is there that information update, belief revision, and preference change among agents occur at their most vivid.[23]

12.6 Literature

There is a huge literature on group behaviour in many fields: cf. Bratman (1992) on shared agency, Anand, Pattanaik & Puppe (2009) on social choice and economics, or Sugden (2003) in social epistemology.

Andréka, Ryan & Schobbens (2002) is our key source for belief and preference merge. The communication dynamics in this chapter comes mainly from van Benthem (2000, 2006c). Van Benthem (2009f) is the first *DEL*-style treatment of belief revision as social choice. But this is just a start, and new logics of group structure are emerging fast.

[23] Our dynamic logics are still far from dealing with subtle procedural phenomena in deliberation, such as rules for speaking, voting, or the dynamics of 'points of order'. Here it may be time to join forces with *Argumentation Theory* and other dialogical traditions, and the work of logicians like Barth, Krabbe, or Gabbay. Van Benthem (2010a) links logical dynamics to the issues of procedure raised in Toulmin (1958).

13 Logical dynamics in philosophy

Logical dynamics is a way of doing logic, but it is also a general stance. Making actions, events, and procedures first-class citizens enriches the ways in which logic interacts with philosophy, and it provides a fresh look at many traditional themes. Looking at logic and philosophy over the last century, key topics like quantification, knowledge, or conditionality have had a natural evolution that went back and forth across disciplines. They often moved from philosophy to linguistics, mathematics, computer science, or economics – and sometimes back to philosophy (cf. van Benthem 2007e). In the same free spirit, this book has studied the logical dynamics of intelligent agency drawing on several disciplines, and the final chapter sequence explores some interfaces in more detail. We start with philosophy, again disregarding boundaries, moving freely between epistemology, philosophy of information, philosophy of action, or philosophy of science.

This chapter discusses a number of philosophical themes that emerge on a dynamic stance, and tries to convey how fresh, but also how radical, logical dynamics can be in its way of dealing with existing issues.[1] We start with a small pilot study on the Fitch Paradox of knowability. Then we move to general epistemology, articulating a view of knowledge as dynamic robustness under various informational processes. This raises a further issue, now in the philosophy of information, namely, what information is, even within the restricted compass of logic. We find several legitimate and complementary notions whose interplay is crucial to rational agency. This leads to the issue of natural mixtures of different notions of information and actions that manipulate them, and we cross over to the philosophy of mathematics, studying intuitionistic logic with dynamic-epistemic tools as a mixture of factual and 'procedural' information. Still, intuitionistic logic and indeed

[1] The reader may want to follow the main line in a first pass, skipping the digressions.

most information and process-oriented logics leave the agents doing all these things out of the picture. We go on to a basic issue in the philosophy of action, and ask ourselves what a rational agent really is, proposing a list of core features on an analogy with the analysis of computation in Turing machines. Having done that, we consider whether the resulting logical dynamics of common sense agency also applies to the more exalted domain of science, discussing some issues in the philosophy of science in this light. Finally, we raise the question of what all this means for logic itself. We compare logical dynamics with logical pluralism, a recent position in the philosophy of logic that reflects on what modern logic has become. We emphasize the characteristic features of the dynamics in this book, and what these entail for our understanding of what the field of logic is, or might become.

These sections are based on the various papers listed in the literature section at the end. The subject of logical dynamics and philosophy clearly deserves a larger book of its own – but hopefully, this chapter will whet the reader's appetite.

13.1 Verificationism and the paradox of the knower

Thinking with the mind set of this book casts many philosophical issues in a new light. It will not necessarily solve them, but it may shift them in interesting ways. We start with a small case study in epistemology, tackling broader themes in the next sections.

Verificationism and the Fitch paradox What is known is evidently true. Now verification-oriented theories of meaning assign truth only to propositions for which we have evidence. But doing that suggests the claim of Verificationism that *what is true can be known*:

$$\varphi \to \Diamond K\varphi \qquad VT$$

Here the K is a knowledge modality of some kind, while the $\Diamond$ is a modality 'can' of feasibility. Now, a surprising argument by Fitch trivializes *VT* on quite general grounds:

FACT The Verificationist Thesis is inconsistent.

Proof Consider the following Moore-style substitution instance for the formula φ in the principle VT:

$$q \wedge \neg Kq \rightarrow \Diamond K(q \wedge \neg Kq)$$

Then we have the following chain of three conditionals:

$$\Diamond K(q \wedge \neg Kq) \rightarrow \Diamond(Kq \wedge K\neg Kq) \rightarrow \Diamond(Kq \wedge \neg Kq) \rightarrow \Diamond\bot \rightarrow \bot$$

Thus, even in a very weak modal logic, a contradiction follows from the assumption $q \wedge \neg Kq$, and we have shown that q implies Kq, making truth and knowledge equivalent. ∎

How disturbing is this? Some paradoxes are just spats on the Apple of Knowledge. But others are signs of deep rot, and cuts are needed to restore consistency. Remedies for the Fitch Paradox fall into two kinds (Brogaard and Salerno 2002; van Benthem 2004b). Some solutions weaken the logic in the above proof. This is like tuning down the volume on your radio so as not to hear the bad news. You will not hear much good news either. Other remedies leave the logic untouched, but weaken the verificationist principle. This is like censoring the news: you hear things loud and clear, but they may not be so interesting. Some choice between these two options is inevitable. But one really wants a non-ad-hoc *new viewpoint* with benefits elsewhere. In our view, its locus is not the Fitch proof as such, but our understanding of the two key modalities involved, either the K or the $\Diamond$, or both.

Acts of coming to know Fitch's substitution instance uses an epistemic Moore-formula $q \wedge \neg Kq$ ('q, but you don't know it') that can be true, but not known, since $K(q \wedge \neg Kq)$ is inconsistent in epistemic logic (cf. Chapter 2). Some truths are fragile, while knowledge is robust. One approach to the paradox weakens *VT* as follows (Tennant 2002):

Truth is knowable only for propositions φ *with* $K\varphi$ *consistent* *CK*

But *CK* provides no exciting account of feasibility. We have put our finger in the dike, and that is all. Indeed, the principle suggests a missing link: we need an informational scenario. We have φ true in an epistemic model $\mathbf{M}$ with actual world s, standing for our current information. Consistency of $K\varphi$ gives truth of $K\varphi$ in some possibly quite different model $\mathbf{N}$, t. The real issue is then the dynamic sense of the feasibility modality $\Diamond$:

What natural step of *coming to know* would take us from $\mathbf{M}$, s to $\mathbf{N}$, t?

The modality $\Diamond$ is sometimes unpacked in terms of *evidence* for φ. We will briefly discuss this view in Section 13.3. But for now, we strike out in an alternative semantic direction:

Announcement dynamics: from paradox to normality We have encountered Moore sentences in this book as true but self-refuting assertions, and so the Fitch Paradox recalls our findings in Chapters 3 and following. In this dynamic setting, *VT* becomes:

A formula φ *that is true may* come to be *known* VT^{dyn}

In terms of public announcement logic *PAL*, this says that there is a true statement or correct observation that changes the current epistemic model **M**, *s* to a submodel *N*, *s* where φ is known. But we already saw that announcing φ *itself*, though an obvious candidate, need not always work for this purpose.[2] What is the import of this analogy? Dynamic-epistemic logic provides definite answers. First, the simple knowledge-generating dynamic law $[!\varphi]K\varphi$ or $[!\varphi]C_G\varphi$ is only valid *factual* statements φ. With epistemic operators present, we found that self-refutation may occur. Still, this was not paradoxical, but quite useful. The Muddy Children puzzle (Chapter 3) got solved precisely because in the last round, uttering a true ignorance statement makes it false, and children learn their status. In Chapter 15, a similar scenario inspires game solution procedures.[3,4]

[2] Going back to the earlier proposal, VT^{dyn} implies *CK*, but here is a counter-example to the converse. The formula $\varphi = (q \wedge \neg Kq) \vee K\neg q$ has $K((q \wedge \neg Kq) \vee K\neg q)$ consistent. (The latter holds in a one-world model with $\neg q$.) Now take an epistemic *S5*-model with just a q- and a $\neg q$-world. Here φ holds in the actual world, but no truthful announcement would ever make us learn that φ. The only relevant update is to a one-world model with q: but there, $K((q \wedge \neg Kq) \vee K\neg q)$ fails.

[3] As a further illustration, Gerbrandy (2005) gives a *PAL* analysis of the *Surprise Exam*. A teacher says the exam will take place on a day next week where the student will not expect it – which flings in the face of a simple Backward Induction argument that there can be no such surprise. Gerbrandy dissolves the perplexity by showing how the teacher's assertion can be true but self-refuting. For instance, with two days, it says (with E_i for 'the exam is on day *i*'): $(E_1 \wedge \neg K_{you} E_1) \vee (E_2 \wedge [!\neg E_1]\neg K_{you}E_2)$. Simple models for *PAL* then clarify variations on the surprise exam. This seems an original track in a well-trodden area, though still ignoring the intuitive self-reference in the teacher's assertion. Baltag & Smets (2010) refine the analysis in terms of iterated soft announcements that change agents' plausibility orderings on worlds (cf. Chapter 7). The iteration limit is a fixed-point model that gets closer to a self-referential effect, creating a final plausibility pattern for the students where beliefs match the intended surprise.

[4] Bonnay & Égré (2007) use dynamic-epistemic techniques coupled with their epistemic 'centering semantics' to analyse Williamson's Margin of Error Paradox.

Dynamic typology These observations do not suggest a ban on self-refuting epistemic assertions – the usual remedy to the Fitch Paradox. They rather call for a typology of *which assertions* are self-fulfilling, becoming common knowledge upon announcement – or self-refuting, making their negation common knowledge. This was the Learning Problem in Chapter 3, and it was just a start. In Chapters 9, 15, we also studied statements that induce common knowledge of themselves or their negations when repeated to a first fixed-point. The Liar Paradox in Kripke's Theory of Truth changed a problem into a gateway to a wonderland of truth predicates reached by approximation procedures. Likewise (though for different reasons), in dynamic-epistemic logic, the Fitch Paradox is not a problem, but an entry point for an exciting area of epistemic effects of a wide array of statements.[5]

More generally, dynamic-epistemic techniques can help provide a typology of paradoxes. For instance, the above features of true but self-refuting assertions, and iteration of announcements to reach stable fixed-points seem common patterns.

Many agents and communication The Fitch Paradox takes on new aspects in social settings with more than one agent. In Chapters 3, 12, we studied communication scenarios that produce common knowledge. These also loosen up the situation here:

Example Fitch with communication.
Consider this model ***M*** with actual world z and a group of agents $\{1, 2\}$:

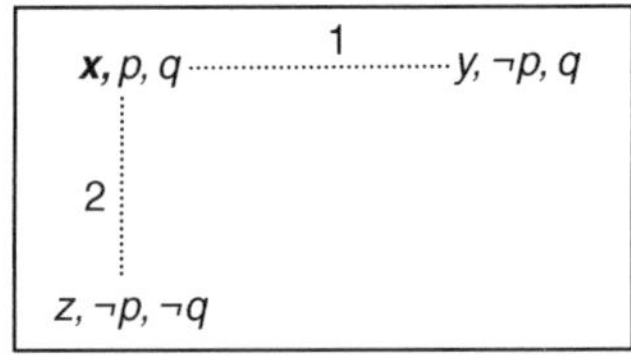

1's saying that q makes 2 know that $p \wedge \neg K_1 p$, which cannot be common knowledge. But the important fact $p \wedge q$ does become common knowledge, even if 2 says that $p \wedge \neg K_1 p$. ■

[5] Van Benthem (2004b) defines three notions of $\varphi \rightarrow \exists A \langle A! \rangle K\varphi$ *(Local Learnability)*, $\exists A$: $\varphi \rightarrow \langle A! \rangle K\varphi$ *(Uniform Learnability)*, and $\varphi \rightarrow \langle !\varphi \rangle K\varphi$ *(Autodidactic Learning)*. He shows that each successive type is stronger than the preceding. In *S5*, all three notions are decidable, but their general theory is open, especially, with *DEL*-style product update (Chapter 4).

We can now have social variants of *VT*, distinguishing between roles of agents, the more common situation in knowledge gathering, where agents can learn facts from, and about others. 'If φ is true, then someone could come to know it' changes the game, witness the consistency of $K_1(q \wedge \neg K_2 q)$. Verificationism might need a take on what we ask of *others*.

This concludes our first dynamic take on an existing philosophical issue. Admittedly, the Fitch Paradox does not go away, Verificationism has not been confirmed or refuted – but all the usual issues are seen in a new light, and with a new thrust.[6]

13.2 Knowledge and dynamic robustness

The preceding case study leads to the general issue of how our logics relate to epistemology. We will now look at this in more generality, focusing on the notion of knowledge.

Worlds apart? In modern surveys, the agenda of epistemology has little to do with logic (cf. Kim and Sosa 2000). Still, epistemic logic started with a view to philosophy, in the book *Knowledge and Belief* (Hintikka 1962). Formulas like $K_i\varphi$ for 'the agent i knows that φ' and $B_i\varphi$ for 'i believes that φ' provided logical forms for philosophical arguments. And the semantics for this language (cf. Chapter 2) was an appealing way of representing what agents know in a world, namely, those propositions that are true in all situations that the agents consider compatible with that world. This was no formal trick, but a philosophical use of the legitimate notion of *information* as a range of alternatives that are still open (Bar-Hillel & Carnap 1953; cf. Chapters 1, 2). Moreover, the modal properties of knowledge validated in this way have been widely discussed. One is

$K_i(\varphi \to \psi) \to (K_i\varphi \to K_i\psi)$ *Distribution Axiom*

Read as a law of omniscience saying that knowledge is closed under known entailments (in particular, logical consequences), this has sparked controversy. Thus, right or wrong, epistemic logic has served at least to focus

6 There are also alternative takes. Chapters 3 and 11 had temporal versions of *PAL*, with a past operator Y for the previous stage. Then $\varphi \to [!\varphi]C_G Y\varphi$ is valid unrestrictedly, yielding a true Verificationist thesis 'Every truth can come to be known *as having been true*.'

philosophical debate on important issues. Of course, in this book, we have rejected the usual terms of this debate. Distribution obviously holds for implicit knowledge $K_i\varphi$, and it equally obviously fails for explicit knowledge $K_i^+\varphi$ that includes syntactic awareness. There is little more to be said at this static level. But Chapter 5 then went further by noting how, in a dynamic-epistemic perspective, the real issue is rather *which epistemic actions a* produce *explicit knowledge*. Thus, the debate should not be about 'omniscience', but about ways of validating the implication

$$K_i(\varphi \to \psi) \to (K_i\varphi \to [a]K_i\psi)$$

Another example of debate fuelled by epistemic axioms is the implication

$K_i\varphi \to K_iK_i\varphi$ *Introspection Axiom*

that raises the issue whether introspection into epistemic states is plausible. Analogous points apply here, and Chapter 5 gave the notion of introspection a similar dynamic twist.

Still, in philosophy, formal notations might just be the last vestiges of a passion long gone – and the role of epistemic logic has clearly diminished: Dretske (1981) ignores it, Barwise & Perry (1983) opposes it, and Willamson (2000) seems neutral at best. But in a remarkable intellectual migration in the 1970s, epistemic logic moved to other disciplines such as economics and computer science. Inspired by this, in the logical dynamics of this book, its agenda has then shifted to the study of information update, communication, and interaction among agents, from humans to machines. Could there now be a return to philosophy? We will illustrate how this might happen in the heartland of epistemology.

Definitions of knowledge Consider the fundamental epistemological issue of what knowledge really *is*. A good starting point is still Plato's Formula

knowledge = justified true belief.

From a logical viewpoint, this intertwines knowledge with *other epistemic attitudes*, viz. belief, while it also highlights *evidence*: sources of knowledge and their certification. Both are major issues in their own right, that have played throughout this book.

But the twentieth century has produced further innovative views of knowledge. Hintikka's take, mentioned above, was *truth throughout a space of possibilities* (called the 'forcing view' in Hendricks 2005), grounding knowledge in semantic information that occurs widely in the study of natural language, but also in much of science. By contrast, Dretske (1981) favoured information theory, defining knowledge as *belief based on reliable correlations* supporting information flow. And another major idea is the 'truth tracking' of Nozick (1981), which gives knowledge of P a counterfactual aspect. In a simplified version, knowledge is

true belief in P, while, if P had not been the case, I would have believed $\neg P$.

On the latter account, knowledge becomes intertwined, not only with static beliefs, but also with dynamic actions of belief *revision* underlying the stated counterfactual.

Clearly, these accounts are richer than standard epistemic logic: the philosophers are ahead of the logicians in terms of imagination. But also, these accounts are still formal, involving connections to belief, evidence, information, or counterfactuals, the very topics modern logicians are interested in. Thus, the distance seems accidental rather than essential.[7]

The right repertoire: epistemic attitudes and actions Our book adds several strands to these developments. Looking at knowledge from our dynamic perspective involves two major methodological points. A first key theme in the logical, computational, and also the psychological literature on agency is this:

Step back and consider which notions belong together in cognitive functioning.

These include knowledge, belief, conditionals, and perhaps even intentions and desires (Cohen & Levesque 1990; Wooldridge 2002). The philosophical point here is this. It cannot be taken for granted that we can explain knowledge per se without tackling the proper cluster of attitudes at the same time – a point also made in Lenzen (1980).[8]

[7] For instance, Nozick's Formula $K_i\varphi \leftrightarrow \varphi \wedge B_i\varphi \wedge (\neg\varphi \Rightarrow B_i\neg\varphi)$ is a logical challenge. Its adoption blocks standard laws of epistemic logic, such as Distribution or Introspection. Are there any inference patterns left? Given some plausible background logic of belief and counterfactuals, what is the complete set of validities of Nozick's K? Arlo-Costa & Pacuit (2006) has a modal formulation with neighborhood models, while Kelly (2002) proposes a learning-theoretic account.

[8] An additional insight from Chapters 2, 11 was that this natural task may be hard. The *complexity* of combined modal logics can go up dramatically from their components, for instance, with epistemic-temporal agents having Perfect Recall. Combination is not a simple matter of adding up.

A second basic insight guiding this book is a Tandem Principle from computer science:

Never study static notions without studying the dynamic processes that use them.

And the two methodological recommendations are connected. The right repertoire of cognitive attitudes will unfold only when we study the right repertoire of epistemic actions. This viewpoint also makes sense in our ordinary use of the term knowledge. Someone only knows φ if she displays dynamic expert behaviour. She should have learnt φ via reliable procedures, but she should also be able to repeat the trick: learn other things related to φ, use φ in new settings, and communicate her knowledge to others.

Knowledge, true belief, and dynamic robustness Rephrasing the above, here is our view on the third ingredient that makes true belief into knowledge. The striking characteristic of knowledge is not some static relationship between a proposition, an agent, and the world, but sustained acts of learning and communication. The quality of what we *have* resides largely in what we *do* – individually, and socially with others. When saying 'I see that φ', I refer to an act of observation or comprehension; when asking a question, I tap into the knowledge of others, and so on with learning, grasping, questioning, or inferring.

The pattern behind all this diversity, also in the above-mentioned philosophical literature, is the *robustness* or *stability of knowledge as informational events occur*.[9] In science, robustness can only be explained if you also give an explicit account of the transformations that can change and disturb a system. In this spirit, the logical dynamics of this book provides a rich account of the informational events that affect knowledge and its relatives. We have seen this at work with the Fitch Paradox and Verificationism. But many more themes in the preceding chapters apply to epistemological issues. In particular, the Tandem View as pursued in Chapter 7 identified three epistemic notions in events of information: belief, knowledge, and a new epistemic attitude intermediate between the two: *stable belief* under learning true facts. Also, the epistemic-temporal logic in Chapters 9, 11 of actions over time turned out to fit well with learning theory (Kelly 1996; Hendricks 2005). Finally, much of our work on multi-agent interaction and

[9] Related ideas occur in Hendricks (2005), and in Roush (2006) on 'truth tracking'.

group knowledge (Chapters 10, 12) fits well with social trends in modern epistemology (cf. Goldman 1999).[10]

Thus, dynamic-epistemic logic presents a much richer picture than older epistemic logics, in terms of interactive events making dynamics and information major themes. We can see a lot more in terms of the new techniques, giving earlier issues new and deeper twists. Things have changed, and contacts between logic and philosophy deserve a new try.

Coda: logic and evidence Before discussing further repercussions of our dynamic stance, we briefly return to Plato's Formula. The latter also contains the notion of justification, or *evidence* that underlies our knowledge. Justification has been explained in many ways, such as proof or information correlation.[11] Evidence seems a natural epistemic notion. That is why van Benthem (1993) proposed a merge of semantics-based epistemic logic with a calculus of evidence. And nowadays there is a candidate for such a merge. Logical proof systems often provide explicit binary type-theoretic assertions of the form

x is a proof for φ[12]

One such calculus extends the provability reading for modal $\Box\varphi$ as saying that φ has a proof. This existential quantifier is unpacked in the Justification Logic of Artemov (1995, 2001, 2008) that includes explicit operations of combination (#), choice (+), and checking (!) on proofs. Then, epistemic axioms can be indexed for the evidence supporting them:[13]

[10] Further logical patterns in epistemology are found in Baltag, van Benthem & Smets (2010), that classifies proposed notions of knowledge in terms of dynamic actions: from observation to contrary-to-fact variation, identifying the sort of stability required.

[11] Note a logical shift. Epistemic logic revolves around a *universal* quantifier: $K_i\varphi$ says that φ is true in all situations that agent *i* considers as candidates for the current world. But the justification quantifier is *existential*: it says that there exists evidence. Now, co-existence of $\forall$ and $\exists$ views is not unheard of in logic. The semantic notion of validity says that a formula φ is true in all models. The syntactic notion says there exists a proof for φ. And Gödel's *completeness theorem* established a harmony: a first-order formula φ is valid if and only if it has a proof. Still, Plato's Formula does not suggest an equivalence, but an additional requirement with bite.

[12] In the labelled deductive systems of Gabbay (1996), the *x* can even be any sort of evidence.

[13] Similar forms of indexing work for epistemological issues such as the Skeptical Argument: *'I know that I have two hands. I know that, if I have two hands, I am not a brain in a vat. So (?): I know that I am not a brain in a vat.'* This is again modal distribution, and it might be analysed as requiring context management: $[c]K_i(\varphi \to \psi) \wedge [c']K_i\varphi \to [c \# c']K_i\psi$.

$[x]K(\varphi \rightarrow \psi) \wedge [y]K\varphi \rightarrow [x\#y]K\psi$ Explicit Omniscience

$[x]K\varphi \rightarrow [!x]KK\varphi$ Explicit Introspection

This is an interesting way to go, but the relation to the logical dynamics of this book is not obvious. Renne (2008) is a first attempt at combining the frameworks. More immediate from the viewpoint of this book are two other approaches. The first would just identify explicit evidence with the information-producing events of *DEL* (Chapter 4). Another way of combining semantic information with evidence is found in Chapter 5. This was the merge of public announcements with dynamic logic of syntactic awareness-raising steps (acts of inference, introspection, 'paying attention'). Van Benthem and Pacuit (2011) offer a compromise. Even so, a systematic treatment of evidence remains a challenge outside the scope of this chapter.

13.3 Varieties of logical information

We have explained knowledge in terms of dynamic robustness under informational actions. But then the issue comes up of *what notions of information* fuel this logical dynamics. There are clear differences, say, between the semantic models of epistemic logic and the syntactic information flow found in proofs, as we have seen in Chapters 1, 2, 5. Thus, information dynamics for knowledge may come in different kinds, more semantic and more syntactic. This diversity is an important topic by itself, and we will now discuss it in more detail. Of course, at the same time, this is a basic issue in the philosophy of information.

Information is a notion of wide use and intuitive appeal. Different paradigms claim it, from Shannon channel theory to Kolmogorov complexity, witness the handbook Adriaans & van Benthem (2008). Information is also a widely used term in logic – but a similar diversity reigns: there are competing accounts, ranging from semantic to syntactic.[14] We will look at the earlier ones as well as others, to get a more complete picture.

[14] Many logicians feel that this diversity is significant. We do not need this notion in the mechanics or the foundations of the discipline. As Laplace once said to Napoleon, who inquired into the absence of God in his *Mécanique Céleste*: 'Sire, je n'avais pas besoin de cette hypothèse.'

Information as range The first logical notion of information is semantic, associated with possible worlds, and we call it *information as range*. This is the main notion in this book, studied in epistemic logic in tandem with the dynamic processes transforming ranges. A concrete illustration were the successive updates for learning first $A \vee B$ and then $\neg A$, starting from an initial situation where all four propositional valuations are still possible:

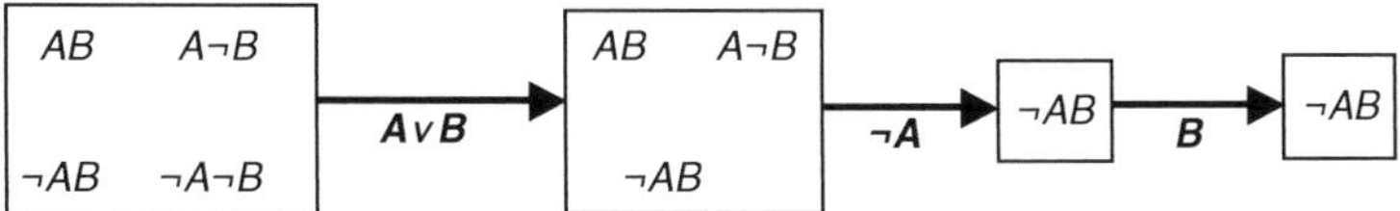

This is the standard text book sense in which valid conclusions 'add no information' to the premises.

Information as correlation But there are other semantic features of information. A second major aspect is that information is about something relevant to us, and so it turns on connections between different situations: my own, and others. *Information as correlation* has been developed in situation theory, starting from a theory of meaning in information-rich physical environments (Barwise & Perry 1983) and evolving to a theory of distributed systems whose parts show dependencies via so-called 'channels' (Barwise & Seligman 1995).

Correlation ties in with inference. One recurrent example in Indian logic runs as follows (Staal 1988). I am standing at the foot of the mountain, and cannot see what is going on there. But I can observe in my current situation. Then, one useful inference is this:

> '*I see smoke right here. Seeing smoke here indicates fire on the mountain.*
> *So, there is a fire on the mountain top.*'[15]

The Aristotelian syllogism is about one situation – while now, inference crosses over. Given the right channels, observations about one situation give reliable information about another.[16] Incidentally, on some of these views,

[15] Cf. the running example in Barwise & Seligman (1995) on seeing a flashlight on the mountain showing that there is a person in distress to some observer safely in the valley.

[16] Barwise & van Benthem (1999) develop the theory of *entailment across a relation* between models, with generalized interpolation theorems allowing for transfer of information.

inference seems a last resort when other processes have failed. If I can see for myself what is happening, that suffices. But if no direct observation is possible, we resort to reasoning.[17] Again, we see the entanglement of different informational processes that guides this book.

Information as code But inference also involves a third major logical sense of information, oriented toward syntax. This is the sense of *information as code* in which valid conclusions do add information to the premises. Its paradigmatic dynamic process is proof or computation, with acts of what may be called elucidation that manipulate representations. Chapter 5 developed systems for this with proofs adding new formulas to our stock of explicit knowledge. For instance, the earlier inference step from $A \vee B$ and $\neg A$ to B, though adding no semantic information to the final semantic information state $\{w\}$, did increase explicit knowledge about the single final world w by adding an explicit property B:

$$\text{from } w, \{A \vee B, \neg A\}, \ \text{ we went to } w, \{A \vee B, \neg A, B\}$$

But insightful representations can also be more graphical. A concrete illustration in Chapter 1 were the stages in the solution of a 3 × 3 'Sudokoid':

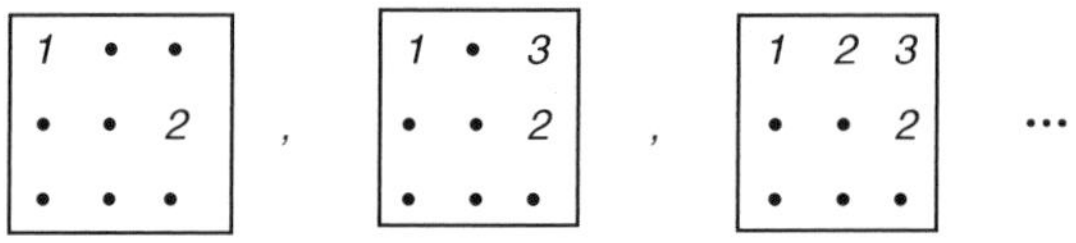

$$\boxed{\begin{array}{ccc} 1 & \bullet & \bullet \\ \bullet & \bullet & 2 \\ \bullet & \bullet & \bullet \end{array}}\ ,\quad \boxed{\begin{array}{ccc} 1 & \bullet & 3 \\ \bullet & \bullet & 2 \\ \bullet & \bullet & \bullet \end{array}}\ ,\quad \boxed{\begin{array}{ccc} 1 & 2 & 3 \\ \bullet & \bullet & 2 \\ \bullet & \bullet & \bullet \end{array}}\quad \ldots$$

Each successive diagram displays more information about the solution.

Diversity and co-existence Summing up, there are three different major notions of information in logic, each with its own dynamic process.[18]

Naturally, this diversity invites comparison as to compatibility:

Case study: co-existence of range and constraint logics We will not look at all possible merges between the three notions, but here is a brief case study on the relation between information as range and information as correlation.

[17] A very Indian traditional example is that of a coiled object in a dark room – using logic, rather than touch, to find out if it is a piece of rope or a cobra.

[18] Later on, we will find a fourth notion of *procedural information* arising from the total process agents are in, beyond single informational steps (Chapters 3, 11).

We first formulate the latter view in a congenial logical framework, for greater ease of comparison:

Example Modal constraint models.
Consider two situations s_1, s_2, where s_1 can have some proposition letter p either true or false, and s_2 a proposition letter q. There are four possible configurations: (a) s_1: p, s_2: q, (b) s_1: p, s_2: $\neg q$, (c) s_1: $\neg p$, s_2: q, and (d) s_1: $\neg p$, s_2: $\neg q$. With all these present, no situation carries information about another, as p and q do not correlate in any way. A significant constraint on the system arises only when we *leave out* some possible configurations. For instance, let the system have just the two states (a) s_1: p, s_2: q, (d) s_1: $\neg p$, s_2: $\neg q$. Now, the truth value of p in s_1 determines that of q in s_2, and vice versa: the constraint s_1: $p \leftrightarrow s_2$: q holds. This motivates defining general *constraint models* $\mathbf{M} = (Sit, State, \mathbf{C}, Pred)$ with a set *Sit* of situations, a set *State* of valuations, a predicate *Pred* recording which atoms hold where, and a constraint C stating which assignments of states to situations can occur.[19] ■

These models support an obvious modal language for reasoning with constraints, with names x for situations ($\mathbf{x}$ denotes a tuple), and atoms $P\mathbf{x}$ for predicates of situations. Next, there is a shift relation $\mathbf{s} \sim_x \mathbf{t}$ *iff* $\mathbf{s}(x) = \mathbf{t}(x)$ *for all* $x \in \mathbf{x}$, that lifts to tuples of situations $\mathbf{x}$ by having equality of $\mathbf{s}$ and $\mathbf{t}$ for all coordinates in $\mathbf{x}$. Thus, there are modalities $\Box_{\mathbf{x}}\varphi$ for each such tuple, which say that the situations in $\mathbf{x}$ settle the truth of φ in the system:[20]

$$\mathbf{M}, \mathbf{s} \models \Box_x \varphi \quad \text{iff} \quad \mathbf{M}, \mathbf{t} \models \varphi \text{ for each global state } \mathbf{t} \sim_x \mathbf{s}$$

This extended modal constraint language has a decidable complete logic for reasoning about constraints and informational correlation.

Correlation and dependence There is more here than meets the eye. Modal constraint logic resembles the decidable first-order logic of *dependent variables* in van Benthem (1996), with modal-style single and polyadic quantifiers as well as substitution operators. Van Benthem (2005c) proves a precise equivalence, making the above correlation view of information and the idea of logical dependence the same topic in different guises.

[19] Constraint models are like the 'context models' in Ghidini & Giunchiglia (2001), and they also resemble the 'local state models' of interpreted systems in Fagin *et al.* (1995).

[20] There is an analogy here with *distributed knowledge* for groups of agents, cf. Chapter 2.

Adding epistemic structure Adding epistemic structure is natural here. A blinking dot on my radar screen is correlated with an airplane approaching. But unless I *know* that there is a blinking dot, the correlation will not help me. That knowledge arises from an event: my observing the screen. This suggests an *epistemic constraint language* for models $\mathbf{M} = (Sit, State, \mathbf{C}, Pred, \sim_i)$ that combine correlation and range. E.g., suppose that $\mathbf{M}$ satisfies the constraint $s_1{:}p \rightarrow s_2{:}q$. Then the agent knows this, as the implication is true in all worlds in $\mathbf{M}$. Now suppose the agent has come to know that $s_1{:}p$. In that case, she also knows that $s_2{:}q$:

$$(K\ s_1:p \wedge K(s_1:\ p \rightarrow s_2:q)) \rightarrow K\ s_2:q$$

Dynamically in our system *PAL*, if the agent *learns* that $s_1{:}p$, she would also know that $s_2{:}q$:

$$[!s_1:p]K\ s_2:q$$

Thus, range and correlation views of information co-exist in obvious ways. This merge is natural, from philosophy to physics and economics.[21,22,23]

Inferential information and realization On the path of comparison, our next task is to draw inferential information into the same circle of ideas. But that is precisely what was done in Chapter 5 of this book, at both static and dynamic levels. The upshot of the analysis was twofold. Models for inferential information flow resemble those of dynamic-epistemic logic, endowing worlds with syntactic access to, or awareness of, propositions. And the update mechanism works for a much wider variety of actions than inference, including acts of introspection – all falling under the common denominator of converting implicit into explicit knowledge.[24] For instance, the axiom

[21] The co-existence extends even further. Situation theory involves event scenarios that 'harness' information, such as the Mousetrap of Israel & Perry (1990). These suggest *dynamic constraint models* $\mathbf{M} = (Sit, State, \mathbf{C}, Pred, Event)$ with standard modal logics.

[22] Similar models with epistemic information, correlations, and informational events occur in Baltag & Smets (2004) on the structure of quantum information states and measurement actions.

[23] Sadzik (2009) connects epistemic logic to correlated equilibria in game theory.

[24] Van Benthem & Martinez (2008) survey many further existing logical approaches to inferential information.

$K(\varphi \rightarrow \psi) \rightarrow (K\varphi \rightarrow K\psi)$ for semantic information told us that $K\psi$ is implicit knowledge. But as we hinted at in the above, the real issue of interest is another one. To those willing to do some epistemic work, the premises also licensed an act of realization producing explicit knowledge:

$$K(\varphi \rightarrow \psi) \rightarrow (K\varphi \rightarrow [\#\psi]Ex\psi)$$

An explicit system like this can help disentangle the notions of information that are run together in some philosophical discussions of omniscience.

Peaceful co-existence or unification? We have demonstrated how the different notions of information identified here live together fruitfully in precise logical settings. Still, peaceful co-existence is not Grand Unification. Several abstract frameworks claim to have unified all notions of logical information. In particular, algebraic and relevant logics have been put forward for this purpose (cf. Dunn 1991; Mares 1996; Restall 2000), while van Benthem (1991) used a categorial Lambek Calculus with dynamic and informational interpretations with a similar intent. Sequoiah-Grayson (2007) is a philosophical defence of the non-associative Lambek calculus as a core system of information structure and information flow.[25]

While all this is appealing, axioms in such systems tend to encode minimal properties of mathematical adjunctions, and these are so ubiquitous that they can hardly be seen as a substantial theory of information (cf. van Benthem 2010b). No Grand Unification has occurred yet that we are aware of. And maybe we do not need one:

Conclusion We have identified three major logical notions of information. We have shown that they are compatible, though we doubt that a grand unification between the three views is possible, or even desirable. They are probably better seen as complementary stances on a rich and complex phenomenon. In particular, we see these stances as a way of sharpening up epistemological definitions of knowledge in terms of three different kinds of information, though we will not pursue this reanalysis here.

Finally, we have not closed the book yet on information diversity in logic. In the following section, in line with the global temporal process

[25] Abramsky (2008) proposes a general category-theoretic framework for information.

perspective of Chapters 4, 11, we will even propose a fourth intuitive stance of 'procedural information'.

13.4 Intuitionistic logic as information theory

Given the entanglement of different notions of information, let us see how this functions in practice. As a concrete test for our analysis, including the combination of observational and access dynamics, we go to the philosophy of mathematics. We bring our dynamic logics to bear on a much older system that has long been connected with both proof and semantic information, viz. *intuitionistic logic*, a famous alternative to epistemic logic. Our aim is to see whether our framework applies, and what new things come to light.

From proof to semantics Intuitionistic logic had its origins in the analysis of constructive mathematical proof, with logical constants acquiring their meanings in natural deduction rules via the Brouwer–Heyting–Kolmogorov interpretation (Dummett 1977; van Dalen 2002). It then picked up algebraic and topological semantic models in the 1930s. In the 1950s, Beth proposed models over trees of finite or infinite sequences, and in line with the proof idea, intuitionistic formulas are true at a node when 'verified' there in some strong intuitive sense. A later simplified version of this semantics, due to Kripke, uses *partially ordered* models $\mathbf{M} = (W, \leq, V)$ with a valuation V, setting:

$\mathbf{M}, s \models p$	iff	$s \in V(p)$
$\mathbf{M}, s \models \varphi \wedge \psi$	iff	$\mathbf{M}, s \models \varphi$ *and* $\mathbf{M}, s \models \psi$
$\mathbf{M}, s \models \varphi \vee \psi$	iff	$\mathbf{M}, s \models \varphi$ *or* $\mathbf{M}, s \models \psi$
$\mathbf{M}, s \models \varphi \rightarrow \psi$	iff	*for all* $t \geq s$, *if* $\mathbf{M}, t \models \varphi$, *then* $\mathbf{M}, t \models \psi$
$\mathbf{M}, s \models \neg\varphi$	iff	*for no* $t \geq s$, $\mathbf{M}, t \models \varphi$

In line with the idea of accumulating certainty, the valuation is *persistent*:

$$\text{if } \mathbf{M}, s \models p, \text{ and } s \leq t, \text{ then also } \mathbf{M}, t \models p$$

The truth definition lifts the persistence to all formulas. In particular, a negation says the formula itself will never become true at any further stage. This makes Excluded Middle $p \wedge \neg p$ invalid, as this fails at states where p is not yet verified though it will later become so. This may happen in several ways: see the black dots in the pictures below, that stand for the start of informational processes unfolding as downward trees:

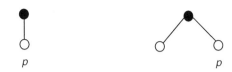

Implicit versus explicit epistemics Information or knowledge enter here, not in the explicit format of epistemic logic, but by re-interpretation of the classical logical constants. Van Benthem (1993) calls this 'meaning loading': new phenomena are absorbed implicitly into an old semantics, rather than getting separate explicit treatment. This is both intriguing, and a source of deep difficulty in comparing with other approaches.

Interpreting the models What logical notion of information is represented by these models? Intuitively, each branching tree describes an *informational process* where an agent learns progressively about the state of the actual world, encoded in a propositional valuation. At end-points of the tree, all information is in, and the agent knows the actual world. Thus, these models seem a sort of temporal alternative to the epistemic models we have had so far.[26]

Procedural information One first striking point is that intuitionistic models immediately extend our analysis of logical information so far, by registering *two* notions:[27]

(a) *factual information* about how the world is; but on a par with this:
(b) *procedural information* about our current investigative process.

How we get our knowledge matters deeply, and while leaves record factual information, the branching structure of our trees itself, with available and missing intermediate points, encodes agents' knowledge of the latter kind. The distinction between factual and procedural information makes sense much more widely, and we saw it already in Chapters 3, 11 that extended dynamic-epistemic logic with protocols stating which histories are admissible in the investigative process. Now we can draw a comparison:

[26] Even so, the technical perspectives of Chapter 2 for epistemic logic apply. For instance, modal bisimulation induces a similar invariance between intuitionistic models.

[27] The following points also apply to modal extensions of intuitionistic logic that allow non-persistent statements in the investigative process (van Benthem 1989).

From intuitionistic to epistemic information How can we model intuitionistic scenarios in dynamic-epistemic logic? Consider the following tree:

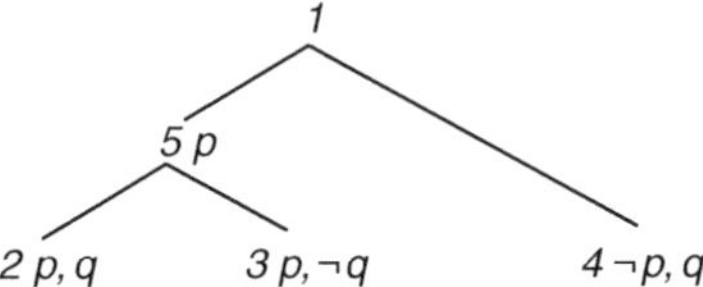

Epistemic logic casts knowledge in terms of worlds representing ways the actual situation might be. At stages of the tree, the obvious candidates are the end-points below, or the *complete histories* of the process. Thus, we can assign epistemic models as follows:

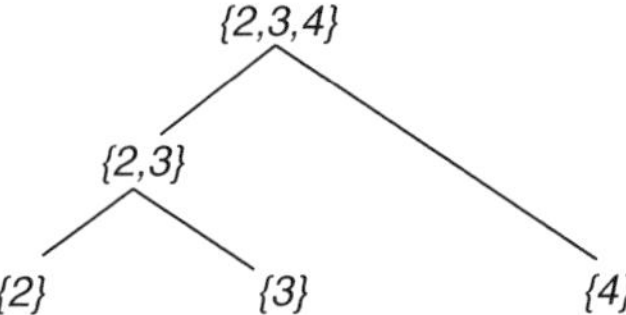

This depicts a family of epistemic models that decrease over time.

An analogy: protocols and games This picture is reminiscent of the epistemic analysis of games in Chapter 10. Indeed, with some give and take,[28] intuitionistic trees convert into the *PAL* protocol models of Chapter 11, relating them to the perspective of this book. But here is a difference with the game scenario: moves are now just anonymous upward inclusion links. So, in the spirit of dynamification, *what are the underlying actions?*

Acts of announcement The first type of action that we see in intuitionistic tree models are *public announcements*. If each end-point is uniquely definable, each step shrinking the set of reachable end-points announces the disjunction of all definitions for end-points that remain. Thus, intuitionistic logic describes effects of observing facts – though without making these acts explicit.

[28] In a game of perfect information, players knew where they were at a node in the game tree, but they did not know what the future history will be. In Section 10.8, we converted global uncertainty about the future into local uncertainty about the present, by taking worlds to be pairs *(h, s)* with *h* any history passing through the tree node *s*. This yielded an epistemic protocol model in the sense of Chapter 11.

Acts of explicit seeing Equally intriguing from our perspective, however, is a second type of informational action, that comes to light in our initial failures of the Double Negation law:

The second model M_2 poses no problems. An epistemic-temporal protocol with announcement actions $!p$ and $!\neg p$ splits the two branches as required. But now consider the first model M_1. A natural *PAL* version would put the set $\{s\}$ at both nodes, as nothing is ruled out. But then no information flows, and, if we think of p as a property of the end-point, we already knew that p at the start.[29] But in intuitionistic models, actually putting p at a stage means more than just its inevitability in the latter sense. What event is taking place then? It is tempting to interpret this as an event of the sort that we studied in Chapter 5, an act $\#p$ of awareness raising, or *explicit seeing*.

Thus, intuitionistic logic combines all notions of information that we have studied, but in new ways.

Conclusion In our perspective, intuitionistic logic as an account of information flow is akin to both semantic dynamic-epistemic logic of observation and syntactic logics of explicit seeing and awareness, merging events of public observation with private acts of realization.[30] Moreover, its semantics in terms of processes of investigation emphasizes the importance of procedural information as a further basic category. Of course, many aspects of intuitionistic logic are not explained in this way, such as its accounts of constructive proof and definition. But our analysis does show that

[29] Knowledge in this sense matches the intuitionistic operator $\neg\neg$ or its modal counterpart $\Box\Diamond$, referring to eventual truth in all reachable end-points.

[30] Van Benthem (2009e) shows how the usual reduction of intuitionistic logic to *S4* is misunderstood as a move to a weaker epistemic logic for *K*, whereas in fact, it mixes factual and procedural information in a new *epistemic-temporal* modality.

intuitionistic logic has a plausible interpretation as a theory of information-driven agency, something that fits its multi-agent game-theoretic interpretation via dialogues (Lorenzen 1955). Thus the logical dynamics of this book might sit well with constructive logic and mathematics.[31]

Open problem: implicit versus explicit approaches once more Still there remains a major conceptual issue that we have not been able to resolve. There is a clear methodological contrast between implicit logical approaches like intuitionism that 'load meanings' of classical constants, and explicit approaches like epistemic logic with new vocabulary for knowledge on a classical base. The contrast occurs wide and far. We also found it with our dynamic logics of questions in Chapter 6 versus 'inquisitive semantics' that provides propositional logic with a new semantics without explicit syntax for question operators. And this is again just an instance of a general contrast between our dynamic logics and *dynamic semantics* for natural language (see Groenendijk & Stokhof 1991, and the survey van Benthem, Muskens & Visser 1996), where meanings of the basic propositional operators and quantifiers change from static to dynamic ones.

How the two approaches compare in general is something we must leave for another study.[32]

13.5 What are rational agents?

Our topics so far were mainly concerned with knowledge, information, and action. But one major topic has been left out of consideration, namely, *who performs* all these activities. What is a rational agent, and what tasks have we set for her as theorists of intelligent interaction? Perhaps surprisingly, this question seldom gets asked in the burgeoning literature on agency. Chapter 1 has stated the main features of rational agency that motivated the logical research programme in this book. But I really think of the underlying issues as deeper, requiring philosophical analysis far beyond what I can present in this book. I will only discuss a few major features, broadly following the programme outlined in Chapter 1:

[31] Intuitionistic logic should also fit then with other aspects of agency, such as belief.

[32] Van Benthem (2010e) is a first broad exploration of this contrast across logic.

Classical foundations To see the point, compare the foundations of mathematics, starting with the logics of Frege, Russell, and others. Hilbert's Program provided appealing *goals*: establish the consistency of formalized mathematics, and where possible completeness, with a logic that is simple, perhaps decidable. This was a programme with panache! The 1930s then discovered its infeasibility with Gödel's incompleteness theorems and the undecidability results of Church and Turing for natural computational and logical problems. Foundational research made exciting refutable claims, and their refutation had positive spin-off. Like Vergilius' Romans after the fall of Troy, the logicians retreating from Paradise founded an empire of recursion theory, proof theory, and model theory, thanks to insights from that turbulent period. In particular, the Universal Turing Machine from the foundational era is still our general model of computation, being a lucid analysis of key features of a human doing sums with pencil and paper.

Now, if the hero of our time is the Generic Rational Agent, what are its key skills and properties, clearly far beyond doing sums on paper? I have asked many colleagues: answers were lively, but diverse. Let me list some issues that I myself find important:

Prelude: idealized or bounded processing powers? But first, we need to set an appropriate level of idealization. The dynamic logics in this book mostly endow agents with unlimited inferential and observational powers, and memory to store their fruits. But I am also attracted by the opposite tendency in cognitive science, stressing the limitations on these powers that human cognition operates under. In that case, the heart of rationality would be optimal performance given heart-breaking constraints.[33] Now, this book does provide tools for this: *DEL* describes limited observational access, its inferential versions describe limited inferential power, and we also had variations on memory. Moreover, we should not confuse (as is sometimes done) the idealization in a logical system with the nature of the agents that it describes: one can have a perfect logic of memory-less agents (cf. Chapter 11) or under-informed players of a game (Chapter 10). Still, we have no systematic account of bounds on agents and their variation.

Which core tasks? But next, powers to what end? A Turing Machine must just compute. Do rational agents have a core business? Is it *reasoning* – as a

[33] Gigerenzer & Todd (1999) gives surprising examples of optimal behaviour even then.

normal form for all other intelligent activities? Reasoning is important, especially when taken in a broad sense.[34] But other crucial abilities such as acumen in perception or other information sources do not reduce to reasoning in any illuminating way.[35] My first key feature is this:

Rational agents intelligently take and combine information from many sources.

Revision and learning But I do not think that informational soundness: being right all the time, is the main hallmark of rational agents. Their peak performances are rather in spotting problems, and trying to solve them. Rationality is constant *self-correction*. This reflects my general take on the foundational collapse in the 1930s. The most interesting issue in science is not guarantees for consistency and safe foundations, but the dynamic ability of repairing theories, and coming up with creative responses to challenges. Thus belief revision and learning are the true tests of rationality, rather than flawless update. Chapter 7 showed how our logics can do justice to this second major feature of agency:

Rational agents crucially correct themselves and respond intelligently to errors.

Evaluation, goals, and commitments We have also seen in Chapters 1, 9 how rational agency is not just blind information processing. It crucially involves a balance between information and *evaluation*. This led to our study of a second stream of preference dynamics behind the flow of information, but this was just a start. There is much more to goals, intentions, and commitments that truly 'make sense' of behaviour.[36]

Rational agents act in balance between their information, preferences, and goals.

[34] Colleagues responding to my request for a core list mentioned making predictions, but even more, a backward-looking talent of explaining and *rationalizing* what has already happened.

[35] Again, the emphasis on reasoning may confuse object-level agency and meta-level theory.

[36] Frankly, the computational literature leads the way in logical studies of action and commitment. Cohen & Levesque (1990), Rao & Georgeff (1991) are major sources for the *BDI* framework ('beliefs, desires, intentions') that still needs to be connected in a serious way to the logical dynamics in this book. Meyer, van der Hoek & van Linder (1999) study commitment with tools from dynamic logic. Dunin-Keplicz and Verbrugge (2002) add connections with cooperative problem solving. Roy (2008) uses ideas from *DEL* to show (partly in cooperation with van Hees) how intentions can facilitate making decisions and even solving games. Closer to the *BDI* framework, Icard, Pacuit & Shoham (2009) develop a dynamic logic of commitment. Shoham (2009) analyses logical theories of intention from a more concrete computational perspective, linking many strands.

Communication, interaction, and groups But even rich single-agent views are still too restricted. The core phenomenon we are after is intelligent interaction. A truly intelligent agent can perform tasks directed toward others: ask the right questions, explain things, convince, persuade, understand strategic behaviour, synchronize beliefs and preferences with other agents, and so on. Almost circularly, I state my final desideratum:

A rational agent is someone who interacts rationally with other agents!

This social perspective showed throughout this book, in logics of mutual knowledge, communication, and games. It has many features that transcend specific logical systems. One is a crucial skill that glues us together: the ability to put yourself in someone else's place. In its bleakest form, this is a role switch in a game. More concretely, it is the ability to see social scenarios through other people's eyes, as in the Categorical Imperative: 'Treat others as you would wish to be treated by them.'

Another social talent is the typical ability of humans to form new entities, viz. intelligent groups with lives of their own. Our identity is made up of layers of belonging to groups, as acknowledged in our logics of common knowledge and group structure. And finally, agents are not all the same, and they form groups whose members have diverse abilities, strategies, and so on. Making room for this diversity is a key task in logic and philosophy. Successful behaviour means functioning well in an environment of agents with different capacities and habits.

Summary I have now identified the four basic features that I consider as constitutive of rational agency. The systems in this book show that these features are not just slogans, since they can be analysed precisely and brought together in one logical framework.

Even so, all this does not add up to one crisp universal model of rational agency. The field of intelligent interaction is still waiting for its Turing.[37] Still, I do think that these broader foundational questions should be asked, much more than they have been so far.

But even then, a difference remains with the foundations of mathematics:

[37] Interestingly, the *Turing Test* in AI is closer to what I am after. An ability to recognize other agents as machines or humans is a typical delicate social rational skill.

'So what': are there refutable claims and goals? Suppose we find one model of agency, what is the agenda providing focus and thrill? Could there be an analogue to Hilbert's Program, setting a worthy heroic goal for interactive logicians to march toward – and fail? Instead of even suggesting an answer, I end with a more modest perspective:

Integrating trends A field may also form around a shared modus operandi. This book has shown trends toward framework integration between dynamic-epistemic logic and game theory (Chapter 10), epistemic temporal logic (Chapter 11), probability theory (Chapter 8), and other areas. An interesting analogy is again with the foundational era. The 1930s saw competing paradigms for defining computation. But gradually it became clear that, at a well-chosen level of input–output behaviour, these all captured the same computable functions. *Church's Thesis* then proclaimed the unity of the field, saying that all approaches described the same notion of computability – despite intensional differences making one or the other better for particular applications. This led to a common field of Recursion Theory, everyone got a place in the joint history, and internal sniping was replaced by external vigour. Something similar might happen in the study of intelligent interaction. Failing a Hilbert or Turing, we might at least have a Church.[38]

13.6 Agency in science and mathematics

Even in the absence of a daring refutable programme for rational agency, we can start with what we have. In particular, the dynamic stance of this book may have something to say about the foundations of science, not just the daily world of common sense activities where most of our motivating examples have come from. Let us look briefly at the philosophy of science to confirm this. In fact, logical dynamics forms a very nice fit:

Common sense and scientific agency First, perhaps, one should remove a barrier. Classics like Nagel (1961) assiduously enumerate all the differences between the two worlds of science and common sense, the habitat of rational agency as studied in this book. I cannot discuss these claims in detail (cf. van

[38] One further unifying force across the area is the *empirical reality* of intelligent interaction, and hence an independent sanity check for whatever theory we come up with.

Benthem 2010d for details, also in what follows), but let me just state my view. Science is the exercise of certain qualities of common sense agency, but taken further in isolation, and also simplified, since some strategic features of intelligent interaction are put out of play.[39] Still, science is clearly a social process – indeed, about the oldest successful one, predating all established empires and religions. And one finds respectable social aspects in science wherever one looks: even traditional logical formalizations emphasized its 'intersubjectivity', a social feature. Thus, there is no barrier, and much in this book immediately applies to our understanding of science.

Diversity of information sources: observation on a par with inference As we have seen from the start, agents manipulate many sources of information, observation, inference, and communication. And this moves us with one great stride from traditional mathematics-centred approaches to the reality of the experimental sciences where observation is a fundamental information source.[40] Indeed, the hard part may be explaining the rationale of mathematical proof. Even though this, too, plays an essential role in science, it is less easy to say just how. In the philosophy of science, this has been a persistent problem: cf. Fitelson (2006) on explaining the informativeness of Einstein's deductions from the General Theory of Relativity. This problem of information gain through inference was our major concern in Chapter 5, and in the above Section 13.3.

Theories, attitudes, and informational actions over time Very typical for our logics of agency was the wide spectrum of attitudes that agents can have towards information, ranging from knowledge and belief to neutral 'entertainment', or even doubt. I feel that all of these are not alien to scientific research, but at the very heart of it. Accordingly, we also need a much richer view of scientific theories in terms of epistemic attitudes. This would start with beliefs, since, as Popper pointed out, *belief revision* as an engine of learning is as essential to science as peaceful accumulation and regurgitation of knowledge. Continuing with this additional structure, there is also the *research agenda* as an object in its own right. What philosophers have said

39 The very term 'research' sounds dynamic. The heart of science seems to lie in its modus operandi: the processes that generate its products, not a museum of certified theories.

40 Newton's *Principia Mathematica* has mainly mathematical axioms, but read his *Optics*, and you will see that experiments, i.e., the voice of Nature, are treated with equal respect.

about that fits well with our interest in questions and issues (Chapter 6). A theory in science, and even in mathematics, is not just the set of its propositions (answers to past questions), but also its current agenda of open problems. Finally, science is a *long-term process* with features that only emerge in the long run. Likewise, our dynamic logics led to epistemic-temporal logics of agency that interface with formal learning theory, a branch of the philosophy of science (cf. Kelly 1996).

The others: social aspects revisited Our third source of information was communication, making conversation and argumentation key topics in logic. But science, too, essentially involves many agents: as we said, it is one of the oldest social activities in history. Various authors have cast experiments in terms of games played by Man and Nature (Giles 1974; Hintikka 1973; Lorenz & Lorenzen 1978; Mittelstaedt 1978). But also, interaction between human agents seems essential to science: for instance, its styles of debate are a major engine of progress. In this book, social agency led to contacts with game theory (Chapter 10). In the philosophy of science, interfaces are often rather with evolutionary games (Skyrms 1990), but the two can meet. And there is more to the social aspect of science. Scientific theories are usually community creations, and their development is a collective group activity. This matches our step toward logics of how groups form and evolve, along with their beliefs and actions (Chapter 12).[41]

Science and values Finally, alongside the dynamics of information flow, there was a second major cognitive system permeating rational action, viz. the dynamics of evaluation. Rational action strives for a balance between information and desire. Now, science is often said to be value-neutral, and only about information. Is that so? Or are we missing crucial aspects by ignoring its goals, perhaps changing over time?

Conclusion The analogies pointed out here amply justify taking a fresh look at the formal philosophy of science in terms of logical dynamics.

[41] Logical dynamics might also overcome the divide between formal approaches in the philosophy of science and historical or sociological views that are supposed to be anti-logical. In our view, the latter rightly stress the importance of revision next to accumulation, and the pervasive role of human agency. But as we have seen, these are not a threat to logic, but rather an enhancement.

I submit that much more of what really makes science tick will come to light in that way.[42]

13.7 What is logic?

From the philosophy of science, it is only one small self-referential turn to the philosophy of logic. Our final topic is the nature of logic itself. As we have said before, textbooks in the philosophy of logic tend to address issues reflecting the state of the art of 50 years ago – and sometimes, they use definitions of the field that were out of date even then, ignoring definability and computability in favour of proof theory. But, where is logic heading? There is a common feeling that it has been broadening its scope and agenda beyond classical foundational issues, and maybe a concern that, like Stephen Leacock's famous horseman, it is 'riding off madly in all directions'. What is the resultant vector?

Dynamics versus pluralism Two broad answers are in circulation today that try to do justice to the real state of the field. One is Logical Pluralism, locating the new scope of the discipline in charting a wide variety of reasoning styles, often marked by non-classical substructural rules of inference (Beall & Restall 2006). This sees the essence of logic as describing consequence relations, a view to which I subscribed in my work on substructural logics around 1990. But I have changed my mind. The Logical Dynamics of this book says that the main issue is not reasoning styles, but the *variety of informational tasks* performed by intelligent agents, of which inference is only one among many, including observation, memory, questions, answers, dialogue, and strategic interaction. Logical systems should deal with all of these, making information-carrying events first-class citizens.

[42] Our discussion may have suggested a certain antagonism between logical dynamics and the role of *mathematics* in thinking about science. This is by no means the case. As we have emphasized in Chapter 1, logical dynamics uses mathematics as its methodology. But also, it can analyse the mathematical activity *itself*. Van Benthem (2010c) is a case study, looking at mathematical theories from all the angles discussed here. In particular, it finds new challenges to our logics beyond the dynamics of inference addressed in Chapter 5. One is the treatment of beliefs in a deductive setting, where the models of Chapter 7 seem inadequate. The other is the crucial role of *definitions* in mathematics, and hence the logical dynamics of language selection and language change.

In this final section, I will contrast the two views. I argue that logical dynamics sets itself the more ambitious goal, including an explanation of *why* substructural phenomena occur, by deconstructing them into classical logic plus an explicit account of the underlying informational events. I see this as a more challenging agenda for logic, and much richer fare for philosophers of logic. However, if I am right in my depiction, this also contains the seeds for a creative co-existence of the two views, and I will show how.

Styles of reasoning Let us start with the basics of the pluralist view, which starts from an appealing abstraction level that uncovers patterns across many different logical systems. Classical consequence $\mathbf{P} \Rightarrow C$ from a finite sequence of premises $\mathbf{P}$ to a conclusion C says that C is true in every situation where all the propositions in $\mathbf{P}$ are true. This relation between premises and conclusions satisfies a number of *structural rules*:

if $\mathbf{P}, Q, R, \mathbf{S} \Rightarrow C$, *then* $\mathbf{P}, R, Q, \mathbf{S} \Rightarrow C$	*Permutation*
if $\mathbf{P}, Q, Q \Rightarrow C$, *then* $\mathbf{P}, Q \Rightarrow C$	*Contraction*
$C \Rightarrow C$	*Reflexivity*
if $\mathbf{P} \Rightarrow Q$ *and* $\mathbf{P}, Q \Rightarrow C$, *then* $\mathbf{P} \Rightarrow C$	*Cut*
if $\mathbf{P} \Rightarrow C$, *then* $\mathbf{P}, Q \Rightarrow C$	*Monotonicity*

Together, these laws encode the basic style of reasoning behind classical consequence. It treats the data that feed into a conclusion as sets (order and multiplicity do not matter), the inferential relation is a pre-order allowing for chaining of conclusions, and overkill does not matter: accumulating more data is not going to endanger earlier conclusions.

The Bolzano Program The 1970s and 1980s saw a wave of different reasoning styles. Relevant logic dropped monotonicity, default logics dropped monotonicity and transitivity, resource logics in categorial grammar and linear logic dropped contraction, and so on. A general theory developed in the work of Gabbay (1996), Dunn (1991), Restall (2000), while Dosen & Schroeder-Heister (1993) coined the term 'sub-structural logics'. Van Benthem (1989) noted an analogy with Bolzano's *Wissenschaftslehre* (1837) that did not focus on logical constants, but on charting properties of different reasoning styles: deductive or probabilistic, in the common sense, or by philosophical standards.[43]

[43] The term I proposed for this enterprise: the *Bolzano Program*, has never caught on, though this Austrian-Italian pioneer continues to exert an appeal to logicians (van Benthem 2003a).

An analysis of consequence in terms of structural rules has an attractive Spartan austerity. Ties to a richer semantic picture are provided by *representation theorems* (van Benthem 1991, 1996) that are close to techniques in algebraic logic (Venema 2006). Here is a simple example, showing the bare bones of classical consequence:

THEOREM A consequence relation $P \Rightarrow C$ satisfies the above five structural rules iff it can be represented by a map sending propositions P to sets $Set(P)$ with $P_1, \ldots, P_n \Rightarrow C$ iff $\cap_{1 \leq i \leq n} Set(P_i) \subseteq Set(C)$.

Proof The proof is simply by setting $Set(B) =_{def} \{A \mid A \Rightarrow B$ in the given relation$\}$, and then checking that the given equivalence holds by an appeal to all the stated structural rules. ■

More sophisticated representation theorems work for further notions of consequence. Hence, the structural analysis of consequence captures abstract essences of reasoning, while still being able to reintroduce the original practice when needed.

Two worries, and the move to logical dynamics Even so, in van Benthem (1989), I voiced two concerns. First, it seemed that structural rules address mere *symptoms* of some underlying phenomenon. Non-monotonicity is like fever: it does not tell you which disease causes it. Thus, I was missing a deeper analysis of underlying phenomena. Matching this was a second worry. Substructural logics give up properties of classical consequence, but they retain the old formal language. Why not be radical *with respect to the language* as well, and reconsider what we want to say?[44] All this is what the logical dynamics of this book is about. We take logic to be a study of all sorts of informational processes in new languages, not just pluralist consequence relations, and design appropriate languages.[45]

The difference in approach may be seen with the logic *PAL* of public announcements in Chapter 3. We did not just look at new 'epistemic

[44] Admittedly, this happened with linear logic and relevant logic. But, it has not happened with circumscription and default logics, and we will return to that issue below.

[45] An exclusive concentration on consequence does not even fit the classical realities in the 1950s, when logic already had model theory and recursion theory in addition to proof theory as its main branches, placing computation, meaning, and expressive power on a par with proof and consequence.

consequence relations', but made it the job of logic to explicitly describe any events *!P* that change agents' information states.

Here is another illustration of the difference: *non-monotonic logics*, a show-piece of pluralism, being a stable practice beyond classical consequence.

Non-monotonic consequence or dynamic logic of belief change Classical consequence from **P** to *C* says that all models of **P** are models for *C*. But McCarthy (1980) showed that problem solving goes beyond this, by zooming in on the 'most congenial' models. A *circumscriptive* consequence from **P** to *C* says that *C* is true in all *minimal models* for **P**. Abstract description levels here include conditional logic (Shoham 1988; Gabbay 1996) and belief revision theory (Gärdenfors 1988), all with representation theorems.

In contrast with this, Chapter 7 represented circumscription in terms of acts of *belief revision*, providing explicit logical systems for agents' belief change under events of hard information *!P*, and also under various sorts of soft information $\Uparrow P$. Again, the underlying informational acts have become first-class citizens of the system.

A compromise: back-and-forth between dynamics and consequence But we cannot break free from our personalities for long. Mine is analogy- and consensus-seeking, and let us now see how this can be done for the case of Pluralism versus Dynamics.

Dynamic consequence for knowledge. A first comparison occurred in Chapter 3. The Restaurant of Chapter 1 had a notion of consequence that fits with public announcement. One first processes the information in the premises, and then checks the conclusion:

DEFINITION A sequent $P_1, \ldots, P_k \Rightarrow \varphi$ is *dynamically valid* if, starting with any epistemic model **M**, *s* whatsoever, successive announcements of the premises result in an updated model where announcement of φ has no further effect: in the model $(\ldots(\mathbf{M}|P_1)\ldots)|P_k, s$, the formula φ was already true everywhere, even before it was announced. ■

Dynamic consequence is expressed by a *PAL* formula making the conclusion *common knowledge*: $[!P_1]\ldots[!P_k]\, C_G\varphi$ (#). For factual formulas $P_1, \ldots, P_k, \varphi$, this holds iff φ follows classically from $P_1, \ldots, P_k$. The reason is that factual formulas do not change their truth values at worlds when passing from **M** to **M**|*P*. This is not true in general, however – and using Moore-type statements, all classical structural rules failed for dynamic validity. But still, *modified structural rules* remain valid, and we do get a new substructural logic:

FACT Dynamic consequence satisfies the following structural rules:[46]

if $\mathbf{P} \Rightarrow C$, *then* $A, \mathbf{P} \Rightarrow C$	*Left-Monotonicity*
if $\mathbf{P} \Rightarrow A$ *and* $\mathbf{P}, A, \mathbf{Q} \Rightarrow C$, *then* $\mathbf{P}, \mathbf{Q} \Rightarrow C$	*Left-Cut*
if $\mathbf{P} \Rightarrow A$ *and* $\mathbf{P}, \mathbf{Q} \Rightarrow C$, *then* $\mathbf{P}, A, \mathbf{Q} \Rightarrow C$	*Cautious Monotonicity*

This abstract substructural analysis extracts the gist of dynamic inference. And this is not just a way of reaching out to consequence-oriented Pluralism, but a provable fact. Call a *super-sequent* $\sum \rightarrow \sigma$ from a set of sequents $\sum$ to sequent σ *update-valid* if all substitution instances with epistemic formulas (reading sequents as the above type (#) formulas) yield valid *PAL*-implications. Then we have (cf. van Benthem 2003a for details):

THEOREM The update-valid structural inferences $\sum \rightarrow \sigma$ are precisely those whose conclusions σ are derivable from their premise sets $\sum$ by the rules of Left-Monotonicity, Left-Cut, and Cautious Monotonicity.

Thus, we have shown how dynamic logics lead to an interesting abstraction level of dynamic consequence relations and its structural rules.[47]

Dynamic consequence for belief. The same ideas returned in Chapter 7. The logic of belief under hard information, too, supports a matching consequence

[46] At the abstraction level of structural rules, we can view propositions A as *partial functions* T_A taking input states to output states. *Transition models* $\mathbf{M} = (S, \{T_A\}_{A \in Prop})$ consist of states in S with a family of transition relations T_A for each proposition A. Here, a sequence of propositions $\mathbf{P} = P_1, \ldots, P_k$ *dynamically implies* conclusion C in $\mathbf{M}$, if any sequence of premise updates starting anywhere in $\mathbf{M}$ ends in a fixed point for the conclusion: *if* $s_1\ T_{P1}\ s_2 \ldots T_{Pk}\ s_{k+1}$, *then* $s_{k+1}\ C\ s_{k+1}$. We then say that *sequent* $P_1, \ldots, P_k \Rightarrow C$ *is true* in the model: $\mathbf{M} \models P_1, \ldots, P_k \Rightarrow C$. Van Benthem (1996) proves a representation theorem: A sequent σ is derivable from a set of sequents $\sum$ by these three rules iff σ is true in all abstract transition models where all sequents in $\sum$ are true.

[47] The leap from *PAL* to a sequent-style analysis is drastic. Another natural abstraction level is a slightly richer *poly-modal language* over transition models. Dynamic validity needs two basic modalities, universal modal boxes for the premise transitions, and a *loop* modality for fixed-points: $\mathbf{M}, s \models (a)\varphi$ iff $R_a ss$ *and* $\mathbf{M}, s \models \varphi$. The modal loop language is decidable and axiomatizable (van Benthem 1996). Reading dynamic sequents $P_1, \ldots, P_k \Rightarrow C$ as modal formulas $[P_1]\ldots[P_k](C)T$, all earlier structural rules become derivable, as well as other properties, and the representation result can be extended. Thus, modal logic itself seems a natural abstract theory of dynamic inference.

relation, saying that processing the premises results in belief, rather than knowledge (or common belief when relevant):

$$[!P_1] \dots [!P_k]B\varphi$$

The latter is the dynamic counterpart to non-monotonic consequence relations like circumscription. Determining its precise substructural rules is an open problem.[48]

But the dynamic setting also suggests *new consequence relations*, whose trigger is *soft* rather than hard information. The lexicographic upgrade $\Uparrow P$ changed the current ordering $\leq$ of worlds as follows: all P-worlds in the current model become better than all $\neg P$-worlds, while, within those two zones, the old plausibility ordering remains. Now we get a new notion of dynamic consequence, more attuned to generic belief revision:

$$[\Uparrow P_1] \dots [\Uparrow P_k]B\varphi$$

This should be grist to the mill of pluralist logicians, since structural rules for this notion are unknown. Further plausibility changes may even suggest more consequence relations.

Thus, in a dynamic setting, both classical reasoning and non-monotonic default reasoning are at heart about epistemic attitudes and responses to information, from hard update to soft plausibility change. Merging things in this way fits with the earlier general conception of agency: processes of inference and self-correction go hand in hand.[49]

Two directions Consequence relations and dynamic logics relate as follows:

Dynamics	$\rightarrow$ *abstraction* $\rightarrow$	Consequence
Consequence	$\rightarrow$ *representation* $\rightarrow$	Dynamics

From dynamics to consequence, the issue is to find abstraction levels capturing significant properties of consequence relations generated by the dynamic activity in the logic. From consequence to dynamics, one looks for representation theorems exhibiting the dynamic practice behind a given

[48] This is because intuitions about circumscription concern only factual statements, whereas dynamic logics think of consequence between all formulas in the language.

[49] It would be nice to do dynamic logics of other reasoning styles, such as *abduction*.

consequence relation. The two directions live in harmony, and we can perform a Gestalt Switch one way or the other. The two directions might also be used to describe historical periods. The avant-garde tendency in applied areas in the 1980s was toward abstraction, and maybe that of the current decade more towards concretization.

Conclusion so far We have contrasted two programmes for legitimizing the diversity of modern logic. Logical Pluralism emphasizes consequence as the locus of research, and finding natural ways to parametrize it. Logical Dynamics emphasizes informational events, from inference to observation, and the processes by which rational agents harness these in order to act. We have shown how to move back and forth between the two perspectives by abstraction and representation, and sometimes get connecting results.

Coda: dynamics is more classical It remains to be noted that this peaceful co-existence still leaves two very different world-views. In particular, logical dynamics remains much closer to classical logic. We saw how a non-monotonic consequence relation relates to a classical dynamic logic of the process that *causes* the non-monotonicity. Thus, when we take the logical analysis one step further, and are willing to choose a language matching this,

monotonic dynamic logic can model non-monotonic consequence.

Thus we can trade in non-classical logics for classical logics, enriched with a dynamic super structure. This observation seems quite general,[50] though I have no proof of its scope.

Once more, what is logic? Finally, in this debate among the avant-garde, we may have lost sight of the larger issue that started our discussion:

[50] More evidence is the classical nature of *PAL* versus non-classical intuitionistic logic, as well as the analysis of quantum logic in terms of classical *PDL* in Baltag & Smets (2008b). As for challenges to this approach, one would want a dynamic take on *paraconsistent* logics in terms of processes that handle inconsistencies, and on *resource-conscious logics* (linear logic, categorial logics) putting management of information pieces explicitly into the logic. Finally, returning to Bolzano after all, who included language as an explicit and crucial parameter in his account of logical consequence, one should find a way to include *language change* into the dynamics.

What is logic?

Many answers are in circulation. Beyond what we have discussed, some people seek the essence in mathematical notions of semantic invariance (Bonnay 2006) or in proof theory (Martin-Löf 1996). I myself have no magic definition to offer. But then, I see no point to essentialist statics of what logic *is*, but rather to the dynamics of what it could *become*. And for that, this book offers a way of broadening the classical agenda of proof and computation towards a wider one of rational agency and intelligent interaction. Of course, there must be some continuity here, or we have just started doing another subject. But I have no specific axioms or invariances that need to be preserved at all cost. Once more in line with the thrust of this book, I see a discipline as a dynamic activity, not as any of its static products: proofs, formal systems, or languages. Logic is a stance, a modus operandi, and perhaps a way of life. That is wonderful enough.

13.8 Conclusion

This concludes our tour of contacts with philosophy. We have seen how Logical Dynamics connects naturally with issues in epistemology, philosophy of information, philosophy of science, and other areas, throwing fresh light on traditional discussions, and suggesting new ways of viewing and using logic. These bridgeheads need strengthening, and many more things will happen – and all it takes is a sensibility to the dynamic stance.

13.9 Literature

The treatment of the Fitch Paradox is largely based on van Benthem (2004b, 2009a). The view of knowledge as dynamic robustness is in van Benthem (2006a). The discussion of logical information comes from van Benthem (2005c) and van Benthem & Martinez (2008). The dynamic-epistemic analysis of intuitionistic logic is in van Benthem (2009e). The general discussion of rational agency follows van Benthem (2008c). Connections with philosophy of mathematics and philosophy of science are taken from van Benthem (2010c, 2010d). The discussion of dynamics and logical pluralism is in van Benthem (2008d). See http://staff.science.uva.nl/~johan/Research for a larger collection of relevant documents.

14 Computation as conversation

The analysis of information and interactive agency in this book crucially involved ideas from computer science. Conversation and games are processes with stepwise changes of information states. And even without a computer around – only hearts and minds – these updates resemble acts of computation: sequential for individuals, and parallel for groups. *Conversation as computation* turned out a fruitful metaphor – and we recall some illustrations below. But ideas also flow the other way. One can take the logical dynamics stance to computer science, inverting the metaphor to *computation is conversation*. We illustrate the latter direction, too, 'epistemizing' and 'gamifying' various algorithmic tasks. This fits with how modern computer systems have moved away from single physical devices to societies of agents engaged in a wide variety of tasks. Thus, in the end, our two directions belong together. Humans and machines mix naturally in logical theory, witness the dynamics of email users in Chapter 4. And joint agency by humans and machines is also a cognitive reality in practice, as we will emphasize in Chapter 16.

14.1 Conversation as computation

Our dynamic logics have adopted many key ideas from computation:

Models Our semantic paradigm were state changes in hard and soft information for individuals and groups, with models forming what might be called abstract distributed computers. Chapters 3–9 sharpened the contours of this, with an ever richer account of information states and triggers for changing them. Indeed, the logical methodology itself reflected a general insight from computer science, the Tandem View that *data and process belong together*. We developed representations best suited for the given informational process, but conversely that process could not work without suitable data structures.

Languages and logics Next, the logics describing the dynamics incorporated a key idea from program logics: event description by preconditions and postconditions. This led to a transparent recursive structure for key axioms of new knowledge or belief after steps of information update, resembling regression procedures in planning (Reiter 2001).

Program structure And another major aspect of computation pervaded conversation, viz. program or process structure (Chapters 3, 10, 11). Like programs, conversation involves control over sequences of base actions. As we saw in Chapter 3, it uses all major constructions of imperative programs: sequential composition;, guarded choice *IF* ... *THEN*... *ELSE*..., and guarded iteration *WHILE*... *DO*... And the Muddy Children's speaking simultaneously added a parallel composition | |. Accordingly, agency involved logics originally developed for analysing programs, like *PDL* and the μ-calculus – all the way to strategies in games (Chapters 10, 11, 15).

Complexity of logical tasks A final theme was difficulty of tasks. While dynamic logics gave a rich account of events that convey information, their expressive power had a computational price. This showed with basic uses of a logic: model checking, model comparison, and validity testing. Here absolute complexity results were not so revealing, but *thresholds* were interesting. While dynamic-epistemic logics tend to be decidable in the vicinity of *Pspace*-complete, further features of conversation can explode this. Notably, the logic PAL^* of public announcement with iteration, that analysed puzzles like Muddy Children (Chapter 3) or game solution methods like Backward Induction (Chapter 10), was undecidable. Single statements have a simple theory, designing conversation is hard.[1] And this is just a beginning. We did not pursue interactive measures like communication complexity (Yao 1979) or computational thresholds in moving to lying and cheating.[2]

[1] Many tasks for agents carry complexity. For instance, a conversation protocol generates a tree of trajectories from some initial model (Chapter 11) where we can ask for *plans* achieving specified effects, say *PDL* programs π that move the initial state to one satisfying a goal proposition φ.

[2] There was also complexity in theories about agents. Chapter 11 showed that, while bounded memory stays easy, agents with Perfect Recall induce grid-like traces on epistemic-temporal trees that make logics Π_1^1*-complete*: a complexity disaster, though a boon in mathematical richness.

14.2 Reversing the direction: computation as conversation

Viewing conversation and interaction as computation, combining ideas from philosophical and computational logic, has been the main thrust of this book. But this very link suggests an inversion. Lower bound proofs for complexity often establish that a problem of known complexity can be reduced to the current one. And such reductions, though technical, often link up two different styles of thinking in a converse direction. Indeed, our analysis of conversation as computation suggests looking for conversation in computation (van Benthem 2008a):

Realizing computation in conversation High complexity for a logic is often seen as bad news. But by the same token, there is good news: the logic encodes significant problems. Consider Cook's Theorem that *SAT* in propositional logic is *NP*-complete. Reversing the perspective, this says that solving just one basic logical task has universal computational power for many practical problems. And the proof even provides an effective translation: logic realizes computation, if you know the key. The same applies to other complexity classes. E.g., *Pspace*-complete is the solution complexity for many games (Papadimitriou 1994; van Emde Boas 2002), and hence being able to solve *SAT* problems for informational scenarios in epistemic logic suffices for solving lots of games.

In this same light, consider the key result from Miller & Moss (2005) on high complexity of PAL^*. What they prove is essentially, that each tiling problem – and hence each significant problem about computability by Turing machines – can be effectively reduced to a *SAT* problem in $PAL + PDL$. I literally take this result to mean the following:

> *Conversation has Universal Computing Power* Any significant computational problem can be realized as one of conversation planning.[3]

14.3 Merging computation and agency: epistemization

But the real benefit of bringing computation and conversation together is not reduction of one to the other. It is creating a merged theory with new questions. In particular, a theory of computation can absorb ideas from the

[3] I was inspired here by Shoham & Tennenholtz (1999) on what a market can compute.

dynamics of information flow. We will discuss this by examples, starting from known algorithmic tasks and adding epistemic structure ('epistemization'). In the next section, we add ideas from social interaction (performing a 'gamification'). At the end of the chapter, we note some general trends.

Epistemizing algorithms Consider Graph Reachability (*GR*), the typical planning problem. Given a graph **G** with distinguished points *x*, *y*, is there a chain of directed arrows leading from *x* to *y*? *GR* can be solved in *Ptime* in the size of **G**: there is a quadratic-time algorithm finding a path (Papadimitriou 1994). The same holds for reachability of a point in **G** that satisfies some specified goal condition φ. The solution algorithm actually performs two related tasks: determining if a route exists at all, and giving us a concrete plan to get from *x* to *y*. We will now consider various ways of introducing knowledge and information.

Knowing that you have made it Suppose you are an agent trying to reach a goal region defined by φ, with only limited observation of the terrain. So, you need not know your precise location. The graph **G** is now a model *(G, R, ~)* with accessibility arrows, but also epistemic uncertainty links between nodes. A first epistemization of *GR* asks for the existence of a plan that will lead you to a point *that you know to be in the goal region* φ. Brafman, Latombe & Shoham (1993) analyse a robot whose sensors do not tell her exactly where she is. They then add a *knowledge test* to the task, inspecting current nodes to see if we are definitely in the goal region: $K\varphi$. In this case, given the *Ptime* complexity of model checking for epistemic logic (Chapter 2), the new search task remains *Ptime*.

Having a reliable plan But what about the plan itself: should we not also *know it to be successful*? Consider the following graph, with an agent at the root trying to reach a φ-point:

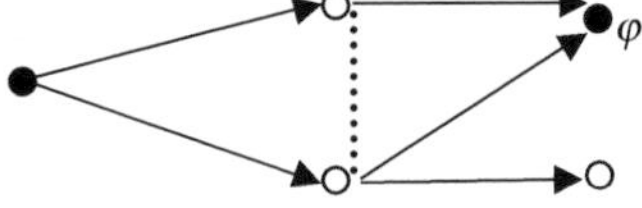

The dotted line says that the agent cannot tell the two intermediate positions apart. A plan that reaches the goal is *Up; Across*. But after the first action, the agent no longer knows where she is, and whether moving *Across* or *Up* will reach the φ-point. Suppose now for simplicity that a plan is a finite sequence ***a*** of arrows. We may require initial knowledge that this will work: $K[\mathbf{a}]K\varphi$. But this is just a start: we also must be sure at intermediate stages that the rest of the plan will still work. This would require truth of all further formulas

$$[a_1]K[a_2]K\varphi \qquad \text{where } a = a_1 ; a_2$$

This *epistemic transparency* of a plan can still be checked in *Ptime*, as the number and size of these requirements only increases polynomially. But this changes with complex *PDL* programs for plans. It is not obvious to how to define the right notion of epistemic reliability – and it may lead to new languages beyond *DEL* and *PDL*.[4]

Different types of agent The agent in the preceding graph has forgotten her first action. Our *DEL* agents with Perfect Recall of Chapters 4, 10, 11 would not be in this mess, since they only have uncertainties about what other agents did. And the commutation law $K[a]\varphi \rightarrow [a]K\varphi$ that holds for them automatically derives intermediate knowledge from initial knowledge $K[a]K\varphi$. But as we have noted in several chapters, there are many kinds of epistemic agent: with perfect memory, finite memory bounds, etc. Thus, epistemized algorithms naturally go together with questions about *what sorts of agents* are supposed to be running them – and the complexity of these tasks-for-agents can vary accordingly.

Epistemic plans In an informational setting, the notion of a plan requires further thought. A plan reacts to circumstances via instructions *IF* φ *THEN do a ELSE b*. The usual understanding of the test condition φ is that one finds out if it holds, and then chooses an action. For this to work, the agent has to be able to perform the test. But in the above graph, the plan '*IF* you went *Up*, *THEN* move *Across ELSE* move *Up*' is of no use, as the agent cannot decide which alternative holds. One solution is the 'knowledge programs' of Fagin *et al.* (1995), making actions dependent on conditions like 'the agent knows that φ' that can always be decided given epistemic introspection.[5] We saw these with uniform strategies in games of imperfect information in Chapter 10. The more general option is to define *executable plans* in epistemic models, making sure that agents can find out whether a test condition holds when needed. But I have found no definition that satisfies me.

[4] Van Benthem (2002a) uses an epistemic μ-calculus to define transparent strategies.

[5] Some of the heuristic algorithms in Gigerenzer & Todd (1999) have this flavour.

Algorithms and explicit informational actions Testing a proposition for truth involves acts of observation or communication, suggesting a move from static to dynamic-epistemic logics. We can model tests α as actions of *asking* whether α holds in models where one can query other agents, or Nature. The logic *DEL* of Chapter 4 was a showcase of such 'dynamification'. It should be well suited for analysing dynamified algorithms, especially in an extension with the dynamic logics of questions in Chapter 6.[6]

14.4 Merging with interaction: gamification

Very often, it is natural to go further, making the epistemics interactive by bringing more agents into a standard algorithm, and then turning traditional tasks into social games:

Multi-agent scenarios and interactive games Reaching a goal and knowing you are there naturally comes with social variants where others should, or should not know where you are. Card games provide many concrete examples. In the Moscow Puzzle (van Ditmarsch 2003), two players must inform each other about the cards they have without letting a third party know the solution. Indeed, games are a general model of computation (Abramsky 2008) and new questions about them abound in logical dynamics (cf. Chapters 10, 15).

Reachability and sabotage Turning algorithms into games involves prying things apart with roles for different agents. Early examples are logic games in the style of Lorenzen, Ehrenfeucht, or Hintikka (cf. the survey van Benthem 2007d). For an algorithmic example, consider again Graph Reachability (van Benthem 2005a). The following picture gives a travel network between two European capitals of logic and computation:

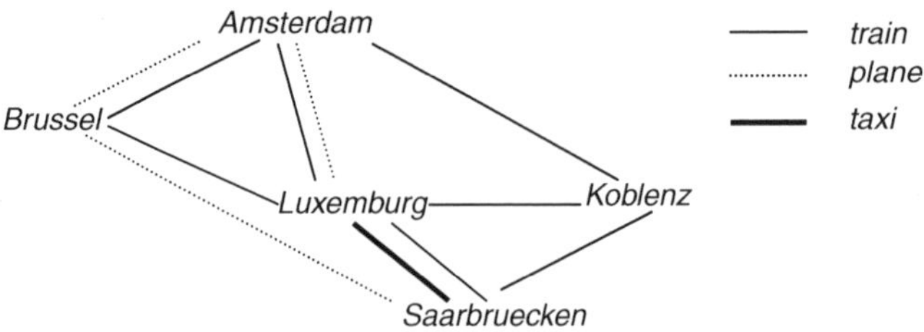

[6] A special case is knowledge games, where moves are asking questions and giving answers, and players go for goals like being the first to know (Agotnes & van Ditmarsch 2009).

It is easy to plan trips either way. But what if transportation breaks down, and a malevolent Demon starts cancelling connections, anywhere in the network? At every stage of our trip, let the Demon first take out one connection, while Traveller then follows a remaining link. This turns a one-agent planning problem into a two-player *sabotage game*, and the question is who can win where. Simple Zermelo-style reasoning (cf. Chapter 10) shows that, from Saarbruecken, a German Traveller still has a winning strategy, while in Amsterdam, Demon has the winning strategy against the Dutch Traveller.[7]

This example suggests a general transformation for any algorithmic task. It becomes a *sabotaged* one when cast as a game with obstructing players. Such a scenario raises many questions. First, there is logic. One can design languages for these games and players' strategies in terms of 'sabotage modalities' on models with accessibility relations R:

$$\boldsymbol{M}, s \models {<}{-}{>}\varphi \text{ iff there is a link } (s,t) \text{ in } R \text{ such that } \boldsymbol{M}[R := R - \{(s,t)\}], s \models \varphi$$

Van Benthem (2005a) has details on these unusual modal logics where intuitively, models change in the process of evaluation.[8] Next, how does the computational complexity of the original task change when we need to solve its game version? For sabotaged Graph Reachability, Rohde (2005) shows that this complexity jumps from *Ptime* to *Pspace*-completeness. This takes a polynomial amount of memory space, like Go or Chess.

Catch Me If You Can But there is no simple prediction when a game is more complex than its algorithmic ancestor. Again consider graphs, but now with another game variant of *GR*. Obstruction could also mean that someone tries to stop me en route:

> Starting from an initial position *(G, x, y)* with me located at *x* and you at *y*, I move first, then you, and so on. I win if I reach my goal region in some finite number of moves without meeting you. You win in all other cases.[9]

This game, too, models realistic situations, such as tactical episodes of warfare, or avoiding people at receptions. The difference with the Sabotage

[7] Link cutting games also have interesting interpretations in terms of learning and teaching. Chapters 1, 10 had a variant where a Teacher tries to trap a Student into reaching a certain state of knowledge.

[8] *Sabotage modal logic* seems related to the 'reactive Kripke models' of Gabbay (2008).

[9] Thus, you win if you catch me before I am in the goal region, if I get stuck, or if the game continues indefinitely. Other ways of casting these winning conditions allow draws.

game is that the graph remains fixed during the game. And indeed, the computational complexity stays lower.[10]

Adding knowledge and observation again Warfare involves limited observation and partial knowledge. If we turn algorithms into games of *imperfect information*, solution complexity may go up again. Jones (1978) is a classic showing how complexity can jump in this setting. Sevenster (2006) is an extensive study of epistemized gamified algorithms, linked with the *IF* logic of Hintikka–Sandu (1997). Things turn out delicate here. Consider the earlier-mentioned game of Scotland Yard. Here an invisible player who tries to avoid getting caught has to reveal her position after every k moves for some fixed k. But then the game turns into one of perfect information by making k-sequences of old moves into new moves. Chapter 10 provides some logical background, merging *DEL* with game theory.

Rephrasing the issues in game theory? The preceding complexity questions ask whether some player has a winning strategy even when hampered by lack of knowledge. But the crucial game-theoretic feature of imperfect information is the existence of more delicate *Nash equilibria in mixed strategies*, letting players play moves with certain probabilities. It is the resulting game values we should be after for gamified algorithms. Thus, gamification as generalized computation should also make us pause and think about what we want. For instance, imperfect information games for algorithms invite explicit events of observation and communication – and hence the logical dynamics of this book (cf. Chapter 10). And then: why two players, and not more? Even for logic games, argumentation, often cast as a tennis match, really needs a Proponent, an Opponent, plus a third player: the *Judge*. Thus, algorithms in a social setting merge computer science, logic, and game theory.

14.5 Toward general theory: transformations and merges

The preceding is just a bunch of examples, but it suggests more systematic topics. What follows is a brief list of further directions and open problems:

[10] The game can be recast over a new graph with positions (G, x, y) with players' moves as before while you get free moves when I am caught or get stuck. You win if you can keep moving forever. Graph games like this can be solved in *Ptime* in the size of the graph.

Epistemizing logics One concern is logical languages for the new structures. This may seem a simple matter of combining dynamic and epistemic logic, but how?[11] Next, relevant tasks for these logics can fall into the cracks of standard notions of complexity. E.g., planning problems are intermediate between model checking and satisfiability. They ask, in a model M with state s, whether some epistemic plan exists that takes us from s to the set of goal states. Thus, epistemizing logics is non-trivial, if done with a heart.

Epistemizing and gamifying algorithms Next, there is the issue of general transformations behind the above examples. Instead of botany, one wants systematic results on defining transformations and resulting changes in the complexity of the original algorithm.[12]

Epistemized process theories and games Adding observation and interaction also suggests epistemizing fundamental theories of computation, such as Process Algebra. As the latter includes explicit communication channels, the link seems congenial.[13] Explicit epistemics also makes sense with game semantics for programming languages (Abramsky 2008). Its usual models achieve non-determinacy by moving to infinite games of perfect information. But as we have seen, non-determinacy reigns in simple finite games with imperfect information (cf. Chapter 10), suggesting natural epistemic versions of the game theory. Also, strategies in game semantics involve switching games, using information about moves in one to play in the other (Abramsky 2006), again an epistemic theme that can be made explicit.

14.6 Conclusion

This chapter has linked the logical dynamics of this book to trends in computer science. Bringing together computation with knowledge and games is in the air. It may be seen with epistemic analyses of communication in Fagin *et al.* (1995), calculi of distributed computing like Milner's *CSP*, and information systems. We have approached this trend from a dynamic logical

[11] Van Benthem (1999a) shows how not even epistemizing propositional dynamic logic is simple, as one can have both *states* and *arrows* with uncertainty relations.

[12] In recursion theory, a pioneering study of this sort is Condon (1988) on Turing machines run by agents with limited observation.

[13] Van Eijck (2005) compares communication scenarios in *DEL* and Process Algebra. Netchitajlov (2000) links logics of parallel games and process algebra.

perspective, raising new issues. This matches hard-core computational logic with the philosophical tradition – including logics of knowledge, and eventually even processes with beliefs and desires. We feel this is a viable marriage. It is rich in theory, and it fits very well with modern computation in societies of interacting agents. Recent slogans like Social Software (Parikh 2002) take this into activist mode, and propose analysing existing social procedures in this style – perhaps even designing better ones.[14,15]

Another way of stating our conclusion is that computation is a fundamental category across the sciences and humanities, if cast in the proper generality. This is much more than computational implementation as the handmaiden of Academia. Our dynamic-epistemic logics show how computational models enrich the study of what used to be the preserves of linguists and philosophers. In the opposite direction, we can epistemize and dynamify existing logics and algorithms in computer science, to get powerful broader theories.

Finally, what about the chapter title *Computation as Conversation?* I think of it as a metaphor that gives scientific inspiration, while enriching one's perception of the ordinary world around us. Imagine all those Amsterdam cafés with swarms of people chatting away, while I sit here writing this book in solitude. That difference in labour is fine with me. But under the right key, human society is also a grand parallel Conversation Machine whose computing power is awesome, if only we harness it to some purpose ...

14.7 Literature

There is a huge literature on the many interfaces of logic and computer science, and several handbooks on logic, computer science, and AI are there to be consulted (Gabbay, Hogger & Robinson 1995; van Leeuwen 1991). Fagin *et al.* (1995) is a key source for epistemic views, and Parikh (2002) is also congenial. The dissertations Sevenster (2006), Kurzen (2010) are Amsterdam-style samples of logics for epistemic algorithms and games. The specifics of this chapter come from van Benthem (2008a).

[14] This is like 'mechanism design' in game theory, now with computational techniques.

[15] One might even throw *physics* into the mix. Dynamic logics are also crossing over to the foundations of quantum information and computation (Baltag & Smets 2004).

15 Rational dynamics in game theory

Games are an arena where all strands of rational agency and interaction come together. In Chapter 10, we showed how extensive games fit well with dynamic logics of knowledge, belief, and preference, adding an account of mid- and long-term agency. In principle, this has made all points about games that are relevant to the main line of this book. For readers who have not had enough of games yet, in this chapter we show how logical dynamics interfaces with game theory, in particular, solution procedures for games in strategic form. Our main tool is iterated public announcement, and our style will be more impressionist.[1]

15.1 Reaching equilibrium as an epistemic process

Iterative solution Solving games often involves an algorithm finding optimal strategies, like Backward Induction for extensive games. Here is a classic for strategic form games:

Example Iterated removal of strictly dominated strategies (SD^{ω}).
Consider the following matrix, with this legend for pairs: *(A-value, E-value)*:

	E	*a*	*b*	*c*
A	*d*	*2,3*	*2,2*	*1,1*
	e	*0,2*	*4,0*	*1,0*
	f	*0,1*	*1,4*	*2,0*

[1] This chapter cannot do justice to the many links between logic and games. Battigalli & Bonanno (1999), Battigalli & Siniscalchi (1999), Stalnaker (1999), Bonanno (2001), Halpern (2003), and Brandenburger & Keisler (2006) represent the state of the art. See also the monograph van Benthem (to appearA).

First remove the dominated right-hand column: E's action *c*. After that, the bottom row for *A*'s action *f* is strictly dominated, and after removal E's action *b* is strictly dominated, and then *A*'s action *e*. The successive removals leave just the Nash equilibrium *(d, a)*. ■

In this example, SD^{ω} reaches a unique equilibrium profile. In general, it may stop at some larger 'solution set' of matrix entries where it performs no more eliminations. There is an extensive literature analysing such game-theoretic solution concepts in terms of epistemic logic (de Bruin 2004 has a survey), defining the optimal profiles in terms of common knowledge or common belief of rationality. Instead, in this chapter we will give a dynamic take:

Solution methods as epistemic procedures Recall the Muddy Children puzzle of Chapter 3. The solution was driven by *repeated announcement* of a single epistemic assertion, viz. the children's ignorance, making the model smaller in each round. Here is an analogy:

Example, continued Updates for SD^{ω} rounds.
Here is the sequence of updates for the rounds of the preceding algorithm:

1	2	3
4	5	6
7	8	9

1	2
4	5
7	8

1	2
4	5

1
4

1

Each box may be viewed as an epistemic model. Each step increases players' knowledge about possible outcomes, until some equilibrium sets in where they know as much as they can. ■

Soon, we shall see which assertion drives these dynamic steps – but we start with the statics of game logic. As always in this book, this is a necessary platform for the dynamic analysis.

15.2 Epistemic logic of strategic game forms

Epistemic game models Consider a strategic game $\mathbf{G} = (I, \{A_j \mid j{\in}I\}, \{P_j \mid j{\in}I\})$ with a player set I, and actions A_j for each $j{\in}I$. A tuple of actions for each player is a *strategy profile* yielding a unique outcome, and each player j has a *preference order* P_j among outcomes.

DEFINITION Game models.

The *full model over game* **G** is an epistemic model **M** whose worlds are all the strategy profiles, and accessibility $\sim_j$ for player j is agreement of profiles in the jth co-ordinate. A *game model* **M** is any submodel of a full game model. ■

Worlds describe decision points for players: they know their own action, not that of the others. More sophisticated models of deliberation exist, but this kind will do for us here. Thus, a game matrix becomes an epistemic model:

Example Matrix game models.

The model for the matrix game in our first example looks as follows:

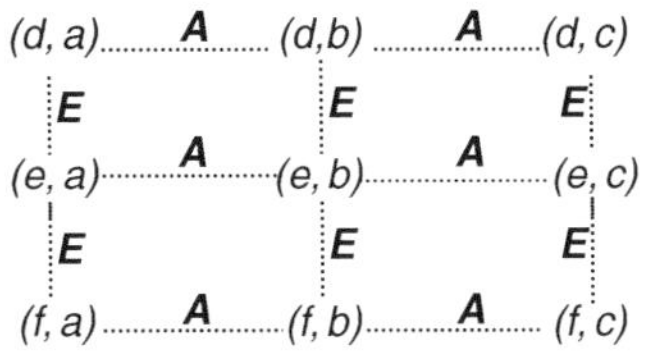

(d, a) ..... A ..... (d, b) ..... A ..... (d, c)
E E E
(e, a) ..... A ..... (e, b) ..... A ..... (e, c)
E E E
(f, a) ..... A ..... (f, b) ..... A ..... (f, c)

E.g., E's relation $\sim_E$ runs along columns: this means that E knows his own action, but not that of A. ■

SD^{ω} changes game models. In the style of Chapter 3, omitting profiles adds knowledge about the actual decision.

A little bit of logic These models are multi-S5 models, and they even provide normal forms for all models of that kind (cf. Chapter 2). They also support an interesting epistemic logic, with laws like $K_A K_E \varphi \rightarrow K_E K_A \varphi$. I know of no complete axiomatization, however, and in terms of Chapter 2, there is a complexity danger lurking, as matrices involve grid structure.[2]

Best response and equilibria To talk about equilibria we need to introduce preferences (cf. Chapter 9):

DEFINITION Expanded game language.

In a game model **M**, player j *performs action* $\omega(j)$ in world ω, while $\omega(j/a)$ is the strategy profile ω with $\omega(j)$ replaced by action a. The *best response proposition* B_j for j says that j's utility cannot improve by changing her action in ω – keeping the others' actions fixed:

[2] A nice new hybrid modal logic of preference, knowledge, and 'freedom' for agents in these same models has been proposed in Seligman (2010).

$$\mathbf{M}, \omega \models B_j \text{ iff } \&_{\{a \in Aj \mid a \neq \omega(j)\}} \ \omega(j/a) \leq_j \omega$$

Nash Equilibrium amounts to the conjunction $NE = \& \ B_j$ of all B_j-assertions. ■

Best response refers to all moves in the game – whether or not occurring in the model **M**. Thus, the proposition letter B_j keeps its truth value constant when models change.

Example Induced models for well-known games.
Consider Battle of the Sexes with its two Nash equilibria. The abbreviated diagram to the right marks the best-response atomic propositions at worlds where they are true:

	E	a	b
A	c	2, 1	0, 0
	d	0, 0	1, 2

(c, a) ····A···· (c, b)
E ⋮ E ⋮
(d, a) ····A···· (d, b)

B_A, B_E	–
–	B_A, B_E

Next, our initial running example yields the following full epistemic game model with 9 worlds:

	a	b	c
d	B_A, B_E	–	–
e	B_E	B_A	–
f	–	B_E	B_A

A (horizontal), E (vertical)

As for the distribution of B_j-atoms, by the above definition, every column in a full game model must have at least one occurrence of B_A, and every row one of B_E. ■

In these models, more complex epistemic assertions can be interpreted:

Example, continued Evaluating epistemic game formulas.
(a) $\langle E \rangle B_E \wedge \langle A \rangle B_A$ says that everyone thinks her current action might be best for her. In our nine-world model, this is true in the six worlds in the *a, b* columns. (b) The same model highlights a key distinction. B_j says that *j*'s current action is *in fact* a best response at ω. But *j* need not *know* that, as she need not know what the other player does. Indeed, $K_E B_E$ is false throughout the model, even though B_E is true at three of the worlds. ■

A fortiori, common knowledge of rationality in its obvious sense usually fails in the full model of a game, even with a unique Nash equilibrium: it will only hold in submodels.

Example Valid game laws for best response.
The following holds in all full game models: $<E>B_A \wedge <A>B_E$. It expresses the basic maths of matrices. We will see further principles later on. ■

A context-dependent best response looks only at profiles *inside* **M** (Apt 2005):

DEFINITION Relative best response.
The *relative best response* proposition B^*_j in a game model **M** is true at only those strategy profiles where *j*'s action is a best response to that of her opponent when the comparison set is all alternative strategy profiles that actually occur inside **M**. ■

Absolute best implies relative best, but not vice versa. With B^*_j, best profiles for *j* may change with **M**. In each one-world model, the one profile is relatively best for all.[3]

Example All models have relative best positions.
To see the difference between the two notions, compare these two models:

$1, 1\ (B_A)$	$0, 2\ (B_E)$
$0, 2\ (B_E)$	$1, 1\ (B_A)$

$1, 1\ (B_A, B^*_E)$	
$0, 2\ (\mathbf{B}_E)$	$1, 1\ (\mathbf{B}_A)$

■

Logics again There are various issues about epistemic preference languages for describing these models, and axiomatizing their validities. Van Benthem (2007f) has more details, but Chapter 9 on preference logic, and the game logics of Chapter 10 are relevant, too.

15.3 Rationality assertions

Playing a best response is not introspective, as we saw: players need not know they are doing it. But conversation scenarios normally assume knowledge, so what *can* players know?

Weak Rationality What players can know is that *there is no alternative action that they* know *to be better*. At least, they are no epistemic fools:

[3] Epistemically, relative best response says that the following is *distributed knowledge* at ω for the group $G-\{j\}$: '*j*'s current action is at most as good for *j* as *j*'s action at ω'.

Definition Weak Rationality.
Weak Rationality at world w in an epistemic-preferential game model $\boldsymbol{M}$ is the assertion that, for each available alternative action, j thinks the current one may be at least as good:

WR_j $\quad \&_{a \neq \omega(j)} <j>$ 'j's current action is at least as good for j as a'

The index set runs over worlds in the current model, like we had with relative best response B^*_j. ■

Weak Rationality WR_j fails exactly at those rows or columns in a two-player general game model that are strictly dominated for j.[4] With a little matrix combinatorics for these logical assertions, one can then show that this property must occur widely:

Fact Every finite game model has worlds with WR_j true for all players j.

For omitted proofs, here and elsewhere, we refer to van Benthem (2007f).

Next, we have:

Fact Weak Rationality is epistemically introspective: $WR_j \rightarrow K_j\, WR_j$ is valid.

This feature makes Weak Rationality suitable for public announcement.

Strong rationality Weak Rationality has the logical form $\&$ $<j>$. A stronger version of rationality inverts the order, making players think that *their actual action may be best*:

Definition Strong Rationality.
Strong rationality for j at a world w in a model $\boldsymbol{M}$ says that

SR_j $\quad <j> \&_{a \neq \omega(j)} <j>$ 'j's current action is at least as good for j as a'

The index set runs over all profiles in the whole game. Thus SR_j is precisely the modal formula $<j>B_j$. Strong Rationality for the whole group (SR) is then the conjunction $\&_j\ SR_j$.[5] ■

By the valid S5-law $<j>\varphi \rightarrow K_j<j>\varphi$, SR_j is something that players j will know if it is true.

Fact SR_j implies WR_j, but not vice versa.

[4] We omit the more general notion of dominance by probabilistic mixtures of rows.

[5] SR is the notion of rationalizability of Bernheim and Pearce (cf. de Bruin 2004; Apt 2005).

Proof In the following game model, WR_E holds everywhere but SR_E does not:

E		a	b	c
A	d	1, 2	1, 0	1, 1
	e	0, 0	0, 2	2, 1

B_E, B_A	B_A	–
–	B_E	B_A

■

SR does not hold in all game models, but it does hold in full models:

FACT Each finite full game model has worlds where Strong Rationality holds.

Again this shows that the logic of these game models has quite some interest.

But now we leave the logical statics, turning to the epistemic dynamics of these assertions. This is the main point of this chapter, though again we remind the reader that we are only trying to whet her appetite. The full story is in the sources mentioned in what follows.

15.4 Iterated announcement of rationality and game solution

Virtual conversation We now cast the earlier iterative solution as a virtual conversation. Take some actual world *s* in a game model **M**. Now, players *deliberate in synch* by thinking of what they know about the scenario. As with Muddy Children, we want generic assertions in the language of knowledge and best response. Our first result recasts SD^{ω}:

THEOREM The following two assertions are equivalent for worlds *s* in full game models *M(G)*:

(a) world *s* is in the SD^{ω} solution zone of *M(G)*,
(b) repeated successive announcement of Weak Rationality for players stabilizes at a submodel *N, s* whose domain is that solution zone.

The general programme The theorem restates what we already knew. But its main point is the style of analysis as such. We can now match games and epistemic logic in two directions:

> *From games to logic* Given some algorithm defining a solution concept, we try to find epistemic actions driving its dynamics.

From logic to games Any type of epistemic assertion defines an iterated solution process that may have independent interest.

Other scenarios: announcing strong rationality Instead of *WR*, we can also announce *SR*. An announcement that j is strongly rational leaves only states s where SR_j holds. But this may eliminate worlds, invalidating SR_k at s for other players, as their existential modalities now lack a witness. Thus, repeated announcement of *SR* makes sense:

Example Iterated announcement of Strong Rationality.
Our running example gives the same model sequence for *SR* as with SD^{ω}. But the Strong Rationality sequence differs from *WR* in the following case:

	E	*a*	*b*	*c*			
A	*d*	2,3	1,0	1,1	B_E	–	–
	e	0,0	4,2	1,1	–	B_E, B_A	–
	f	3,1	1,2	2,1	B_A	B_E	B_A

WR does not do anything, *SR* removes the top row and rightmost column. ■

Example Ending in '*SR*-loops'.
In this model, successive announcement of *SR* gets stuck in a 4-loop:

B_E	–	B_A
–	B_A	B_E
B_A	–	B_E

B_E	B_A
–	B_E
B_A	B_E

B_E	B_A
B_A	B_E

■

Nash equilibria in an initial game survive into the limit of *SR*. But this may not be all that remains:

Example Battle of the Sexes revisited.
In this game, announcing *SR* gets stuck: $\langle E \rangle B_E \wedge \langle A \rangle B_A$ holds everywhere. ■

Finally, repeated announcement of *SR* is *self-fulfilling* (cf. Chapters 3, 10):

FACT Strong Rationality becomes common knowledge on game models.

Other rational things to say If the initial game has a Nash equilibrium, and players have decided on one, they can also keep announcing something stronger than *SR*:

$<E>NE \wedge <A>NE$ *Equilibrium Announcement*

Its limit model leaves the Nash equilibria plus all worlds that are both $\sim_E$ and $\sim_A$ related to one.

Further logical aspects Many properties of our dynamic-epistemic logics make sense here. For instance, dynamic-epistemic procedure involves *scheduling*. Order of assertions may matter in game solution (recall the sequential Muddy Children in Chapter 3), reflecting the role of procedure in social scenarios. Another interesting logical fact is that different game models may be epistemically *bisimilar* (cf. Chapter 2), inviting simplification:

Example Bisimulation contractions of game models.
The epistemic-preferential model to the left contracts to that on the right:

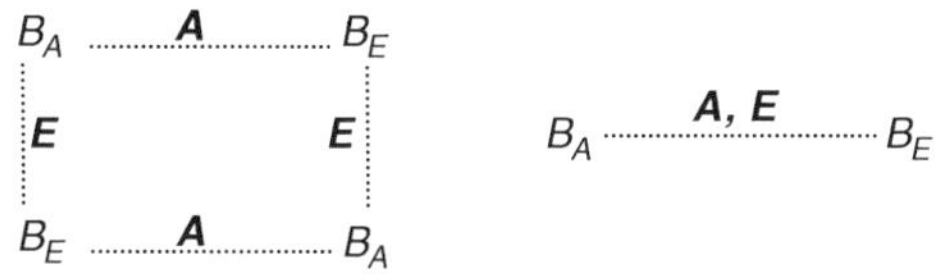

B_A ...A... B_E
E ⋮ ⋮ E
B_E ...A... B_A

B_A ...A, E... B_E

■

Conclusion This concludes our brief description of the *PAL*-style deliberative view of game solution. It offers a radical dynamic alternative to the usual epistemic foundations of game theory. Common knowledge of rationality is not *presupposed*, but *produced* by the logic.

15.5 Logical background: from epistemic dynamics to fixed-point logic

Now we present some further background to the preceding analysis:

Program structure and iterated announcement limits Here is a link with Chapters 3, 11. Consider any formula φ in our epistemic language. For each model $\boldsymbol{M}$ we can keep announcing it, retaining just those worlds where φ holds. This yields a sequence of nested decreasing sets. In infinite models, we take the sequence across limit ordinals by taking intersections of all stages so far. This process always reaches a first *fixed-point*, a submodel where taking just the worlds satisfying φ no longer changes things:

DEFINITION Announcement limits in a model.
For any model **M** and formula φ, the *announcement limit #(φ, **M**)* is the first submodel in the iteration sequence where announcing φ has no further effect. If *#(φ, **M**)* is non-empty, φ has become common knowledge, and we call φ *self-fulfilling* in **M**. In the other case, φ is *self-refuting* in **M**, and its negation has become common knowledge. ■

The assertions for games in Section 15.4 were self-fulfilling, resulting in common knowledge of rationality. The joint ignorance statement of the muddy children was self-refuting.

New problems in dynamic-epistemic logic Determining syntactic shapes of self-fulfilling or self-refuting formulas generalizes the semantic Learning Problem stated in Chapter 3. As for axiomatization, one can add announcement limits explicitly to our language: $\mathbf{M}, s \models \#(\psi)$ iff s belongs to $\#(\mathbf{M}, \psi)$. Is *PAL* with this operator # still decidable?

Equilibria and fixed-point logic To motivate our next step, here is a different take on the SD^{ω} algorithm. The original game model need not shrink, but we can compute a new *property of worlds* in stages, zooming in on a new subdomain that is a fixed-point. Fixed-point operators can be added to first-order logic (Moschovakis 1974; Ebbinghaus & Flum 1995), but we use an epistemic version of the modal *μ-calculus* (Chapters 2, 4, 10):

DEFINITION Positive formulas and fixed-point operators.
Formulas $\varphi(p)$ with only positive occurrences of the proposition letter p define a monotonic set transformation in any epistemic model **M**:

$$F_{\varphi}(X) = \{s \in M \mid \mathbf{M}[p := X], s \models \varphi\}$$

The formula $\mu p\bullet\ \varphi(p)$ defines the smallest fixed point of F_{φ}, starting from the empty set as a first approximation, and $\nu p\bullet\ \varphi(p)$ defines the greatest fixed point, starting from the whole **M** as a first approximation. Both exist for monotone F by the Tarski–Knaster theorem. ■

The smallest fixed-point of a monotone function F is the intersection of all subsets X with $F(X) \subseteq X$, the greatest fixed-point is the union of all X with $X \subseteq F(X)$. Here is a telling connection:

THEOREM The limit set of worlds for repeated announcement of *SR* is defined inside the full game model by $\nu p\bullet\ (\langle E\rangle(B_E \wedge p) \wedge \langle A\rangle(B_A \wedge p))$.

Announcement limits as inflationary fixed points Now for a general analysis behind this. An announcement limit *#(φ, **M**)* arises by continued application of the following map:

$$F^*_{\mathbf{M},\varphi}(X) = \{s \in X \mid \mathbf{M}|X, s \models \varphi\}, \quad \text{with } \mathbf{M}|X \text{ the restriction of } \mathbf{M} \text{ to } X.$$

The function F^* need not be monotone with respect to set inclusion. The reason is this: when $X \subseteq Y$, an epistemic φ may change its truth value from $\mathbf{M}|X$ to the larger model $\mathbf{M}|Y$. We do not compute in a fixed model, as with *vp•* $\varphi(p)$, but in ever smaller ones, changing the range of the modal operators in *j* all the time. Still, announcement limits can be defined in *inflationary fixed-point logic* (Ebbinghaus & Flum 1995):

THEOREM The iterated announcement limit is a deflationary fixed point.

Proof Take any φ, and relativize it to a fresh proposition letter *p*, yielding $(\varphi)^p$. Here, *p* need not occur positively (e.g., it becomes negative when relativizing box modalities), and no μ-calculus fixed-point operator is allowed. But a *Relativization Lemma* applies (cf. Chapter 3). Let *P* be the denotation of *p* in **M**. Then for all *s* in *P*:

$$\mathbf{M}, s \models (\varphi)^p \text{ iff } \mathbf{M} \mid P, s \models \varphi$$

Therefore, the above definition of $F^*_{\mathbf{M},\varphi}(X)$ as $\{s \in X \mid \mathbf{M}|X, s \models \varphi\}$ equals

$$\{s \in M \mid \mathbf{M}[p := X], s \models (\varphi)^p\} \cap X$$

But this computes a greatest fixed point of a generalized sort. Consider any formula $\varphi(P)$, without restrictions on the occurrences of *P*. Now define

$$F^{\#}_{\mathbf{M},\varphi}(X) = \{s \in \mathbf{M} \mid \mathbf{M}[p := X], s \models \varphi\} \cap X$$

This map need not be monotone, but it always yields subsets. Thus, it still finds a greatest *inflationary fixed-point* starting with **M**, and iterating, taking intersections at limit ordinals. If the function $F^{\#}$is monotonic, this coincides with the usual fixed point procedure. ■

Dawar, Graedel & Kreutzer (2004) show that we cannot do better: announcement limits go beyond the epistemic μ-calculus – while the extension with inflationary fixed points is undecidable. Fortunately, however, special types of announcement are better behaved:

Monotone fixed points after all The announcement limit for *SR* was definable in the epistemic μ-calculus. The reason is that the update function $F_{M,\ SR}\ (X)$ is monotone for set inclusion. This has to do with a special syntactic form. *Existential modal formulas* are built with only existential modalities, proposition letters or their negations, conjunction and disjunction. Semantically, they are preserved under *extensions of models*. Now we have

THEOREM $F^{\#}{}_{M,\varphi}\ (X)$ is monotone for *existential* modal formulas φ.

Proof The map $F^{\#}{}_{M,\varphi}$ is monotone by the preservation under extensions: if $X \subseteq Y$, and $M|X, s \models \varphi$, then also $M|Y, s \models \varphi$. Therefore, a μ-calculus greatest fixed point works here. ■

The ignorance announcements of the Muddy Children were existential, too, and hence reasoning about this puzzle lives in the epistemic μ-calculus.[6]

Next, consider the earlier-mentioned order dependence. With monotonic assertions, this can often be avoided. For instance, it can be shown that the announcement limit of SR_E ; SR_A is the same as that of *SR*. A final application of our main theorem uses the decidability of the μ-calculus:

COROLLARY Dynamic-epistemic logic plus $\#(\psi)$ for existential ψ is decidable.

Thus reasoning about Strong Rationality or Muddy Children is simple.

15.6 Variations and extensions: worries, enabling, beliefs, and syntax

Muddy Children revisited The ignorance assertion driving the puzzle suggests *self-defeating game solution methods*. Players might say 'My action may turn out badly' until their worries dissolve. Also, Muddy Children has a crucial *enabling action*, viz. the Father's announcement. Internal communication only

[6] An application of this syntax to games is the fact that, for any epistemic model M, $\#(SR, M) \subseteq \#(WR, M)$. With non-existential formulas, however, even when φ implies ψ, the limit model $\#(\varphi, M)$ need not be included in $\#(\psi, M)$.

reaches the goal after external information has broken the symmetry of the diagram. This, too, makes sense in games:

Example With a little help from my friends.
Some equilibria are reached only after external information removes strategy profiles, say, breaking the symmetry of the '*SR*-loops' that witness Strong Rationality in games:

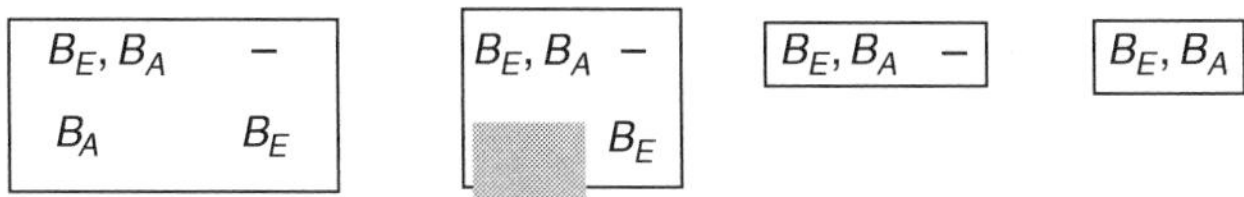

In the first model, nothing is eliminated by *SR*. But after an announcement ruling out the bottom-left world, updates take the resulting 3-world model to its Nash equilibrium. ■

The art here is to find well-motivated enabling statements that set, or keep conversation going: Roy (2008) uses players' *intentions* for this purpose.

Changing beliefs and plausibility Many logical analyses of games use players' *beliefs* instead of knowledge. The main results in this chapter then generalize, using the models of Chapter 7 like we did with Backward Induction in Chapter 10. For instance, with the SD^ω algorithm, in the limit players *believe* they are in the solution set. In addition to hard information about beliefs, one can now also produce *soft information*, changing plausibility patterns over strategy profiles. Zvesper (2010) explores these issues further.

15.7 Recent developments in dynamic-epistemic logic and games

We conclude with some recent work on the dynamic-epistemic logic of games that fits with the broad lines drawn in this chapter:

Agreement and hard update Dégrémont & Roy (2009) use iterated *PAL* statements to analyse the 'agreeing to disagree' results of Aumann (1976). They find curious scenarios:

Example Stated disagreements in beliefs can switch.
Consider two models **M**, **N** with actual world **s**. Accessibility runs as indicated: e.g., in **M**, $\{s, t, u\}$, $\{v\}$ are epistemic equivalence classes for agent 2

ordered by plausibility. For agent 1, the (ordered) epistemic classes are $\{s, v\}$, $\{t\}$, $\{u\}$. Now $B_1\neg p \wedge B_2 p$ is true at worlds s and t only in **M**. Announcing it updates **M** to **N**, in whose actual world $B_1 p \wedge B_2\neg p$ is true:

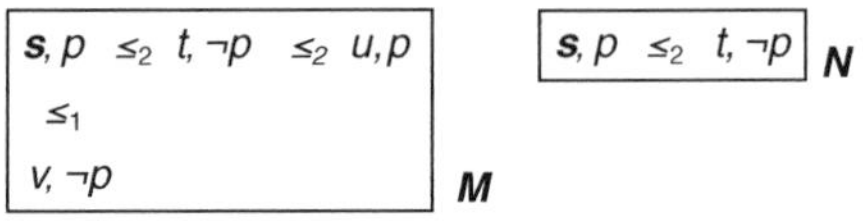

$\mathbf{s}, p \leq_2 t, \neg p \leq_2 u, p$
$\leq_1$
$v, \neg p$

M

$\mathbf{s}, p \leq_2 t, \neg p$

N

■

This leads to a dynamic-epistemic analysis of game-theoretic results from Geanakoplos & Polemarchakis (1982). Any dialogue where players keep stating whether or not they believe that formula φ is true at the current stage leads to agreement in the limit. If agents share a well-founded plausibility order at the start (though their hard information may differ), then in the first fixed-point, they all believe, or all do not believe that φ. Dégrémont (2010) links these results to the earlier issue of syntactic definability in epistemic fixed-point logics.[7]

Iterated soft update Baltag & Smets (2009a) analyse limit behaviour of soft announcements, including the radical $\Uparrow\varphi$ and conservative $\uparrow\varphi$ of Chapter 7. Again, surprises occur:

Example Cycling radical upgrades.
Consider a one-agent plausibility model with proposition letters as shown:

$s, p \leq t, q \leq \mathbf{u}, r$

Here $\mathbf{u}$ is the actual world. Now make the following soft announcement: $\Uparrow(r \vee (B^{\neg r}q \wedge p) \vee (B^{\neg r}p \wedge q))$. This formula is true in worlds s, u only, and hence the new pattern becomes:

$t, q \leq s, p \leq \mathbf{u}, r$

In this new model, the formula $r \vee (B^{\neg r}q \wedge p) \vee (B^{\neg r}p \wedge q)$ is true in t, u only, so radical upgrade yields the original model, starting a cycle:

$s, p \leq t, q \leq \mathbf{u}, r$

■

Note that the final world 3 stays in place. This stability is significant. Among many relevant results, Baltag & Smets (2009a) prove that every truthful

[7] Demey (2010a) extends this style of analysis to a probabilistic setting.

iterated sequence of radical upgrades stabilizes all *simple* non-conditional beliefs. This work is connected to learning theory in Baltag, Gierasimczuk & Smets (2010) (cf. Chapter 11).

Fixed-point logics and abstract game solution Finally, Zvesper (2010) is a belief-based study of game solution using dynamic-epistemic logics, linking up with the framework of Apt (2005). The setting is abstract Level-Three models for games (cf. Chapter 10), with belief operators $B(X)$ sending sets of worlds X to the worlds where the agent finds X most plausible, and optimality operators $O(X)$ selecting those worlds from a set X that are best for the agent in an absolute global sense.[8] The counterpart of the solution concept SD^{ω} is then the greatest fixed-point O^{∞} arising from repeatedly applying the optimality operator O to the model. In this setting, rationality *(rat)* denotes the set $O(\mathbf{B})$ where agents do their best given their most plausible belief worlds $\mathbf{B}$. These sparse models suggest a modal fixed-point language that can define quite a few basic epistemic notions. For instance, the proposition $CB\varphi$ (common belief in φ) is defined by the greatest fixed-point formula $\nu p \bullet \wedge_i B_i(\varphi \wedge p)$.

This syntax supports very perspicuous formal derivations, in a minimal modal logic plus the basic inference rules of the modal μ-calculus for greatest fixed-points:

$\nu p \bullet \psi(p) \rightarrow \psi(\nu p \bullet \psi(p))$	unfolding axiom
if $\vdash \alpha \rightarrow \psi(\alpha)$, *then* $\vdash \alpha \rightarrow \nu p \bullet \psi(p)$	inclusion rule

Example Proving an epistemic characterization theorem syntactically.
We will leave out agent indices for convenience. First, for all formulas φ,

(a) $rat \rightarrow (B\varphi \rightarrow O\varphi)$

since $B\varphi$ implies that $\mathbf{B} \subseteq [[\varphi]]$ – and using the monotonicity of O plus the definition of *rat*. Simplifying $CB\varphi$ to the notation $\nu p \bullet B(\varphi \wedge p)$, one fixed-point unfolding yields:

(b) $CB\ rat \rightarrow B(rat \wedge CB\ rat)$

[8] Absolute best (cf. our earlier atoms B_A, B_E) makes optimality *upward monotonic*: if $X \subseteq Y$, then $O(X) \subseteq O(Y)$. By contrast, context-dependent 'best' cookies need not be best foods.

From this, using observation (a) with $\varphi = rat \wedge CB\ rat$, we get

(c) $rat \wedge CB\ rat \rightarrow O(rat \wedge CB\ rat)$

Finally, by the introduction rule for greatest fixed-points,

(d) $rat \wedge CB\ rat \rightarrow \nu q \bullet Oq\ (= O^{\infty})$ ■

Thus, a few lines of modal proof capture the essence of a famous characterization theorem in the epistemic foundations of game theory.[9]

Moreover, we can inspect proofs for new interesting features. Indeed, in the above, little logic is used: only monotonicity of belief and optimality modalities, not distribution over conjunction. Hence the preceding analysis also works on the general topological or neighbourhood models of Chapter 2. These are closer to the syntactic models of Chapter 5, suggesting that the core of game theory might also apply to agents that are bounded in their syntactic logical powers of inference.[10]

15.8 Conclusion

Acts of deliberation underlie much of game theory. Now, in physics, an equilibrium only becomes intelligible with a dynamic account of the forces producing it. Likewise, epistemic equilibrium is best understood with an account of the actions leading to it. For this purpose, we used update scenarios for virtual communication, in a dynamic-epistemic logic that changes game models. To us, this approach also fits better with the intuitive notion of rationality, not as static harmony between utilities and expectations, but as a *quality of action*. In that sense, rationality resides in the procedure being followed. Summarizing, solving a game involves dynamic-epistemic procedures, and game-theoretic equilibria are their greatest fixed points. This suggests a general study of game solution, instead of separate characterization theorems. We found technical results on dynamic-epistemic

[9] De Bruin (2004) has further examples of the virtues of modal proof for game theory.

[10] Zvesper defines *DEL*-style dynamic actions on neighbourhood models (cf. Chapters 2, 4, 5). This is extended to a general evidence dynamics in van Benthem & Pacuit (2011).

fixed-point logics that may help. Of course, our models were still crude, and missed some sophisticated distinctions. And we do not claim that logical dynamics is a miracle cure for all cracks in the foundations of game theory. But it does add a new way of looking at things, and our scenarios seem fun to explore, extend, and play with.

16 Meeting cognitive realities

The themes in this book paint a more realistic picture than traditional views of logic as a study of abstract consequence relations. Agency, information flow, and interaction seem close to human life, suggesting links between logical dynamics and empirical cognitive science. But how much of this is genuine? Our systems are still mathematical in nature, and they do not question the normative stance of logic toward inference: they rather extend it to observations, questions, and other informational acts. Still, many logicians today are intrigued and inspired by empirical facts about cognition: pure normativity may be grand, but it can be lonely, and even worse: boring. In this chapter, we therefore present a few thoughts on the interface of logical dynamics and cognitive science.

16.1 Self-imposed borders

Logic has two historical sources: argumentation in the dialectical tradition, and axiom-based proof patterns organizing scientific inquiry. Over the centuries, the discipline turned mathematical. Is logic still about human reasoning? Or is it, as Kant and Bolzano said, an abstraction in the realm of pure ideas? Then logical consequence is an eternal relationship between propositions, firmly cleansed of any stains, smells, or sounds that human inferences might have – and hence also of their colours, and tantalizing twists and kinks. Do empirical facts about human reasoning matter to logic, or should we just study proof patterns, and their armies called formal systems, in an eternal realm where the sun of Pure Reason never sets? Most logicians think the latter. Universities should just hire logicians: in Tennyson's words, 'theirs not to reason why'. If pressed, philosophers might say that logic is normative; it describes *correct* reasoning. People would be wise to follow its advice – but so much the worse for them, if they do not. The butler Gabriel Betteredge said it all (Wilkie Collins, *The Moonstone*, 1868):

> *'Facts?' he repeated. 'Take a drop more grog, Mr. Franklin, and you'll get over the weakness of believing in facts! Foul play, sir!'*

The divide between logic and human reasoning is enshrined in Frege's famous doctrine of Anti-Psychologism: human practice can never tell us what is *correct* reasoning. I find this demarcation line a sign of poverty, rather than splendid isolation. Frege wrote in the days when modern psychology was taking off, in the pioneering work of Helmholtz on the origin of mathematical concepts, or of Wundt on perception. Frege's Thesis blocks creative impulses from best practices of humans, and psychology as their scientific chronicler. Anti-Psychologism is still defended with fervour today. My own view is less definite. Logic is certainly not psychology, and it has goals of its own. And a logical theory is not useless if people do not quite behave according to it. But the boundary is delicate. If logical theory were really totally disjoint from actual reasoning, then it would be no use at all. And in fact, it is not. In this chapter, I chart relations between theory and human practice in logic.[1]

16.2 Logical systems and human behaviour

An inference pattern like Modus Tollens: *from $A \to B$ and $\neg B$, it follows that $\neg A$*, is valid whether we want it or not. Not even a *UN* referendum on our planet would change that. But so what? Humans are designed by Nature according

[1] Many disciplines steer a course between normativity and description. For the delicate status of game theory, cf. the entries by Camerer and Rubinstein in Hansen & Hendricks (2007). Probability theory, too, has a tension between correct rules and failures when humans reason with uncertainty, such as converting conditional probabilities. But it also acknowledges that a core calculus cannot be tested separately from choices made by subjects of a *probabilistic model* for their situation. An analogue for logic is a deeper study of choosing and managing qualitative models, as done for the Wason Card Task in van Lambalgen & Stenning (2007). Probability theory is unlike logic in having internalized the tension, studying objective frequency-based probability in tandem with subjective degrees of belief. Already Johan de Witt's famous treatise *Waerdije* from 1671, in terms of betting odds, measures a subjective notion by observable transactions. But also philosophy and especially epistemology have tensions between theory and empirical claims. We are all in the same boat.

to its objective laws, so there is at least no a priori conflict between what we do and logical norms of correctness.

Indeed, if observed practice diverges from a logical norm, what does that mean? Before we can reason at all, we must have a *representation* of the relevant facts and events with the information they convey. This modelling phase can be conscious, or hard-wired in our neural nets, but it is indispensable. Next, on these models, there is a choice of processes for the task at hand: classical or constructive inference in mathematics, classical or non-monotonic in problem solving, and so on. Given the freedom in this tandem scheme of *representation* + *transformation*, most practices seem consistent with logical theory. And there is even one more degree of freedom in this book that blurs the confrontation with reality. Our dynamic logics parametrized agents' powers of observation or policies for belief revision by devices like multi-agent event models and agent-dependent plausibility orders. While these are appealing conceptually, as a side-effect, much experimentally observed behaviour can be absorbed into setting such parameters, immunizing the logic.

Thus, our worry should not be divergence, but lack of it. Our modus operandi may be too good a protective belt for logic versus practice: it can fit anything. But I would rather let logic learn from confrontation with the facts. To do so, one must aim for direct accounts of stable informational practices. And this is exactly what is happening. Before spinning further a priori thoughts,[2] let us take a look at some real developments.

16.3 Logical theory already listens to practice

For a start, much of Philosophical Logic has always been about representing structures in language and thought beyond the minimum provided by first-order logic.[3] Prior's work in the 1950s on time and temporal reasoning introduced temporal structure into logic to account for the tenses and other modalities of actual use. And in the 1960s, Lewis and Stalnaker appealed to actual conditional reasoning when introducing their new logical notion $\varphi \Rightarrow \psi$ saying that the minimal φ-worlds in the ordering also have ψ true.

[2] Books on the philosophy of logic mostly ignore developments in modern logic, making them defenders of the status quo of at best the 1950s – and often not even that.

[3] The handbook Gabbay & Guenthner (1983–1999) shows this in great detail.

And there are many other influential examples. Benchmarks for correctness of such logical accounts were diverse: faithful analysis of philosophical arguments in natural language, a priori conceptual intuitions, but definitely also ordinary usage.[4] For instance, conditional reasoning is close to the irrealis mode where we consider situations beyond the here and now. Next came the logical semantics of natural language, studying expressions beyond standard logic such as generalized quantifiers, moving closer to actual reasoning. This resulted in rich accounts of information states for real language users, with discourse representation theory as a leading example.[5] Finally, modelling realistic common sense reasoning has been a hallmark of logics in Artificial Intelligence – in Clausewitzian terms: a continuation of philosophy by other means, in this case, an influx of reality.

Even though none of this involved actual psychological experiments, pioneers like Prior, Lewis, or Hintikka did appeal to ordinary language use – not just unverifiable a priori intuitions of gentlemen of leisure. One might summarize this as a trend toward a rich logical modelling machinery beyond standard systems, with families of worlds, temporal perspective, minimization along orderings, and at the level of syntax, systems for anaphoric text coherence and incremental construction of representations.[6]

In using these richer representations, logical theory has again been influenced by practice. In fact, the idea that there is just one notion of consequence was absent in the pioneering work of Bolzano, who charted many different consequence relations, depending on the reasoning task at hand. A similar view is found in the work of Peirce on the triad of deduction, induction, and abduction that is still quite alive today. All this got further impetus in the 1980s with non-monotonic logics for default reasoning in AI, that model human problem solving or planning in the presence of errors and corrections. The important thing here is not a catalogue of inference rules (non-monotonicity is a symptom, not a diagnosis), but underlying mechanisms that might have cognitive reality.[7]

4 Philosophical logic is sometimes taken to be the mind police of formal language philosophy – whereas 'natural language philosophy' was more enamoured of actual behaviour. But facts have played a crucial role in both paradigms.

5 The *Handbook of Logic and Language* (van Benthem & ter Meulen 1997, updated in 2010) is an extensive source for the topics mentioned here, and many more.

6 Another strand is graphical reasoning, as in Barwise & Etchemendy (1991), Kerdiles (2001).

7 Another idea closer to human practice is the role of *resources* in substructural logics.

In recent years, the influence of facts has gone even further. For instance, recent work at the interface of natural language, logic, and game theory (Gärdenfors & Warglien 2007; van Rooij 2004) has a clear cognitive slant. Moreover, formal semanticists are beginning to feel the pull of statistical corpus research, i.e., records of what language users actually do. If the issues logicians have worried about hardly occur at all (Sevenster 2006 has a telling example about the scarcity of branching quantifiers), while others do, this is now cause for unease, where a decade ago, one would just dismiss this as inferior to theorists' intuitions.[8] We will see the important role of large records again later.

Finally, as we said at the beginning, this book suggests greater realism by virtue of its broader range of informational processes, starting from varieties of inference, but also including observation, communication, and strategic interactions between agents.

16.4 From cognitive science to logic, and back

Admittedly, the link to reality in the above is not with experimental psychology or neuroscience. Love of facts can be very Platonic. Logicians analysing natural language, or computer scientists modelling common sense, tend to go by their own intuitions, anecdotal evidence from colleagues, statistically non-significant surveys of sometimes surprising naiveness, and other procedures that avoid the labours of careful experimental design. But even so, experimental evidence is relevant, in that these theories can be, and sometimes are, modified under pressure of evidence from actual usage. There is a growing literature linking logical research with experimental psychology:

Things are happening Semantics of natural language has old links with psychology, witness the chapter by Steedman in van Benthem & ter Meulen (1997), the natural logic in Geurts (2003), or Hamm and van Lambalgen (2004) on default reasoning and linguistic processing. Other contacts include the monograph Stenning (2001) on visual and symbolic reasoning, Castelfranchi & Paglieri (2005), Lorini & Castelfranchi (2007) on psychological models for

[8] I am sensitive to this worry of scarcity myself, but fortunately, one finds logical words and function words in general very high up in frequency lists of natural languages. We are on to something real.

revision of beliefs and goals, and Dunin-Keplicz & Verbrugge (2002) on formation and maintenance of collective intentions. Yet another interface runs between psychology and non-monotonic logics, witness the 2005 *Synthese* volume edited by Leitgeb & Schurz on Non-Monotonic and Uncertain Reasoning in Cognition, including links between default logics and neural nets. A review of links with neuroscience is Baggio, van Lambalgen & Hagoort (2007).

From cognitive science to logic What I find intriguing about our cognitive behaviour is a number of features that call for a richer notion of a logical system. One is the *situatedness* of reasoning that has been noted by philosophers and cognitive scientists. It involves the embodied nature of cognition, based on stable correlations between human bodies and their physical environment (Barwise & Perry 1983). In line with this, reasoning mixes inference with observation of the current situation. Recall our Restaurant of Chapter 1 with questions and inferences: the answers are the situated observations. Accounting for this mixture was one of the key themes in this book. As a second major issue, in cognitive reality, there are other information sources, in particular, our *memory*. Modern linguistic theories take this crucial capacity seriously (cf. the 'data-oriented parsing' of Bod 1998). With experience, we accumulate a stock of understood sentences and comprehended situations (a 'mental corpus'). The same is true for reasoning with a stock of solved problems. When confronted with a new task, *two* processes kick in: memory search for related solved problems, and supplementary rule-based analysis. This seems a much better account of reasoning, both in the common sense and in science, than traditional proof search from a tabula rasa. The interplay of memory and inference in problem solving is an intriguing challenge to the systems of this book. I expect logics based on this architecture to lead to exciting new mathematical theory, merging classical proof systems with probabilistic memory.

Next, I find two further phenomena of particular interest. While most logics model mature performance in steady state, perhaps the most striking cognitive process is *learning*. How do we come to learn logical inference, a skill that comes in stages (Piaget 1953)? Should learnability be a constraint on logics? The other feature that strikes me is *diversity* of cognitive agents. People do differently on tasks, and yet we manage to cooperate quite well – even roping in dumb cognitive partners like machines.[9]

[9] Even professional talent shows great diversity in skills, not excepting talent in logic.

This broader agenda is not hostile to traditional logic. Logical systems are a good focus for research, if we see them for what they are: *models* for cognitive activities.[10]

From logic to cognitive science There are also influences from logic on cognitive science. The early psychology of reasoning (Wason & Johnson-Laird 1972) seemed to show that people do not think according to the rules of logical calculi. But as we have noted, human behaviour in experiments needs to be interpreted, and claimed divergences may be questionable.[11] Also, experiments in the psychology of reasoning and their follow-up are just coral reefs in an ocean of practice. Logical theories have turned out welcome as a way of deriving testable predictions from a more systematic viewpoint. The new logics of inference, update, and interaction in this book all suggest hypotheses about behaviour. I even think that logical dynamics can contribute to understanding of how humans form and maintain representations, turning model theory into *theory of modelling*.[12]

16.5 Cognition as interaction

Let us now move to the perspective of this book, viz. Logical Dynamics.

Multi-tasking and agent diversity Some of the above themes are central to logical dynamics. From the start we emphasized how intelligent behaviour involves multi–tasking: not just inference, but also asking questions, and getting information out of events. To do so, we modelled agents' information, and kept it attuned to the truth through updates. Thus, both situatedness of reasoning and mixes of logical tasks drive our dynamic-epistemic

[10] This point has been made forcefully in Barwise & Etchemendy (1991). For a discussion of architectural issues in the design of logical systems, cf. van Benthem (1999c).

[11] A psychologist once confessed to me that despite all deviations from logic in short-term inferences like the Wason Card Task, he had never met an experimental subject who did not understand the logical solution when it was explained to him, and then agreed it was correct. Why should the latter longer-term fact be cognitively less significant than the short-term one?

[12] Evidence for optimism is the *Topoi* issue 'Logic and Psychology' (Hodges, Hodges & van Benthem 2007), with papers on belief revision, default reasoning, numerical reasoning, natural language interpretation, conditional reasoning, and cognitive evolution, with links between logic, linguistics, game theory, cognitive psychology, and brain research.

logics. And other ideas about agents percolated, too, including a richer set of cognitive attitudes, as well as the dynamics of belief, preferences, and goals. With belief revision in particular, we had our first connection with learning. And we encountered diversity in styles and talents for inference, computation, belief revision, or memory capacity. Thus, our dynamic logics provide a systematic take on many of the earlier themes.[13]

Social interaction But most of all, we have emphasized a feature of cognition that goes beyond the usual emphasis on performance of individuals: whether their bodies, minds, or brains. When King Pyrrhus of Epirus, one of the foremost well-educated generals of his age, crossed over to Italy for his famous expedition, the first reconnaissance of a Roman camp near Tarentum dramatically changed his earlier perception of his enemies (Plutarch, *Pyrrhus*, Penguin Classics, Harmondsworth, 1973):

> Their discipline, the arrangement of their watches, their orderly movements, and the planning of their camp all impressed and astonished him – and he remarked to the friend nearest him: 'These may be barbarians; but there is nothing barbarous about their discipline'.

It is intelligent social life that shows human cognitive abilities at their best, just as competitive games lead to the most beautiful exhibitions of graceful movement. Yet much of cognitive science emphasizes powers of single agents: reasoning, perception, memory, or learning. And this focus persists in neuroscience, as single brains are easiest to put in a scanner. By contrast, this book has emphasized the social side of cognition, with multi-agent activities from language and communication to games.[14] To some, lonesome proof is the cognitive peak experience. I find a heart-to-heart conversation, a committee meeting, or a military campaign more striking as displays of what makes us intelligent.[15]

Other partners Admittedly, other disciplines than logic lead the way in empirical contacts. Experimental game theory developed in the 1990s, and mixtures of games, computation, and real behaviour have been studied even

[13] As a challenge, one would like to find logical reasons for the success of the simple knowledge- and ignorance-based algorithms in Gigerenzer, Todd & the ABC Group (1999).

[14] Similar trends underlie modern studies of argumentation (Gabbay & Woods 2004).

[15] Verbrugge (2009) is a masterful survey of current research on social cognition where cognitive psychology and computational models for the brain meet with logic.

earlier. Consider the above theme of diversity. When interacting with other agents, we need to understand their processing capacities and their strategies. Famously, Axelrod (1984) studied the simple automaton strategy of *Tit-for-Tat* that cooperates if you did in the preceding round, and defects if you defected there. Axelrod showed how, in social encounters, *Tit-for-Tat* wins out against much more complex ways of punishing and rewarding past behaviour. Still, *Tit-for-Tat* is essentially a logical *Copy-Cat* strategy (it is central to interactive computation, cf. Abramsky 2008), so we do not lose the link with logic in studies of real behaviour.

16.6 Logic, cognition, and computation

So, what is the proper relation between logic and cognition? Should we endorse a New Psychologism? This book has a richer conception of logic than the traditional one, inviting contacts with the empirical facts. But all this should be taken in moderation:

Modest psychologism First, as said before, the dynamic logics of this book are normative: they describe how information flows in some ideal sense. And even their parameters for agents' attitudes and processing abilities can be seen as an a priori buffer, as we saw. Still, our discussion showed that logic can absorb psychological insights, and inspire cognitive science. In what sense is this psychologism? Compare the term 'physicalism'. Nowadays, that is not the eighteenth-century claim that everything is just moving particles and their collisions, but the idea that mental behaviour can be described using sophisticated abstract concepts of modern physics. Likewise, human behaviour in cognitive science is a hierarchy of levels, ranging from plain observable facts to sophisticated higher-order descriptions. And then, the fit with logical theory becomes much more plausible, in both directions.[16] I conclude with two aspects in this view of logic that I see as crucial.

[16] Will this prespective dissolve logic as we know it? My view, stated in Chapter 13, is that there remains one logic, but not in any particular definition of logical consequence, or favoured logical system. The unity of logic resides in the modus operandi of its practitioners: i.e., its style of agency.

The balance between correctness and correction My modest psychologism does shift away from the classical foundational turn. Ordinary reasoning becomes the key topic, with mathematical proof a special case. Now there is an undeniable otherworldly beauty to the fundamentalist view that logic should be concerned only with scientific truth. But I would posit another conception, much more dynamic. Frege was obsessed with providing security, once and for all, for mathematical reasoning. To me, however, the key issue is not the static notion of *correctness*, but dynamic *correction*, and the balance between the two. The most striking feature of human intelligence is not being right all the time, but an amazing competence in getting things on the right track once beliefs or plans have gone astray.[17] The same is also true in natural language. We do not have to be completely successful in every step of communication, because we have the striking ability to deal with problems of misunderstanding as they arise. Semantics, too, needs mechanisms of revision, and the balance between correctness and correction is the central phenomenon to grasp. Logic should understand this dynamic behaviour, which involves revision mechanisms beyond inference, as we have amply demonstrated in this book. And on that view, logic is not the static guardian of correctness, as most textbooks have it, but rather the much more dynamic, and much more inspiring, *immune system of the mind*!

A triangle: logic, cognition, and computation Here is one more key fact about the match between logic and reality. There is a third partner. Logic provides models for human reasoning, but they are idealized and have creative divergences. The latter are important in suggesting *new practices*, that often come in computational guises, witness areas like model checking or automated theorem proving.[18] Indeed, one striking feature about human cognition are the ways in which we manage to integrate formal practices into our human behaviour. This *insertion* of designed practice into behaviour happens all the time. Puzzles and games become a natural repertoire, and so do new media like email, enhancing our communicative capabilities – or mathematics itself, enhancing normal reasoning with new tools. Going beyond the status

[17] As Joerg Siekmann once said, the peak moment in a mathematics seminar is not when someone presents a well-oiled proof, but when she sees a mistake and recovers on the spot.

[18] Many reasons for the congeniality of logic and computation were discussed in Chapter 14. Indeed, the 'Turing Test' of Turing (1950) is a powerful historical example of how computation and cognition have long interacted.

quo creates new behaviour that works, and that creative role of logic, too, is a cognitive reality all around the dynamic systems in this book.[19]

Facts to investigate In logic, much professed love of cognition or common sense is mere rhetoric. We still have to go out and confront our theories with cognitive practice. What information do people take out of assertions? How do plans and strategies really work? How do people cope with agent diversity? How do they learn skills in intelligent interaction? Experimental projects in my immediate vicinity include research on theory of mind and iterated knowledge, relating computational complexity to difficulty in games, and stages in which young children learn to play games. Still, in addition to controlled experiments, we should also keep looking at the world around us. Just watch your students play information games on their *GSMs* (epistemic puzzles come as accessories), or give people games with modified rules, say, changing Chess to *Kriegsspiel*, where one does not see the opponent's pieces. After lectures on this book, audiences told me how they play modified information games, e.g., complicating *Clue* to make it more interesting. Society around us is a free cognitive lab, where experimental subjects enjoy what they are doing.

16.7 Conclusion

Logic and cognitive science interface in many ways today, to mutual benefit. A Barrier Thesis like Frege's Anti-Psychologism may have kept the faithful together at a distance from other flocks – but reality always seeps through the cracks. Moreover, since abstract theory influences reality, not just by being right about the status quo, but also through design of new intelligent practices with successful insertions into our lives, the interface between logic and practice is much more diverse than a simplistic normative/descriptive rift would ever allow us to see. And thus, logic can be much more than what it is now.

[19] We communicate with more machines than fellow humans, and virtual reality is all around us. This triangle affair between logical theory, experimental facts, and computational design reflects a deep insight. The very same phenomenon may live in the models of a logical theory, be embodied in human cognition, or exist in the design of some computational process running on a machine.

16.8 Literature

There are no standard textbooks or handbooks on logic and cognitive science. The *Topoi* issue Hodges, Hodges & van Benthem (2007) collects a few current strands, Baggio, van Lambalgen & Hagoort (2007) references other streams of work, and so does Verbrugge (2009), including many interfaces with computer science and neuroscience. Leitgeb (2008) has a number of philosophical papers reappraising the status of psychologism in logic. This chapter is based on the two position papers van Benthem (2007a, 2008b).

17 Conclusion

This book has presented a logical theory of agents that receive and exchange information. It shows how the programme of Logical Dynamics can deal with inference, observation, questions, communication, and other informational powers in one uniform methodology, putting the relevant actions and processes explicitly in the logic, and making use of all major kinds of information: semantic and syntactic. The agents involved in all this have knowledge, beliefs, and preferences – and thus, our systems incorporated knowledge update by observation and inference, varieties of belief revision and learning, and even acts of evaluation and preference change providing a richer motivating dynamics behind the pure kinematics of information. Moreover, from the start, a focus on single agents was not enough: interaction with others is of the essence in argumentation, communication, or games. Our logics dealt with multi-agent scenarios that led to longer-term interaction over time in larger groups, linking up eventually with temporal logic, game theory, and social choice theory. The resulting systems are perspicuous in design, and the dynamic stance also raises many new problems for technical research and philosophical reflection.

More concretely, this book contains a coherent series of technical results on logics of agency, and it generates many further open problems now that the basic theory is in place. Many of these problems are enumerated at the end of chapters. Some are just interesting further paths where readers might stroll, while others represent more urgent tasks given the current limitations of the framework. Of the latter, I would mention *dynamic logics over neighbourhood models* as more fine-grained representations of information, *predicate-logical versions* of our propositional dynamic logics that can deal with objects, and also, the logical dynamics of *changing languages* and concepts. But also, there are challenges in building further bridges between dynamic-epistemic logic and mathematical logic, as well as areas of mathematics like probability theory and dynamical systems. A final challenge is putting together our separate systems for informational and evaluative actions into richer logical models for cognitive agents. Questions and inference were natural companions right at the start of this book, inference and corrective acts of belief revision were another natural pair, and we studied even more complex kinds of entanglement adding preference and action, in the setting of games. But we are far from understanding the total behaviour of the appropriate *combined systems* and of the factors that determine complexity as subsystems are linked.

Next, there is the link with empirical reality, suggested by the very term 'agency'. Our logics normatively describe correct information flow, but they offer more footholds for reality than standard ones. One area where this comes up is natural language, our major vehicle of communication, where one single sentence can serve many cognitive purposes at the same time: informational, directive, evaluative, and so on. Here, logical dynamics meets the *dynamic semantics* of linguistic expressions, and an interesting encounter might happen between the two 'cousins'. Our dynamic logics tend to separate different acts of information and evaluation in their syntax, whereas dynamic semantics executes all of them 'implicitly' with the same linguistic code. A second relevant empirical area that calls for logical study in our sense is the daily practice of deliberation and *argumentation*, since so many of our topics in this book occur in that concrete multi-agent setting. Just think of the fact that one of the most frequent triggers for belief revision and preference change is not bare signals, but contacts with other people: say, through confrontation in debate.

While we find our dynamic-epistemic logic attractive, it is by no means the only method for studying agency. We think it is good for a field to foster different perspectives that complement each other – and we would be satisfied if this book added a stance that is simple and yet revealing. Indeed, we do look at connections with other paradigms, such as probabilistic methods and temporal logics. Our feeling is that, despite occasional ill-humour between schools, at some level of rational agency, there is a Church Thesis about equivalence, or at least complementarity of all major approaches. But at this stage, we are far from even having a consensus on what a rational agent is supposed to be and do.

This book also sits at the interface of several disciplines: philosophy, computer science, game theory, and the social sciences, to name a few. All share a trend of moving away from isolated activities to multi-agent interaction, and logic itself is no exception. Logic in this book is about all sorts of informational processes: observation, inference, communication, and others – whose proper setting are many-mind problems: multi-agent scenarios of language use, argumentation, and games. We think this is close to the historical origins of the discipline in rational argumentation, rather than a new-fangled fad. And also, it is one of the enduring virtues of logic that it sits at an academic crossroads between the sciences and humanities, allowing us to see new developments in many fields, their analogies, and deep structure. We have seen in a number of chapters what things become visible when we take the theory of this book to the disciplines that meet at the logician's doorstep.

We even think that the perspective of Logical Dynamics is worth teaching at early levels, since it makes logic very different from the formalist enterprise that it should try not to be. But in addition to teachable skills, Logical Dynamics is a general intellectual stance that one can take in many settings. Hopefully, our chapters on philosophy, computer science, cognitive science, and game theory have shown by example how fresh that stance can be.

This book is just a beginning, and there are further things on the horizon. In particular, I am completing a companion volume on *Logic and Games*, highlighting that interface in much more detail than has been done here. Likewise, there are connections with computer science, where dynamic-epistemic logic ties in naturally with studies of agency and security. But the old connection with *philosophy* is closest to my heart right now, as may be seen in the length of the relevant chapter. I hope that logic will have a second chance in its mother area. A monograph on the interface of dynamic-epistemic logic and epistemology is in the making, together with Alexandru Baltag and Sonja Smets.

Finally, I repeat some points from the Preface. Being explicit about one's motivations, as I have tried to be, may be helpful, but at the same time a hindrance to readers who do not share the author's ideology. I hope that many notions and results in this book are appealing per se, even if you have no wish to reform logic, and lose no sleep over the daily affairs of rational agents. And if there is pleasure in reading, it is bound to mean something in the end.

References

S. Abramsky (2006) 'Socially Responsive, Environmentally Friendly Logic', in T. Aho & A.-V. Pietarinen, eds., *Truth and Games: Essays in Honour of Gabriel Sandu*, Acta Philosophica Fennica, 17–45.

(2008) 'Information, Processes and Games', in P. Adriaans & J. van Benthem, eds., *Handbook of the Philosophy of Information*, Amsterdam, Elsevier Science Publishers, 483–549.

S. Abramsky & R. Jagadeesan (1992) 'Games and Full Completeness for Multiplicative Linear Logic', *Journal of Symbolic Logic* 59:2, 543–574.

P. Adriaans & J. van Benthem, eds. (2008) *Handbook of the Philosophy of Information*, Amsterdam, Elsevier.

M. d'Agostino & L. Floridi (2007) 'The Enduring Scandal of Deduction. Is Propositional Logic Really Uninformative?', First Oxford Workshop on the Philosophy of Information. Final version in *Synthese* 167, 2009, 271–315.

T. Ågotnes, J. van Benthem, H. van Ditmarsch & S. Minica (2010) 'Question Answer Games', *Proceedings LOFT* 2010, University of Toulouse.

T. Ågotnes & H. van Ditmarsch (2009) 'What Will They Say? Public Announcement Games', University of Bergen & University of Seville. Presented at: Logic, Game Theory and Social Choice 6, Tsukuba, Japan. To appear in *Synthese (Knowledge, Rationality and Action).*

T. Ågotnes, V. Goranko & W. Jamroga (2007) 'Alternating-Time Temporal Logics with Irrevocable Strategies', in D. Samet, ed., *Proceedings of TARK IX*, Univ. Saint-Louis, Brussels, 15–24.

M. Aiello, I. Pratt & J. van Benthem, eds. (2007) *Handbook of Spatial Logics*, Dordrecht, Springer Academic Publishers.

R. Alur, T. Henzinger & O. Kupferman (1997) 'Alternating-Time Temporal Logic', *Proceedings of the 38th IEEE Symposium on Foundations of Computer Science*, Florida, October 1997, 100–109.

P. Anand, P. Pattanaik & C. Puppe, eds. (2009) *The Handbook of Rational and Social Choice*, Oxford University Press.

H. Andréka, J. van Benthem & I. Németi (1998) 'Modal Logics and Bounded Fragments of Predicate Logic', *Journal of Philosophical Logic* 27, 217–274.

H. Andréka, M. Ryan, & P.-Y. Schobbens (2002) 'Operators and Laws for Combining Preference Relations', *Journal of Logic and Computation* 12, 13–53.

K. Apt (2005) 'The Many Faces of Rationalizability', CWI Amsterdam. Final version in *Berkeley Electronic Journal of Theoretical Economics* 7, 2007.

K. Apt & R. van Rooij, eds. (2007) *Proceedings KNAW Symposium on Games and Interaction*, Texts in Logic and Games, Amsterdam University Press.

K. Apt, A. Witzel & J. Zvesper (2009) 'Common Knowledge in Interaction Structures', *Proceedings TARK XII*, Stanford, 4–13.

H. Arlo-Costa & E. Pacuit (2006) 'First-Order Classical Modal Logic', *Studia Logica* 84, 171–210.

S. Artemov (1994) 'Logic of Proofs', *Annals of Pure and Applied Logic* 67, 29–59.

(2001) 'Explicit Provability and Constructive Semantics', *Bulletin of Symbolic Logic* 7, 1–36.

(2007) 'Justification Logic', Technical Report TR-2007019, CUNY Graduate Center, New York.

G. Aucher (2004) 'A Combined System for Update Logic and Belief Revision', Master of Logic Thesis, ILLC, University of Amsterdam.

(2008) 'Perspectives on Belief and Change', Dissertation, IRIT, Toulouse.

(2009) 'BMS Revisited', in A. Heifetz, ed., *Proceedings of Theoretical Aspects of Rationality and Knowledge (TARK 2009)*, Stanford, 24–33.

R. Aumann (1976) 'Agreeing to Disagree', *The Annals of Statistics* 4:6, 1236–1239.

R. Axelrod (1984) *The Evolution of Cooperation*, New York, Basic Books.

F. Bacchus (1990) *Representing and Reasoning with Probabilistic Knowledge, A Logical Approach to Probabilities*, Cambridge (Mass.), The MIT Press.

G. Baggio, M. van Lambalgen & P. Hagoort (2007) 'Language, Linguistics and Cognition', to appear in M. Stokhof & J. Groenendijk, eds., *Handbook of the Philosophy of Linguistics*, Amsterdam, Elsevier.

P. Balbiani, A. Baltag, H. van Ditmarsch, A. Herzig, T. Hoshi & T. de Lima (2008) 'Knowable as Known after an Announcement', *Review of Symbolic Logic* 1, 305–334.

P. Balbiani, A. Herzig & N. Troquard (2007) 'Alternative Axiomatics and Complexity of Deliberative STIT Theories', IRIT, Toulouse.

A. Baltag (2001) 'Logics for Insecure Communication', in J. van Benthem, ed., *Proceedings TARK Siena 2001*, 111–122.

(2002) 'A Logic for Suspicious Players: Epistemic Actions and Belief Update in Games', *Bulletin of Economic Research* 54, 1–46.

A. Baltag, J. van Benthem & S. Smets (2010) *A Dynamic-Logical Approach to Epistemology*, Universities of Oxford, Amsterdam, and Groningen.

A. Baltag, H. van Ditmarsch & L. Moss (2008), 'Epistemic Logic and Information Update', in P. Adriaans & J. van Benthem, eds., *Handbook of the Philosophy of Information*, Amsterdam, Elsevier Science Publishers, 361–456.

A. Baltag, N. Gierasimczuk & S. Smets (2010), 'For Tracking the Truth, Keep Revising Your Beliefs', Invited lecture at *NASSLLI 2010*, Bloomington.

A. Baltag & L. Moss (2004) 'Logics for Epistemic Programs', *Synthese: Knowledge, Rationality, and Action* 2, 165–224.

A. Baltag, L. Moss & S. Solecki (1998) 'The Logic of Public Announcements, Common Knowledge and Private Suspicions', *Proceedings TARK 1998*, 43–56, Morgan Kaufmann Publishers, Los Altos.

A. Baltag & S. Smets (2004) 'The Logic of Quantum Programs', *Proceedings of the 2nd International Workshop on Quantum Programming Languages*, TUCS General Publication No. 33, Turku Center for Computer Science. Extended version '*LQP*: The Dynamic Logic of Quantum Information', *Mathematical Structures in Computer Science* 16, 2006, 491–525.

(2006) 'Dynamic Belief Revision over Multi-Agent Plausibility Models', in G. Bonanno, W. van der Hoek, M. Wooldridge, eds., *Proceedings LOFT '06*, Department of Computing, University of Liverpool, 11–24.

(2007a) 'From Conditional Probability to the Logic of Doxastic Actions', in *Proceedings TARK XI*, UCL Presses Universitaires de Louvain, 52–61.

(2007b) 'Probabilistic Dynamic Belief Revision', in J. van Benthem, S. Ju & F. Veltman, eds., *A Meeting of the Minds*, Proceedings LORI Beijing 2007, London, College Publications, 21–39. Extended version in *Synthese* 165, 2008, 179–202.

(2008a) 'A Qualitative Theory of Dynamic Interactive Belief Revision', in G. Bonanno, W. van der Hoek, M. Wooldridge, eds., *Texts in Logic and Games Vol. 3*, Amsterdam University Press, 9–58.

(2008b) 'A Dynamic-Logical Perspective on Quantum Behavior', *Studia Logica* 89, 185–209.

(2009a) 'Group Belief Dynamics under Iterated Revision: Fixed Points and Cycles of Joint Upgrades', *Proceedings TARK XII*, Stanford, 41–50.

(2009b) 'Learning by Questions and Answers: From Belief-Revision Cycles to Doxastic Fixed Points', in M. Kanazawa, H. Ono & R. de Queiroz, eds., *LNAI Lecture Notes in Computer Science*, vol. 5514, 124–139.

(2009c) 'Talking Your Way into Agreement: Belief Merge by Persuasive Communication', Proceedings of the Second Multi-Agent Logics, Languages, and Organisations Federated Workshops (FAMAS) Turin, Italy. *CEUR Workshop Proceedings* 494, 129–141.

(2010) 'Surprise?! An Answer to the Hangman, or How to Avoid Unexpected Exams!', University of Oxford and University of Groningen.

A. Baltag, S. Smets & J. Zvesper (2008) 'When All is Done but not (yet) Said: Dynamic Rationality in Extensive Games', in J. van Benthem & E. Pacuit, eds., *Proceedings of the Workshop on Logic and Intelligent Interaction*, ESSLLI 2008, 58–73. Extended version 'Keep 'Hoping' for Rationality: a Solution to the Backward Induction Paradox', *Synthese* 169, 2009, 301–333.

A. Baltag, N. Gierasimczuk & S. Smets (2010) 'For Tracking the Truth, Keep Revising Your Beliefs', Invited Lecture at NASSLLI 2010, Indiana University, Bloomington.

Y. Bar-Hillel & R. Carnap (1953) 'Semantic Information', *The British Journal for the Philosophy of Science* 4:14, 147–157.

J. Barwise (1985) 'Three Theories of Common Knowledge', in *Proceedings TARK II*, San Francisco, Morgan Kauffman, 365–379.

J. Barwise & J. van Benthem (1999) 'Interpolation, Preservation, and Pebble Games', *Journal of Symbolic Logic* 64, 881–903.

J. Barwise & J. Etchemendy (1991) 'Visual Information and Valid Reasoning', in *Visualization in Teaching and Learning Mathematics*, Washington, DC, Mathematical Association of America, 9–24.

J. Barwise & L. Moss (1996) *Vicious Circles: On the Mathematics of Non-Wellfounded Phenomena*, Stanford, CSLI Publications.

J. Barwise & J. Perry (1983) *Situations and Attitudes*, Cambridge (Mass.), The MIT Press.

J. Barwise & J. Seligman (1995) *Information Flow, the Logic of Distributed Systems*, Cambridge University Press.

P. Battigalli & G. Bonanno (1999) 'Recent Results on Belief, Knowledge and the Epistemic Foundations of Game Theory', *Research in Economics* 53, 149–225.

P. Battigalli & M. Siniscalchi (1999) 'Hierarchies of Conditional Beliefs and Interactive Epistemology in Dynamic Games', *Journal of Economic Theory* 88, 188–230.

J. Beall & G. Restall (2006) *Logical Pluralism*, Oxford University Press.

N. Belnap, M. Perloff & M. Xu (2001) *Facing the Future*, Oxford University Press.

N. Belnap & Th. Steele (1976) *The Logic of Questions and Answers*, New Haven, Yale University Press.

J. van Benthem (1984) 'Correspondence Theory', in D. Gabbay and F. Guenthner, eds., *Handbook of Philosophical Logic*, vol. II, Dordrecht, Reidel, 167–247. (Reprint with addenda in D. Gabbay, ed., 2001, *Handbook of Philosophical Logic*, vol. III., Dordrecht, Kluwer, 325–408.)

(1989) 'Semantic Parallels in Natural Language and Computation', in H.-D. Ebbinghaus *et al.*, eds., *Logic Colloquium. Granada 1987*, Amsterdam, North-Holland, 331–375.

(1991) *Language in Action: Categories, Lambdas and Dynamic Logic*, Amsterdam, North-Holland & Cambridge (Mass.), The MIT Press.

(1993) 'Reflections on Epistemic Logic', *Logique et Analyse* 34, 5–14.

(1996) *Exploring Logical Dynamics*, Stanford, CSLI Publications.

(1997) 'Dynamic Bits and Pieces', ILLC, University of Amsterdam.

(1999a) *Logic in Games*, lecture notes, ILLC, University of Amsterdam.

(1999b) 'Update as Relativization', ILLC, University of Amsterdam.

(1999c) 'Wider Still and Wider: Resetting the Bounds of Logic', in A. Varzi, ed., *The European Review of Philosophy*, Stanford, CSLI Publications, 21–44.

(2000) 'Update Delights', Invited Lecture, ESSLLI Summer School, University of Birmingham.

(2001) 'Games in Dynamic Epistemic Logic', *Bulletin of Economic Research* 53, 219–248.

(2002a) 'Extensive Games as Process Models', *Journal of Logic, Language and Information* 11, 289–313.

(2002b) 'Invariance and Definability: Two Faces of Logical Constants', in W. Sieg, R. Sommer & C. Talcott, eds., *Reflections on the Foundations of Mathematics. Essays in Honor of Sol Feferman*, ASL Lecture Notes in Logic 15, 426–446.

(2003a) 'Conditional Probability Meets Update Logic', *Journal of Logic, Language and Information* 12, 409–421.

(2003b) 'Is there still Logic in Bolzano's Key?', in E. Morscher, ed., *Bernard Bolzanos Leistungen in Logik, Mathematik und Physik*, Bd.16, Academia Verlag, Sankt Augustin 2003, 11–34.

(2003c) 'Logic and the Dynamics of Information', *Minds and Machines* 13, 503–519.

(2003d) 'Structural Properties of Dynamic Reasoning', in J. Peregrin, ed., *Meaning: the Dynamic Turn*, Amsterdam, Elsevier, 15–31. Final version 'Inference in Action', *Publications de l'Institut Mathématique*, Nouvelle Série 82, 2008, Beograd, 3–16.

(2004a) *Update and Revision in Games*, lecture notes, ILLC Amsterdam & Philosophy Stanford.

(2004b) 'What One May Come to Know', *Analysis* 64, 95–105.

(2005a) 'An Essay on Sabotage and Obstruction', in D. Hutter, ed., *Mechanizing Mathematical Reasoning, Essays in Honor of Jörg Siekmann on the Occasion of his 60th Birthday*, Springer, 268–276.

(2005b) 'Guards, Bounds, and Generalized Semantics', *Journal of Logic, Language and Information* 14, 263–279.

(2005c) 'Information as Correlation versus Information as Range', Research Report, ILLC, University of Amsterdam. To appear in L. Moss, ed., *Logic and Cognition*, Memorial Volume for Jon Barwise.

(2006a) 'Epistemic Logic and Epistemology: the State of their Affairs', *Philosophical Studies* 128, 49–76.

(2006b) 'One is a Lonely Number: on the Logic of Communication', in Z. Chatzidakis, P. Koepke & W. Pohlers, eds., *Logic Colloquium '02*, ASL & A.K. Peters, Wellesley MA, 96–129.

(2006c) 'Open Problems in Update Logic', in D. Gabbay, S. Goncharov & M. Zakharyashev, eds., *Mathematical Problems from Applied Logic I*, Springer, New York & Novosibirsk, 137–192.

(2007) 'Modeling Protocols in Temporal Models', ILLC, University of Amsterdam.

(2007a) 'Cognition as Interaction', in G. Bouma, I. Krämer & J. Zwarts, eds., *Cognitive Foundations of Interpretation*, KNAW Amsterdam, 27–38.

(2007b) 'Dynamic Logic of Belief Revision', *Journal of Applied Non-Classical Logics* 17, 129–155.

(2007c) 'In Praise of Strategies', ILLC, University of Amsterdam. To appear in J. van Eijck & R. Verbrugge, eds., *Games, Actions, and Social Software*, London, College Publications.

(2007d) 'Logic Games, From Tools to Models of Interaction', in A. Gupta, R. Parikh & J. van Benthem, eds., *Logic at the Crossroads*, Mumbai, Allied Publishers, 283–317.

(2007e) 'Logic in Philosophy', in D. Jacquette, ed., *Handbook of the Philosophy of Logic*, Amsterdam, Elsevier, 65–99.

(2007f) 'Rational Dynamics', *International Game Theory Review* 9:1, 2007, 13–45. Erratum reprint, Volume 9:2, 377–409.

(2007g) 'Rationalizations and Promises in Games', *Philosophical Trends*, 'Supplement 2006' on logic, Chinese Academy of Social Sciences, Beijing, 1–6.

(2008a) 'Computation as Conversation', in B. Cooper, B. Löwe & A. Sorbi, eds., *New Computational Paradigms: Changing Conceptions of What is Computable*, New York, Springer, 35–58.

(2008b) 'Logic and Reasoning: Do the Facts Matter?', *Studia Logica* 88, 67–84.

(2008c) 'Logic, Rational Agency, and Intelligent Interaction', in C. Glymour, W. Wei & D. Westerståhl, eds., *Logic, Methodology and Philosophy of Science XIII Beijing 2007*, London, College Publications, 137–161.

(2008d) 'Logical Pluralism Meets Logical Dynamics', *Australasian Journal of Logic* 6, 1–28.

(2008e) 'Merging Observational and Access Dynamics in Logic', *Journal of Logic Studies*, 1:1, 1–16, Institute of Logic and Cognition, Soon Yat-Sen University, Guangzhou.

(2008f) 'Tell it Like It Is', *Journal of Peking University*, Humanities and Social Science Edition, No. 1, 80–90.

(2009a) 'Actions that Make Us Know', in J. Salerno, ed., *New Essays on the Knowability Paradox*, Oxford University Press, 129–146.

(2009b) 'Decisions, Actions, and Games: a Logical Perspective', in R. Ramanujam & S. Sarukkai, eds., *Proceedings of the Third Indian Conference on Logic and its Applications ICLA 2009*, Springer LNAI 5378, 1–22.

(2009c) 'For Better of for Worse: Dynamic Logics of Preference', in T. Grüne-Yanoff & S.-O. Hansson, eds., *Preference Change*, Dordrecht, Springer, 57–84.

(2009d) 'McCarthy Variations in a Modal Key', ILLC, University of Amsterdam, *Artificial Intelligence*, 175:1 (2010), 428–439.

(2009e) 'The Information in Intuitionistic Logic', *Synthese* 167, 251–270.

(2009f) 'Update as Social Choice', ILLC, University of Amsterdam. To appear in P. Girard, M. Marion & O. Roy, eds., *Proceedings Dynamics Workshop Montreal 2007*, Springer.

(2010a) 'A Logician Looks at Argumentation Theory', *Cogency*, vol. 1:2, Universidad Diego Portales, Santiago de Chili.

(2010b) 'Categorial versus Modal Information Theory', *Linguistic Analysis*, 36: 1–4, 533–544.

(2010c) 'Logic, Mathematics, and General Agency', in P. E. Bour, M. Rebuschi & L. Rollell, eds., *Construction*, London, College Publications, 281–300.

(2010d) 'The Logic of Empirical Theories Revisited', to appear in *Studia Logica*.

(2010e) 'Two Kinds of Dynamics in Logic', lecture at Kampfest September, University of Stuttgart.

(to appearA), *Logic in Games*, ILLC, University of Amsterdam and Texts in Logic and Games, Heidelberg, Springer.

(to appearB), *Modal Logic for Open Minds*, Stanford, CSLI Publications.

J. van Benthem & P. Blackburn (2006) 'Modal Logic, a Semantic Perspective', in J. van Benthem, P. Blackburn & F. Wolter, eds., *Handbook of Modal Logic*, Amsterdam, Elsevier, 1–84.

J. van Benthem & C. Dégrémont (2008) 'Multi-Agent Belief Dynamics: bridges between dynamic doxastic and doxastic temporal logics', Paper presented at the 8th Conference on Logic and the Foundations of Game and Decision Theory, Amsterdam & Workshop on Intelligent Interaction, *ESSLLI* Hamburg. To appear in G. Bonanno, W. van der Hoek & B. Löwe, eds., *Postproceedings Lecture Notes in Computer Science Vol. 6006, LOFT 08*, Texts in Logic and Games, Heidelberg, Springer.

J. van Benthem, J. van Eijck & A. Frolova (1993) *Changing Preferences*, Report CS-93-10, Centre for Mathematics & Computer Science, Amsterdam.

J. van Benthem, J. van Eijck & B. Kooi (2006) 'Logics of Communication and Change', *Information and Computation* 204, 1620–1662.

J. van Benthem, J. Gerbrandy, T. Hoshi & E. Pacuit (2007) 'Merging Frameworks for Interaction', *Proceedings TARK 2007*, University of Namur. Final version in the *Journal of Philosophical Logic* 38, 2009, 491–526.

J. van Benthem, J. Gerbrandy & B. Kooi (2006) 'Dynamic Update with Probabilities', ILLC Prepublication PP-2006-21, University of Amsterdam. Final version in *Studia Logica* 93, 2009, 67–96.

J. van Benthem & A. Gheerbrant (2010) 'Game Solution, Epistemic Dynamics, and Fixed-Point Logics', *Fundamenta Informaticae*, 100, 19–41.

J. van Benthem, P. Girard & O. Roy (2009) 'Everything Else Being Equal. A Modal Logic Approach to Ceteris Paribus Preferences', *Journal of Philosophical Logic* 38, 83–125. To appear in 'The Philosopher's Annual 2009', www.pgrim.org/philosophersannual/index.html

J. van Benthem, D. Grossi & F. Liu (2010) 'Deontics = Betterness + Priority', in G. Governatori & G. Sartor, eds., Proceedings *Deontic Logic in Computer Science*, DEON 2010, Fiesole, Italy. Lecture Notes in Computer Science 6181, Springer, 50–65.

J. van Benthem & D. Ikegami (2008) 'Modal Fixed-Point Logic and Changing Models', in A. Avron, N. Dershowitz & A. Rabinovich, eds., *Pillars of Computer Science: Essays Dedicated to Boris (Boaz) Trakhtenbrot on the Occasion of his 85th Birthday*, Berlin, Springer, 146–165.

J. van Benthem & B. Kooi (2004) 'Reduction Axioms for Epistemic Actions', *Proceedings Advances in Modal Logic 2004*, Department of Computer Science, University of Manchester. Report UMCS-04 9-1, Renate Schmidt, Ian Pratt-Hartmann, Mark Reynolds, Heinrich Wansing, eds., 197–211.

J. van Benthem & F. Liu (2004) 'Diversity of Logical Agents in Games', *Philosophia Scientiae* 8:2, 163–178.

(2007) 'Dynamic Logic of Preference Upgrade', *Journal of Applied Non-Classical Logics* 17, 157–182.

J. van Benthem & M. Martinez (2008) 'The Stories of Logic and Information', in P. Adriaans & J. van Benthem, eds., *Handbook of the Philosophy of Information*, Amsterdam, Elsevier Science Publishers, 217–280.

J. van Benthem & S. Minica (2009) 'Toward a Dynamic Logic of Questions', in X. He, J. Horty & E. Pacuit, eds., *Logic, Rationality, and Interaction: Proceedings LORI II Chongqing*, Springer Lecture Notes in Artificial Intelligence, 27–41.

J. van Benthem & A. ter Meulen, eds. (1997) *Handbook of Logic and Language*, Amsterdam, Elsevier Science Publishers.

J. van Benthem, R. Muskens & A. Visser (1996) 'Dynamics', in J. van Benthem & A. ter Meulen, eds., *Handbook of Logic and Language*, Amsterdam, Elsevier, 587–648.

J. van Benthem, S. van Otterloo & O. Roy (2006) 'Preference Logic, Conditionals, and Solution Concepts in Games', in H. Lagerlund, S. Lindström & R. Sliwinski, eds., *Modality Matters*, University of Uppsala, 61–76.

J. van Benthem & E. Pacuit (2006) 'The Tree of Knowledge in Action', *Proceedings Advances in Modal Logic*, ANU Melbourne, 87–106.

J. van Benthem & E. Pacuit (2007). 'Modeling Protocols in Temporal Models', ILLC, University of Amsterdam.

J. van Benthem & E. Pacuit, eds. (2010) 'Temporal Logics of Agency', special issue of the *Journal of Logic, Language and Information*, 19:4, with editorial on pp. 1–5.

J. van Benthem & E. Pacuit (2011) 'The Dynamic Logic of Evidence', to appear in *Studia Logica*.

J. van Benthem & D. Sarenac (2005) 'The Geometry of Knowledge', in J.-Y. Béziau, A. Costa Leite & A. Facchini, eds., *Aspects of Universal Logic*, Centre de Recherches Sémiologiques, Université de Neuchatel, 1–31.

J. van Benthem & F. Velázquez-Quesada (2009) 'Inference, Promotion, and the Dynamics of Awareness', ILLC, Amsterdam. Appeared in *Knowledge, Rationality and Action*, 177:1, 5–27.

J. Bergstra, A. Ponse & S. Smolka, eds. (2001) *Handbook of Process Algebra*, Amsterdam, Elsevier.

P. Blackburn, J. van Benthem & F. Wolter, eds. (2006) *Handbook of Modal Logic*, Amsterdam, Elsevier.

P. Blackburn, M. de Rijke & Y. Venema (2000) *Modal Logic*, Cambridge University Press.

O. Board (1998) 'Belief Revision and Rationalizability', *Proceedings TARK 1998*, 201–213.

R. Bod (1998) *Beyond Grammar: An Experience-Based Theory of Language*, Stanford, CSLI Publications.

G. Boella, G. Pigozzi & L. van der Torre (1999) 'Normative Framework for Normative System Change', in C. Sierra *et al.*, eds., *8th International Joint Conference on Autonomous Agents and Multiagent Systems*, Volume 1, 169–176.

B. Bolzano (1837) *Wissenschaftslehre, Seidelsche Buchhandlung, Sulzbach*. Translated as *Theory of Science* by R. George, Berkeley & Los Angeles, University of California Press, 1972.

G. Bonanno (2001) 'Branching Time, Perfect Information Games, and Backward Induction', *Games and Economic Behavior* 36, 57–73.

G. Bonanno (2004) 'Memory and Perfect Recall in Extensive Games', *Games and Economic Behaviour* 47, 237–256.

G. Bonanno (2007) 'Axiomatic Characterization of the *AGM* theory of Belief Revision in a Temporal Logic', *Artificial Intelligence* 171, 144–160.

D. Bonnay (2006) *What is a Logical Constant?*, Ph.D. thesis, École Normale Supérieure, Paris.

D. Bonnay & P. Égré (2007) 'Knowing One's Limits; An Analysis in Centered Dynamic Epistemic Logic', University of Paris. To appear in P. Girard, M. Marion & O. Roy, eds., *Proceedings Dynamics Montreal 2007*, Springer.

C. Boutilier (1994) 'Conditional Logics of Normality; A Modal Approach', *Artificial Intelligence* 68, 87–154.

C. Boutilier & M. Goldszmidt (1993) 'Revision by Conditional Beliefs', *Proceedings AAAI 11*, Washington DC, Morgan Kaufmann, 649–654.

J. Bradfield & C. Stirling (2006) 'Modal μ-Calculi', in P. Blackburn, J. van Benthem & F. Wolter, eds., 721–756.

R. Bradley (2007) 'The Kinematics of Belief and Desire', *Synthese* 156:3, 513–535.

R. Brafman, J.-C. Latombe & Y. Shoham (1993) 'Towards Knowledge-Level Analysis of Motion Planning', *Proceedings AAAI 1993*, 670–675.

A. Brandenburger & H. J. Keisler (2006) 'An Impossibility Theorem on Beliefs in Games', *Studia Logica* 84, 211–240.

M. Bratman (1992) 'Shared Cooperative Activity', *The Philosophical Review* 101:2, 327–341.

J. Broersen (2009) 'A *STIT*-Logic for Extensive Form Group Strategies', *Proceedings 2009 IEEE/WIC/ACM International Joint Conference on Web Intelligence and Intelligent Agent Technology*, 484–487.

J. Broersen, A. Herzig & N. Troquard (2006) 'A *STIT*-Extension of *ATL*', *Proceedings JELIA* 2006, 69–81.

B. Brogaard & J. Salerno (2002) 'Fitch's Paradox of Knowability', *Stanford Electronic Encyclopedia of Philosophy*, http://plato.stanford.edu/entries/fitch-paradox/.

B. de Bruin (2004) *Explaining Games: on the Logic of Game Theoretic Explanations*, Dissertation, ILLC, University of Amsterdam. Also in *Synthese Library*, Springer, 2010.

J. Burgess (1981) 'Quick Completeness Proofs for some Logics of Conditionals', *Notre Dame Journal of Formal Logic* 22:1, 76–84.

R. Carnap (1952) *The Continuum of Inductive Methods*, University of Chicago Press.

C. Castelfranchi & F. Paglieri (2007) 'The Role of Beliefs in Goal Dynamics: Prolegomena to a Constructive theory of Intentions', *Synthese* 155, 237–263.

B. ten Cate & Ch-ch Shan (2002) 'The Partition Semantics of Questions, Syntactically', In *Proceedings of the ESSLLI-2002 student session*, Malvina Nissim, ed., 55–269. 14th European Summer School in Logic, Language and Information.

P. Cohen & H. Levesque (1990) 'Intention is Choice with Commitment', *Artificial Intelligence* 42, 213–261.

A. Condon (1988) *Computational Models of Games*, Dissertation, Computer Science Department, University of Washington.

I. Cornelisse (2010) 'Dynamic Doxastic Probability Logic', Dynamics Seminar, ILLC, University of Amsterdam.

D. van Dalen (2002) 'Intuitionistic Logic', in D. Gabbay & F. Guenthner, eds., *Handbook of Philosophical Logic*, vol. 5 (2nd edn), Dordrecht, Kluwer, 1–114.

A. Dawar, E. Grädel & S. Kreutzer (2004) 'Inflationary Fixed Points in Modal Logic', *ACM Transactions on Computational Logic*, vol. 5, 282–315.

F. Dechesne & Y. Wang (2007) 'Dynamic Epistemic Verification of Security Protocols', in J. van Benthem, S. Ju & F. Veltman, eds., *A Meeting of the Minds*, Proceedings LORI Beijing 2007, London, College Publications, 129–143.

C. Dégrémont (2010) *The Temporal Mind: Observations on Belief Change in Temporal Systems*, Dissertation, ILLC, University of Amsterdam.

C. Dégrémont & O. Roy (2009) 'Agreement Theorems in Dynamic Epistemic Logic', in A. Heifetz, ed., *Proceedings TARK 2009*, Stanford, 91–98.

L. Demey (2010a) *Agreeing to Disagree in Probabilistic Dynamic Epistemic Logic*, Master's Thesis, ILLC, University of Amsterdam.

L. Demey (2010b) 'Some Remarks on the Model Theory of Epistemic Plausibility Models', Dynamics Seminar, ILLC, University of Amsterdam.

H. van Ditmarsch (2000) *Knowledge Games*, Dissertation, ILLC University of Amsterdam & Department of Informatics, University of Groningen.

(2003) 'The Russian Cards Problem', *Studia Logica* 75, 31–62.

H. van Ditmarsch & T. French (2009) 'Awareness and Forgetting of Facts and Agents', in P. Boldi, G. Vizzari, G. Pasi & R. Baeza-Yates, eds., *Proceedings of the WI-IAT Workshops 2009*, IEEE Press, 478–483.

H. van Ditmarsch, A. Herzig & T. de Lima (2007) 'Optimal Regression for Reasoning about Knowledge and Actions', in G. Bonanno, J. Delgrande, J. Lang & H. Rott, eds., *Formal Models of Belief Change in Rational Agents*, Dagstuhl Seminar Proceedings 07351, Schloss Dagstuhl, Germany 2007.

H. van Ditmarsch, W. van der Hoek & B. Kooi (2007) *Dynamic-Epistemic Logic*, Synthese Library 337, Berlin, Springer.

H. van Ditmarsch & B. Kooi (2006) 'The Secret of My Success', *Synthese* 151, 201–232.

H. van Ditmarsch & B. Kooi (2008) 'Semantic Results for Ontic and Epistemic Change', in G. Bonanno, W. van der Hoek & M. Wooldridge, eds., *Proceedings LOFT VII*, Texts in Logic and Games, Amsterdam University Press, 87–117.

K. Dosen & P. Schroeder-Heister, eds. (1993) *Substructural Logics*, Oxford University Press.

J. Doyle & M. Wellman (1994) 'Representing Preferences as Ceteris Paribus Comparatives', in *Decision-Theoretic Planning: Papers from the 1994 Spring {AAAI} Symposium*, Menlo Park (Calif.), AAAI Press, 69–75.

F. Dretske (1981) *Knowledge and the Flow of Information*, University of Chicago Press.

M. Dummett (1977) *Elements of Intuitionism*, Oxford University Press.

R. Dunbar (1998) *Grooming, Gossip, and the Evolution of Language*, Cambridge (Mass), Harvard University Press.

P. Dung (1995) 'An Argumentation-Theoretic Foundation for Logic Programming', *Journal of Logic Programming* 22, 151–177.

B. Dunin-Keplicz & R. Verbrugge (2002) 'Collective Intentions', *Fundamenta Informaticae* 51, 271–295.

M. Dunn (1991) 'Gaggle Theory: An abstraction of Galois connections and Residuation, with applications to Negation, Implication, and various Logical Operators', in J. van Eijck, ed., *Logics in AI (Amsterdam, 1990)*, Berlin, Springer, 31–51.

H.-D. Ebbinghaus & J. Flum (1995) *Finite Model Theory*, Berlin, Springer.

J. van Eijck (2005) 'DEMO - A Demo of Epistemic Modelling', Augustus de Morgan Workshop, King's College, London. Final version in J. van Benthem, D. Gabbay & B. Loewe *et al.*, eds., 2007, *Interactive Logic*, Amsterdam University Press, 305–363.

J. van Eijck, J. Ruan & T. Sadzik (2006) 'Action Emulation', CWI, Amsterdam.

J. van Eijck & F. Sietsma (2009) 'Multi-Agent Belief Revision with Linked Plausibilities', CWI Amsterdam, in *Proceedings LOFT VIII*, Amsterdam.

J. van Eijck, F. Sietsma & Y. Wang (2009) 'Logic of Information Flow on Communication Channels', CWI, Amsterdam, posted at http://loriweb.org/

P. van Emde Boas (2002) *Models for Games and Complexity*, lecture notes, ILLC, Amsterdam.

U. Endriss & J. Lang, eds. (2006) *Proceedings of the First International Workshop on Computational Social Choice*, ILLC, University of Amsterdam. Available at www.illc.uva.nl/~ulle/COMSOC-2006/

R. Fagin, J. Halpern, Y. Moses & M. Vardi (1995) *Reasoning about Knowledge*, Cambridge (Mass.), The MIT Press.

R. Fagin & J. Halpern (1993), 'Reasoning about Knowledge and Probability', *Journal of the ACM* 41 (2), 340–367.

R. Fagin, J. Halpern & M. Vardi (1990) 'A Nonstandard Approach to the Logical Omniscience Problem', in *Theoretical Aspects of Reasoning about Knowledge: Proceedings of the Third TARK Conference*, Los Altos, Morgan Kaufmann, 41–55.

Y. Feinberg (2007) 'Meaningful Talk', in J. van Benthem, S. Ju & F. Veltman, eds., *A Meeting of the Minds*, Proceedings LORI Beijing 2007, London, College Publications, 41–54.

B. Fitelson (2006) 'Old Evidence, Logical Omniscience & Bayesianism', Lecture ILLC Workshop Probability and Logic, Department of Philosophy, University of California at Berkeley.

T. French & H. van Ditmarsch (2008) 'Undecidability for Arbitrary Public Announcement Logic', in C. Areces & R. Goldblatt, eds., *Proceedings Advances in Modal Logic VII*, London, College Publications, 23–42.

D. Gabbay (1996) *Labeled Deductive Systems*, vol. 1, Oxford, Clarendon Press.

(2008) 'Reactive Kripke Semantics and Arc Accessibility', in A. Avron, N. Dershowitz & A. Rabinovich, eds. *Pillars of Computer Science: Essays dedicated to Boris (Boaz) Trakhtenbrot on the occasion of his 85th birthday*, LNCS, vol. 4800, Berlin, Springer, 292–341.

D. Gabbay & F. Guenthner, eds. (1983–1999) *Handbook of Philosophical Logic*, Dordrecht, Kluwer Academic Publishers.

D. Gabbay, C. Hogger & J. Robinson, eds. (1995) *Handbook of Logic in Artificial Intelligence and Logic Programming*, Oxford University Press.

D. Gabbay & J. Woods, eds. (2004) *Handbook of Logic and Argumentation*, Amsterdam, Elsevier Science Publishers.

P. Gärdenfors (1988) *Knowledge in Flux*, Cambridge (Mass.), Bradford Books/The MIT Press.

P. Gärdenfors & H. Rott (1995) 'Belief Revision', in D. M. Gabbay, C. J. Hogger & J. A. Robinson, eds., 35–132.

P. Gärdenfors & M. Warglien (2007) 'Semantics, Conceptual Spaces, and the Meeting of Minds', LUCS Cognitive Science Centre, University of Lund.

J. Geanakoplos (1992) 'Common Knowledge', *The Journal of Economic Perspectives* 6:4, 53–82.

J. Geanakoplos & H. Polemarchakis (1982) 'We Can't Disagree Forever', *Journal of Economic Theory* 28, 192–200.

J. Gerbrandy (1999a) *Bisimulations on Planet Kripke*, Dissertation, ILLC, University of Amsterdam.

(1999b) 'Dynamic Epistemic Logic', in L. S. Moss, J. Ginzburg & M. de Rijke, eds., *Logic, Language and Computation*, vol. 2, Stanford, CSLI Publications, 67–84.

(2005) 'The Surprise Examination in Dynamic Epistemic Logic', Department of Informatics, University of Torino. Final version in *Synthese* 155, 2007, 21–33.

J. Gerbrandy & W. Groeneveld (1997) 'Reasoning about Information Change', *Journal of Logic, Language and Information* 6, 147–169.

B. Geurts (2003) 'Reasoning with Quantifiers', *Cognition* 86, 223–251.

A. Gheerbrant (2010) *Fixed-Point Logics on Trees*, Ph.D. Thesis, ILLC, University of Amsterdam.

C. Ghidini & F. Giunchiglia (2001) 'Local Model Semantics, or Contextual Reasoning = Locality + Compatibility', *Artificial Intelligence* 127, 221–259.

N. Gierasimczuk (2009) 'Bridging Learning Theory and Dynamic Epistemic Logic', *Synthese* 169, 371–384.

N. Gierasimczuk (2010) *In the Limits of Knowledge: Logical Analysis of Inductive Inference*, Dissertation, ILLC, University of Amsterdam.

N. Gierasimczuk & D. de Jongh (2009) 'On the Minimality of Definite Tell-tale Sets in Finite Identification of Languages', ILLC, University of Amsterdam.

N. Gierasimczuk, L. Kurzen & F. Velázquez-Quesada (2009) 'Learning and Teaching as a Game: A Sabotage Approach', in X. He, J. Horty & E. Pacuit, eds., *Proceedings LORI 2009*, LNAI 5834, 119–132.

G. Gigerenzer, Peter M. Todd, and the ABC Research Group (1999) *Simple Heuristics That Make Us Smart*, Oxford University Press.

R. Giles (1974) 'A Non-Classical Logic for Physics', *Studia Logica* 33, 399–417.

J. Ginzburg (2009) *The Interactive Stance: Meaning for Conversation*, Department of Computer Science, King's College, London.

P. Girard (2008) *Modal Logic for Belief and Preference Change*, Dissertation, Department of Philosophy, Stanford University & ILLC Amsterdam.

P. Gochet (2006) 'La Formalisation du Savoir-Faire', Lecture at Pierre Duhem Colloquium IPHRST Paris, Philosophical Institute, Université de Liege.

A. Goldman (1999) *Knowledge in a Social World*, Oxford University Press.

V. Goranko & G. van Drimmelen (2006) 'Complete Axiomatization and Decidability of Alternating-Time Temporal Logic', *Theoretical Computer Science* 353, 93–117.

J. Groenendijk (2008) 'Inquisitive Semantics: Two possibilities for disjunction', ILLC, University of Amsterdam. Also in P. Bosch, D. Gabelaia & J. Lang, eds., 2009, *Proceedings Seventh Tbilisi Symposium on Language, Logic and Computation*, Springer Lecture Notes in Artificial Intelligence, 80–94.

J. Groenendijk & M. Stokhof (1985) *Studies in the Semantics of Questions and the Pragmatics of Answers*, Dissertation, Philosophical Institute, University of Amsterdam.

(1991) 'Dynamic Predicate Logic', *Linguistics and Philosophy* 14:1, 39–100.

(1997) 'Questions', in J. van Benthem & A. ter Meulen, eds., *Handbook of Logic and Language*, Amsterdam, Elsevier, 1055–1124.

D. Grossi (2007) *Designing Invisible Handcuffs. Formal Investigations in Institutions and Organizations for Multi-Agent Systems*, Dissertation, Department of Computer Science, Utrecht University.

D. Grossi (2009), 'Doing Argumentation Theory in Modal Logic', Technical Report PP-2009-24, ILLC, University of Amsterdam.

D. Grossi & F. Velázquez-Quesada (2009) 'Twelve Angry Men: A Study on the Fine-Grain of Announcements', in X. He, J. Horty & E. Pacuit, eds., *Logic, Rationality and Interaction*, Proceedings LORI II Chongqing, Springer Lecture Notes in Artificial Intelligence, 147–160.

A. Grove (1988) 'Two Modelings for Theory Change', *Journal of Philosophical Logic* 17, 157–170.

T. Gruene-Yanoff & S.-O. Hanson, eds. (2008) *Preference Change*, Dordrecht, Springer.

P. Grunwald & J. Halpern (2003) 'Updating Probabilities', *Journal of Artificial Intelligence Research* 19, 243–278.

A. Gupta, R. Parikh & J. van Benthem, eds. (2007) *Logic at a Cross-Roads: Logic and its Interdisciplinary Environment*, Mumbai, Allied Publishers.

Y. Gurevich & S. Shelah (1986) 'Fixed-Point Extensions of First-Order Logic', *Annals of Pure and Applied Logic* 32, 265–280.

J. Halpern (1997) 'Defining Relative Likelihood in Partially-Ordered Preferential Structure', *Journal of Artificial Intelligence Research* 7, 1–24.

(2003) 'A Computer Scientist Looks at Game Theory', *Games and Economic Behavior* 45(1), 114–131.

(2003) *Reasoning about Uncertainty*, Cambridge (Mass.), The MIT Press.

J. Halpern, R. van der Meyden & M. Vardi (2004) 'Complete Axiomatizations for Reasoning about Knowledge and Time', *SIAM Journal of Computing* 33:2, 674–703.

J. Halpern & M. Tuttle (1993) 'Knowledge, Probability, and Adversaries', *Journal of the ACM* 40, 917–962.

J. Halpern & M. Vardi (1989) 'The Complexity of Reasoning about Knowledge and Time, I: Lower Bounds' *Journal of Computer and System Sciences* 38, 195–237.

Y. Hamami (2010) *The Interrogative Model of Inquiry Meets Dynamic Epistemic Logics*, Master's thesis, ILLC, University of Amsterdam.

F. Hamm & M. van Lambalgen (2004) *The Proper Treatment of Events*, Oxford, Blackwell Publishers.

H. Hansen, C. Kupke & E. Pacuit (2008) 'Neighbourhood Structures: Bisimilarity and Basic Model Theory', in D. Kozen, U. Montanari, T. Mossakowski & J. Rutten, eds., *Logical Methods in Computer Science* 15, 1–38.

P. Hansen & V. Hendricks, eds. (2007) *Five Questions on Game Theory*, Roskilde, Automatic Press.

S. O. Hanson (1995) 'Changes in Preference', *Theory and Decision* 38, 1–28.

(2001) 'Preference Logic', in D. Gabbay & F. Guenthner, eds., *Handbook of Philosophical Logic IV*, 319–393, Dordrecht, Kluwer.

B. Hansson (1969) 'An Analysis of some Deontic Logics', *Noûs* 3, 373–398.

D. Harel (1985) 'Recurring Dominoes: Making the Highly Undecidable Highly Understandable', *Annals of Discrete Mathematics* 24, 51–72.

D. Harel, D. Kozen & J. Tiuryn (2000) *Dynamic Logic*, Cambridge (Mass.), The MIT Press.

P. Harrenstein (2004) *Logic in Conflict*, Dissertation, Institute of Computer Science, University of Utrecht.

H. Helmholtz (1878) *The Facts of Perception*, Middletown (Conn.), Wesleyan University Press.

V. Hendricks (2003) 'Active Agents', in J. van Benthem & R. van Rooij, eds., special issue on Information Theories, *Journal of Logic, Language and Information* 12, 469–495.

(2005) *Mainstream and Formal Epistemology*, Cambridge University Press.

A. Herzig & E. Lorini (2010) 'A Dynamic Logic of Agency I: STIT, Capabilities and Powers', *Journal of Logic, Language and Information* 19, 89–121.

J. Hintikka (1962) *Knowledge and Belief*, Ithaca, Cornell University Press.

(1973) *Logic, Language Games and Information*, Oxford, Clarendon Press.

J. Hintikka, I. Halonen & A. Mutanen (2002) 'Interrogative Logic as a General Theory of Reasoning', in D. Gabbay, R. Johnson, H. Ohlbach & J. Woods, eds., *Handbook of the Logic of Argument and Inference*, Amsterdam, Elsevier, 295–338.

J. Hintikka & G. Sandu (1997) 'Game-Theoretical Semantics', in J. van Benthem & A. ter Meulen, eds., *Handbook of Logic and Language*, 361–410.

H. Hodges, W. Hodges & J. van Benthem, eds. (2007) 'Editorial Logic and Psychology', *Topoi* 26, 1–2.

I. Hodkinson & M. Reynolds (2006) 'Temporal Logic', in P. Blackburn, J. van Benthem & F. Wolter, eds., *Handbook of Modal Logic*, Amsterdam, Elsevier, 655–720.

W. van der Hoek, B. van Linder & J.-J. Meijer (1999) 'Group knowledge is Not Always Distributed (neither is it Always Implicit)', *Mathematical Social Sciences* 38, 215–240.

W. van der Hoek & J.-J. Meijer (1995) *Epistemic Logic for AI and Computer Science*, Cambridge University Press.

W. van der Hoek & M. Pauly (2006) 'Modal Logic for Games and Information', in P. Blackburn, J. van Benthem & F. Wolter, eds., 1077–1148.

W. van der Hoek & M. Wooldridge (2003) 'Cooperation, Knowledge, and Time: Alternating-Time Temporal Epistemic Logic and Its Applications', *Studia Logica* 75, 125–157.

J. Hofbauer & K. Sigmund (1998) *Evolutionary Games and Population Dynamics*, Cambridge University Press.

M. Hollenberg (1998) *Logic and Bisimulation*, Dissertation, Philosophical Institute, University of Utrecht.

W. Holliday (2009) 'Dynamic Testimonial Logic', in X. He, J. Horty & E. Pacuit, eds., *Logic, Rationality and Interaction*, Proceedings LORI II Chongqing, Springer Lecture Notes in AI, 161–179.

W. Holliday & Th. Icard (2010) 'Moorean Phenomena in Epistemic Logic', *Proceedings Advances in Modal Logic 2010*, Steklov Mathematical Institute, Moscow.

J. Horty (2001) *Agency and Deontic Logic*, Oxford University Press.

T. Hoshi (2009) *Epistemic Dynamics and Protocol Information*, Ph.D. thesis, Department of Philosophy, Stanford University (ILLC-DS-2009-08).

T. Hoshi & A. Yap (2009) 'Dynamic Epistemic Logic with Branching Temporal Structure', *Synthese* 169, 259–281.

Th. Icard, E. Pacuit & Y. Shoham (2009) 'Intention Based Belief Revision', Departments of Philosophy and Computer Science, Stanford University.

D. Israel & J. Perry (1990) 'What is Information?', in P. Hanson, ed., *Information, Language and Cognition*, Vancouver, University of British Columbia Press, 1–19.

M. Jago (2006) *Logics for Resource-Bounded Agents*, Dissertation, Department of Philosophy, University of Nottingham.

J. Jaspars (1994) *Calculi for Constructive Communication*, Ph.D. Thesis, University of Tilburg, ITK & ILLC Dissertation series.

R. Ji (2004) *Exploring the Update Universe*, Master's Thesis, ILLC, University of Amsterdam.

N. Jones (1978) 'Blindfold Games are Harder than Games with Perfect Information', *Bulletin EATCS* 6, 4–7.

D. de Jongh & F. Liu (2006) 'Optimality, Belief, and Preference', in S. Artemov & R. Parikh, eds., *Proceedings of the Workshop on Rationality and Knowledge*, ESSLLI Summer School, Malaga.

K. Kelly (1996) *The Logic of Reliable Inquiry*, Oxford University Press.

(1998) 'The Learning Power of Belief Revision', *Proceedings TARK VII Evanston Illinois*, Morgan Kaufmann, San Francisco, 111–124.

(2002) 'Knowledge as Reliable Inferred Stable True Belief', Department of Philosophy, Carnegie Mellon University, Pittsburgh.

G. Kerdiles (2001) *Saying it with Pictures*, Dissertation, ILLC, University of Amsterdam.

J. Kim & E. Sosa, eds. (2000) *Epistemology: An Anthology*, Malden (Mass.), Blackwell.

B. Kooi (2003) 'Probabilistic Dynamic Epistemic Logic', *Journal of Logic, Language and Information* 12, 381–408.

S. Kramer (2007) 'The Meaning of a Cryptographic Message via Hypothetical Knowledge and Provability', in J. van Benthem, S. Ju & F. Veltman, eds., *A Meeting of the Minds*, Proceedings LORI Beijing 2007, London, College Publications, 187–199.

S. Kreutzer (2004) 'Expressive Equivalence of Least and Inflationary Fixed-Point Logic', *Annals of Pure and Applied Logic* 130, 61–78.

L. Kurzen (2007) *A Logic for Cooperation, Actions and Preferences*, Master's Thesis, ILLC, University of Amsterdam.

(2010) *Cooperation and Complexity of Interaction*, Ph.D. thesis in progress, ILLC, University of Amsterdam.

G. Lakemeyer (2009) 'The Situation Calculus: A Case for Modal Logic', *Journal of Logic, Language and Information* 19:4, 431–450.

M. van Lambalgen & K. Stenning (2007) *Human Reasoning and Cognitive Science*, Cambridge (Mass.), The MIT Press.

J. Lang, L. van der Torre & E. Weydert (2003) 'Hidden Uncertainty in the Logical Representation of Desires', *Proceedings IJCAI XVIII*, 685–690.

J. Lang & L. van der Torre (2008) 'From Belief Change to Preference Change', *Proceedings ECAI 2008: 18th European Conference on Artificial Intelligence*, Amsterdam, IOS Press, 351–355.

J. van Leeuwen, ed. (1991) *Handbook of Theoretical Computer Science*, Amsterdam, Elsevier.

H. Leitgeb, ed. (2008) 'Psychologism in Logic', Special issue, *Studia Logica* 88:1.

H. Leitgeb & G. Schurz, eds. (2005) 'Non-Monotonic and Uncertain Reasoning in Cognition', *Synthese* 146, 1–2.

H. Leitgeb & K. Segerberg (2007) 'Dynamic Doxastic Logic: why, how, and where to?', *Synthese* 155, 167–190.

W. Lenzen (1980) *Glauben, Wissen und Wahrscheinlichkeit*, Wien, Springer Verlag, Library of Exact Philosophy.

D. Lewis (1969) *Convention*, Oxford, Blackwell.

(1973) *Counterfactuals*, Oxford, Blackwell.

(1988) 'Desire as Belief', *Mind* 97, 323–332.

K. Leyton-Brown & Y. Shoham (2008) *Essentials of Game Theory: A Concise Multidisciplinary Introduction*, Cambridge University Press.

C. List & Ph. Pettit (2004) 'Aggregating Sets of Judgments. Two Impossibility Results Compared', *Synthese* 140, 207–235.

C. List & R. Goodin (2006) 'Conditional Defense of Plurality Rule: Generalizing May's Theorem in a Restricted Informational Environment', *American Journal of Political Science* 50:4, 940–949.

F. Liu (2005) *Diversity of Agents and their Interaction*, Master's Thesis, ILLC, University of Amsterdam.

(2008) *Changing for the Better: Preference Dynamics and Preference Diversity*, Dissertation DS-2008-02, ILLC, University of Amsterdam. Book version to appear in *Synthese Library*, Springer Science Publishers, 2011.

(2009) 'Diversity of Agents and their Interaction', *Journal of Logic, Language and Information* 18, 23–53.

K. Lorenz & P. Lorenzen (1978) *Dialogische Logik*, Darmstadt, Wissenschaftliche Buchgesellschaft.

P. Lorenzen (1955) *Einfuhrung in die Operative Logik und Mathematik*, Berlin, Springer Verlag.

E. Lorini & C. Castelfranchi (2007) 'The Cognitive Structure of Surprise: Looking for Basic Principles', *Topoi* 26, 133–149.

C. Lutz (2006) 'Complexity and Succinctness of Public Announcement Logic', *Proceedings of the Fifth International Conference on Autonomous Agents and Multiagent Systems (AAMAS06)*, 137–143.

E. Mares (1996) 'Relevant Logic and the Theory of Information', *Synthese* 109, 345–360.

P. Martin-Löf (1996) 'On the Meanings of the Logical Constants and the Justifications of the Logical Laws', *Nordic Journal of Philosophical Logic*, 1, 11–60.

M. Marx (2006) 'Complexity of Modal Logics', in P. Blackburn, J. van Benthem & F. Wolter, eds., *Handbook of Modal Logic*, 139–179.

M. Maynard-Reid II & Y. Shoham (1998) 'From Belief Revision to Belief Fusion', *Proceedings of LOFT-98*, 1998, Torino.

J. McCarthy (1963) 'Situations, Actions, and Causal Laws', Technical report, Stanford University. Reprinted in M. Minsky, ed. (1968) *Semantic Information Processing*, Cambridge (Mass.), The MIT Press, 410–417.

(1980) 'Circumscription – A Form of Non-Monotonic Reasoning', *Artificial Intelligence* 13, 27–39.

R. van der Meijden (1996) 'The Dynamic Logic of Permission', *Journal of Logic and Computation* 6, 465–479.

J.-J. Meyer, W. van der Hoek & B. van Linder (1999) 'A Logical Approach to the Dynamics of Commitments, *Artificial Intelligence* 113, 1–41.

J. Miller & L. Moss (2005) 'The Undecidability of Iterated Modal Relativization', *Studia Logica* 97, 373–407.

R. Milner (1999) *Communicating and Mobile Systems: The Pi Calculus*, Cambridge (Mass.), The MIT Press.

S. Minica (2010) *Dynamic Logic of Questions*, Dissertation in progress, ILLC, University of Amsterdam.

P. Mittelstaedt (1978) *Quantum Logic*, Dordrecht, Reidel.

R. Moore (1985) 'A Formal Theory of Knowledge and Action', in J. Hobbs & R. Moore, eds., *Formal Theories of the Commonsense World*, Ablex Publishing Corp, 319–358.

Y. Moschovakis (1974) *Elementary Induction on Abstract Structures*, Amsterdam, North-Holland.

L. Moss & J. Seligman (1997) 'Situation Theory', in J. van Benthem & A. ter Meulen, eds., *Handbook of Logic and Language*, Amsterdam, North Holland, 239–309.

E. Nagel (1961) *The Structure of Science*, Indianapolis, Hackett.

Y. Netchitajlov (2000) *An Extension of Game Logic with Parallel Operators*, Masters Thesis, ILLC, Amsterdam.

R. Nozick (1981) *Philosophical Explanations*, Cambridge (Mass.), Harvard University Press.

M. Osborne & A. Rubinstein (1994) *A Course in Game Theory*, Cambridge (Mass.), The MIT Press.

S. van Otterloo (2005) *A Strategic Analysis of Multi-Agent Protocols*, Dissertation DS-2005-05, ILLC, University of Amsterdam & University of Liverpool.

E. Pacuit & R. Parikh (2007) 'Reasoning about Communication Graphs', in *Interactive Logic, Proceedings of the 7th Augustus de Morgan Workshop*, J. van Benthem, D. Gabbay, and B. Löwe, eds. London, King's College Press.

E. Pacuit and O. Roy (2006) 'Preference Based Belief Dynamics', *Proceedings of The 7th Conference on Logic and the Foundations of Game and Decision Theory (LOFT 2006)*, Computer Science Department, University of Liverpool.

(2010) *Interactive Rationality*, Department of Philosophy, University of Groningen and University of Tilburg.

C. Papadimitriou (1994) *Computational Complexity*, Reading, Addison-Wesley.

R. Parikh (1985) 'The Logic of Games', *Annals of Discrete Mathematics* 24, 111–140.

(1991) 'Monotonic and Non-Monotonic Logics of Knowledge', *Fundamenta Informaticae* XV, 255–274.

(2002) 'Social Software', *Synthese* 132, 187–211.

R. Parikh & R. Ramanujam (2003) 'A Knowledge-Based Semantics of Messages', *Journal of Logic, Language and Information* 12, 453–467.

M. Pauly (2001) *Logic for Social Software*, dissertation DS-2001-10, ILLC, University of Amsterdam.

S. Peters & D. Westerståhl (2006) *Quantifiers in Language and Logic*, Oxford University Press.

J. Piaget (1953) *The Origins of Intelligence in Children*, London, Routledge and Kegan Paul.

J. Plaza (1989) 'Logics of Public Communications', *Proceedings 4th International Symposium on Methodologies for Intelligent Systems*, 201–216.

G. Priest (1997) 'Impossible Worlds–Editor's Introduction', *Notre Dame Journal of Formal Logic* 38, 481–487.

R. Ramanujam (2008) 'Some Automate Theory for Epistemic Logic', Invited lecture at Workshop on Intelligent Interaction, ESSLLI Summer School, August 11–15, Hamburg.

(2010) 'Memory and Logic: A Tale from Automata Theory', 'Logic and Philosophy Today', J. van Benthem & A. Gupta, eds., special issue, *Journal of the Indian Council of Philosophical Research*, 27:1–2, 305–337.

A. Rao & M. Georgeff (1991) 'Modeling Rational Agents within a BDI-Architecture', in R. Fikes & E. Sandewall, eds., *Proceedings of Knowledge Representation and Reasoning (KR&R-91)*, San Mateo, Morgan Kaufmann, 473–484.

R. Reiter (2001) *Knowledge in Action*, Cambridge (Mass.), The MIT Press.

B. Renne (2008) *Public Communication in Justification Logic*, Dissertation, CUNY Graduate Center, New York.

G. Restall (2000) *An Introduction to Substructural Logics*, London, Routledge.

B. Rodenhäuser (2001) *Updating Epistemic Uncertainty*, Master Thesis MoL-2001-07, ILLC, University of Amsterdam.

R. Rodriguez (2006) 'Notes on Topological Updates', paper Dynamics Seminar, ILLC, University of Amsterdam.

F. Roelofsen (2006) *Distributed Knowledge*, Master's Thesis, ILLC, University of Amsterdam.

Ph. Rohde (2005) *On Games and Logics over Dynamically Changing Structures*, Dissertation, Rheinisch-Westfälische Technische Hochschule Aachen.

J. W. Romeijn (2009) 'Meaning Shifts and Conditioning', Philosophical Institute, University of Groningen. Available at http://irs.ub.rug.nl/dbi/4aeee1a7af4a0.

R. van Rooij (2003) 'Quality and Quantity of Information Exchange', *Journal of Logic, Language and Information* 12, 423–451.

(2004) 'Signalling Games Select Horn Strategies', *Linguistics and Philosophy* 27, 493–527.

(2005) 'Questions and Relevance', in *Questions and Answers, Proceedings 2nd CoLogNET ElsNET Symposium*, ILLC, Amsterdam, 96–107.

H. Rott (2001) *Change, Choice and Inference*, Oxford University Press.

(2006) 'Shifting Priorities: Simple Representations for 27 Iterated Theory Change Operators', in H. Lagerlund, S. Lindström & R. Sliwinski, eds., *Modality Matters: Twenty-Five Essays in Honour of Krister Segerberg*, Uppsala Philosophical Studies 53, 359–384.

(2007) 'Information Structures in Belief Revision', in P. Adriaans & J. van Benthem, eds., *Handbook of the Philosophy of Information*, Amsterdam, Elsevier Science Publishers, 457–482.

Sh. Roush (2006) *Tracking Truth: Knowledge, Evidence and Science*, Oxford University Press.

O. Roy (2008) *Thinking before Acting: Intentions, Logic, and Rational Choice*, Dissertation, ILLC, University of Amsterdam.

J. Sack (2008) 'Temporal Language for Epistemic Programs', *Journal of Logic, Language and Information* 17, 183–216.

(2009) 'Extending Probabilistic Dynamic Epistemic Logic', *Synthese* 169, 241–257.

M. Sadrzadeh & C. Cirstea (2006) 'Relating Algebraic and Coalgebraic Logics of Knowledge and Update', in G. Bonanno & W. van der Hoek, eds., *Proceedings LOFT 2006*, Department of Computer Science, University of Liverpool. Appeared as 'Coalgebraic Epistemic Update without Change of Model', *Proceedings of the 2nd International Conference on Algebra and Coalgebra in Computer Science*, Bergen, Norway, 187, 158–172.

T. Sadzik (2005) 'Exploring the Iterated Update Universe', Department of Economics, Stanford University. Appeared as ILLC Research Report 2006, ILLC, University of Amsterdam, 2006.

(2009) 'Beliefs Revealed in Bayesian-Nash Equilibrium', Department of Economics, New York University. Talk presented at *9th SAET Conference on Current Trends in Economics*, Ischia, Italy.

D. Sarenac (2009). 'Modal Logic for Qualitative Dynamics', Department of Philosophy, Colorado State University, Fort Collins.

J. Searle & D. Vanderveken (1985) *Foundations of Illocutionary Logic*, Cambridge University Press.

K. Segerberg (1995) 'Belief Revision from the Point of View of Doxastic Logic', *Bulletin of the IGPL* 3, 534–553.

(1999) 'Default Logic as Dynamic Doxastic Logic', *Erkenntnis* 50, 333–352.

J. Seligman (2010) 'Using Hybrid Logic to Analyze Games', Department of Philosophy, University of Auckland. Invited Lecture at Workshop *'A Door to Logic'*, Beijing, May 2010.

S. Sequoiah-Grayson (2007) 'Information Gain from Inference', Philosophical Institute, Oxford. LogKCA-07 (2007), X. Arrazola and J. Larrazabal, eds., University of Basque Country Press, 351–368. Final version: 'A Positive Information Logic for Inferential Information', *Synthese* 167, 2009, 409–431.

(2009) 'Mono-Agent Dynamics', in X. He, J. Horty & E. Pacuit, eds., *LORI 2009*, LNAI 5834, 321–323.

M. Sergot (2008) 'Temporal Logic of Events and Preference', Department of Computing, Imperial College, London.

M. Sevenster (2006) *Branches of Imperfect Information: Logic, Games, and Computation*, Dissertation DS-2006-06, ILLC Amsterdam.

Y. Shoham (1988) *Reasoning About Change: Time and Change from the Standpoint of Artificial Intelligence*, Cambridge (Mass.), The MIT Press.

(2009) 'Logical Theories of Intention and the Database Perspective', *Journal of Philosophical Logic* 38, 633–647.

Y. Shoham & K. Leyton-Brown (2008) *Multiagent Systems: Algorithmic, Game Theoretic and Logical Foundations*, Cambridge University Press.

Y. Shoham & M. Tennenholtz (1999) 'What Can a Market Compute and at What Expense?', Department of Computer Science, Stanford University.

B. Skyrms (1990) *The Dynamics of Rational Deliberation*, Cambridge (Mass.), Harvard University Press.

W. Spohn (1988) 'Ordinal Conditional Functions: A Dynamic Theory of Epistemic States', in W. Harper *et al.*, eds., *Causation in Decision, Belief Change and Statistics II*, Dordrecht, Kluwer, 105–134.

F. Staal (1988) *Universals: Studies in Indian Logic and Linguistics*, University of Chicago Press.

R. Stalnaker (1978) 'Assertion', in P. Cole, ed., *Syntax and Semantics* 9, New York, Academic Press, 315–332.

(1999) 'Extensive and Strategic Form: Games and Models for Games', *Research in Economics* 53, 293–291.

K. Stenning (2001) *Seeing Reason*, Oxford University Press.

R. Sugden (2003) 'The Logic of Team Reasoning', *Philosophical Explorations* 6, 165–181.

Y.-H. Tan & L. van der Torre (1999) 'An Update Semantics for Deontic Reasoning', in P. McNamara & H. Prakken, eds., *Norms, Logics and Information Systems*, IOS Press, 73–90.

N. Tennant (2002) 'Victor Vanquished', *Analysis* 62, 135–142.

W. Thomas (1992) 'Infinite Trees and Automaton Definable Relations over Omega-Words', *Theoretical Computer Science* 103, 143–159.

L. van der Torre & Y.-H. Tan (1999) 'Contrary-to-Duty Reasoning with Preference-Based Dyadic Obligations', *Annals of Mathematics and Artificial Intelligence* 27, 49–78.

(2001) 'Dynamic Normative Reasoning Under Uncertainty: How to Distinguish Between Obligations Under Uncertainty and Prima Facie Obligations', in D. Gabbay & Ph. Smets, eds., *Handbook of Defeasible Reasoning and Uncertainty Management Systems*, vol. 6: *Agents, Reasoning and Dynamics*, Dordrecht, Kluwer, 267–297.

S. Toulmin (1958) *The Uses of Argument*, Cambridge University Press.

A. M. Turing (1950) 'Computing Machinery and Intelligence', *Mind* 59, 433–460.

J. Väänänen (2007) *Dependence Logic*, Cambridge University Press.

F. Velázquez-Quesada (2008) 'Inference and Update', ILLC, University of Amsterdam, presented at Workshop on Logic and Intelligent Interaction, ESSLLI Summer School, Hamburg. Appeared in *Synthese (Knowledge, Rationality, and Action)* 169:2, 283–300.

(2010) *Small Steps in the Dynamics of Information*, Dissertation, ILLC, University of Amsterdam.

F. Veltman (1985) *Logics for Conditionals*, Dissertation, Philosophical Institute, University of Amsterdam.

(1996) 'Defaults in Update Semantics', *Journal of Philosophical Logic* 25, 221–261. Also appeared in *The Philosopher's Annual*, 1997.

Y. Venema (2006) 'Algebras and Co-Algebras', in P. Blackburn, J. van Benthem & F. Wolter, eds., *Handbook of Modal Logic*, Amsterdam, Elsevier, 331–426.

R. Verbrugge (2009) 'Logic and Social Cognition', *Journal of Philosophical Logic* 38, 649–680.

P. Wason & P. Johnson-Laird (1972) *The Psychology of Reasoning*, Cambridge (Mass.), Harvard University Press.

Y. Wang (2010) *Epistemic Modelling and Protocol Dynamics*, Dissertation, CWI Amsterdam.

T. Williamson (2000) *Knowledge and its Limits*, Oxford University Press.

A. Wisniewski (1995) *The Posing of Questions*, Dordrecht, Kluwer.

M. Wooldridge (2002) *An Introduction to Multi-Agent Systems*, Colchester, John Wiley.

G. H. von Wright (1963) *The Logic of Preference*, Edinburgh University Press.

T. Yamada (2006) 'Acts of Commanding and Changing Obligations', in K. Inoue, K. Satoh & F. Toni, eds., *Computational Logic in Multi-Agent Systems CLIMA VII*. Also in *Lecture Notes in AI*, 4371 (2007), 1–19, Berlin, Springer Verlag, 2007.

A. C. Yao (1979) 'Some Complexity Questions Related to Distributed Computing', *Proceedings of the 11th STOC*, 209–213.

A. Yap (2006) 'Product Update and Temporal Modalities', Department of Philosophy, University of Victoria. To appear in P. Girard, M. Marion & O. Roy, eds., *Proceedings Workshop on Dynamics*, University of Montreal.

B. Zarnic (1999) 'Validity of Practical Inference', ILLC Research report PP-1999-23, University of Amsterdam.

J. Zhang & F. Liu (2007) 'Some Thoughts on Mohist Logic', in J. van Benthem, S. Ju & F. Veltman, eds., *A Meeting of the Minds*, Proceedings LORI Beijing 2007, London, College Publications, 85–102.

M. Xu (2010) 'Combinations of *STIT* and Actions', in 'Temporal Logics of Agency', special issue of the *Journal of Logic, Language and Information*, on-line version January 2010.

T. Yamada (2008) 'Logical Dynamics of some Speech Acts that Affect Obligations and Preferences', *Synthese* 165:2, 295–315.

J. Zvesper (2010) *Playing with Information*, Dissertation, ILLC, University of Amsterdam.

Index

Made in the USA
Middletown, DE
18 January 2018